Edgar G. Goodaire
Department of Mathematics and Statistics
Memorial University
St. John's, Newfoundland, Canada A1C 5S7

Linear Algebra I

Course Notes for Mathematics 2050

Kendall Hunt
publishing company

Kendall Hunt
publishing company

www.kendallhunt.com
Send all inquiries to:
4050 Westmark Drive
Dubuque, IA 52004-1840

ISBN 978-1-4652-0898-9

Printed in Canada
10 9 8 7 6 5 4 3 2

Contents

To My Students

From aviation to the design of cellular phone networks, from data compression (CDs and jpegs) to oil and gas exploration, from computer graphics to Google, linear algebra is indispensable. With relatively little emphasis on sets and functions, linear algebra is "different." Most students find it enjoyable and a welcome change from calculus. Be careful though. The answers to most calculus problems are numerical and easily confirmed through text messages to a friend. The answers to many problems in these notes are **not** numerical, however, and require explanations as to **why** things are as they are. So let me begin this note to you with a word of caution: this is a set of notes you **must read**.

For many of you, linear algebra will be the first course where many exercises ask you to explain, to answer why or how. It can be a shock to discover that there are mathematics courses (in fact, most of those above first year) where words are more important than numbers. Say good-bye to the days when the solution to a homework problem lies in finding an identical worked example in the text. Homework is now going to require some critical thinking!

Many years ago, a student came into my office one day to ask for help with a homework question. When this happens with a book of which I am an author, I am always eager to discover whether or not I have laid the proper groundwork in the section so that the average student could be expected to make a reasonable attempt at an exercise. From your point of view, a homework exercise should be very similar to a worked example. Right? In the instance I am recalling, I went through the section of these notes with the student in question page by page until we found such an example. In fact, we found precisely the question I had assigned, with the complete solution laid out as an example that I had forgotten to delete when I transferred it to the exercises! The student felt a little sheepish while I was completely shocked to be reminded, once again, that some students don't read their textbooks.

It is always tempting to start a homework problem right away, without preparing yourself first, but this approach isn't going to work very well here. You will find it imperative to read a section from start to finish before attempting an exercise based on that section.

A lot of students don't understand what is being asked because they don't know what the words mean. If this happens to you, go to the glossary at the end of these notes where most technical terms are defined and where you will often find an example of something that fits the definition and an example of something that doesn't. Many students complain they have no idea how to write a "proof." If this is you, my first comment is "Fine. You're in the same boat with lots of other students." A proof is just a convincing argument. If you can explain "why" to a friend, and she is convinced, then you have just given a proof. Hey, one has to **learn** how to write a good proof, and you will over time. Pay attention in class where you will see a proof almost every class. There's an appendix called "Show and Prove" at the back of these notes. With time, you'll get the hang of a proof.

In a short note entitled "Things I Must Remember," also at the back, I have included many important ideas that my students have helped me to collect over the years. You will also find there some ideas that are often just what you need to solve homework problems.

I hope that you like my writing style, that you discover you like linear algebra, and that you soon surprise yourself with your ability to write a good clear mathematical proof. I hope that you do well in your linear algebra courses and all those other courses where linear algebra plays an important role. Let me know what you think of this book. I like receiving comments—good, bad and ugly—from anyone.

Edgar G. Goodaire
`edgar@mun.ca`
St. John's, Newfoundland, Canada
July 2012

Week 1

UNIT ONE: EUCLIDEAN n-SPACE

Vectors and Arrows

A 2-*dimensional vector* is a column of two numbers surrounded by brackets. For example,

$$\begin{bmatrix}1\\3\end{bmatrix}, \begin{bmatrix}2\\4\end{bmatrix}, \begin{bmatrix}-2\\3\end{bmatrix}, \text{ and } \begin{bmatrix}0\\0\end{bmatrix}$$

are 2-dimensional vectors. Different people use different notation for vectors. When typesetting, my preference is for boldface type, so if v is the name of a vector, you'll see v in these notes. When writing by hand, I underline, like this, $\underline{v}$.

The *components* of the vector $\mathsf{v} = \begin{bmatrix}a\\b\end{bmatrix}$ are the numbers a and b. Two vectors are *equal* if and only if they have the same corresponding components. Thus, if

$$\begin{bmatrix}a-3\\2b\end{bmatrix} = \begin{bmatrix}-1\\6\end{bmatrix},$$

then $a - 3 = -1$ and $2b = 6$, so that $a = 2$ and $b = 3$.

A vector can be pictured with an arrow, actually many arrows. To picture $\mathsf{v} = \begin{bmatrix}3\\2\end{bmatrix}$, for instance, start at any point $A(x_0, y_0)$, go right 3 units and up 2, putting the arrow head at $B(x_0 + 3, y_0 + 2)$. Then $\mathsf{v} = \overrightarrow{AB} = \begin{bmatrix}3\\2\end{bmatrix}$. The notation $\overrightarrow{AB}$ means the vector that is pictured by the arrow from A to B.

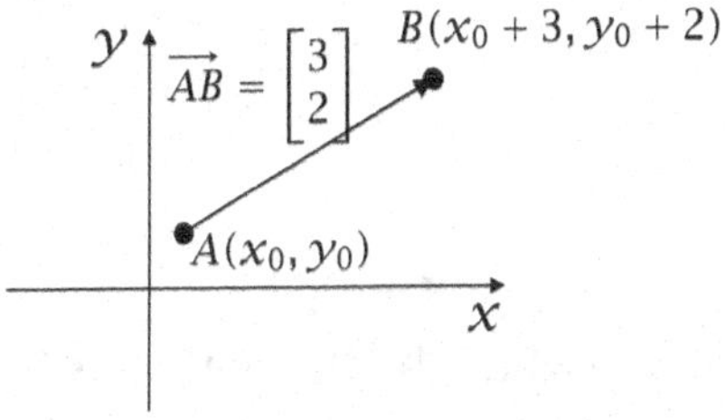

Notice how to get the components 3, 2, from the coordinates of A and B; just subtract the coordinates of A from the coordinates of B. For general points $A(x_0, y_0)$ and $B(x_1, y_1)$, the vector $\overrightarrow{AB} = \begin{bmatrix}x_1 - x_0\\y_1 - y_0\end{bmatrix}$.

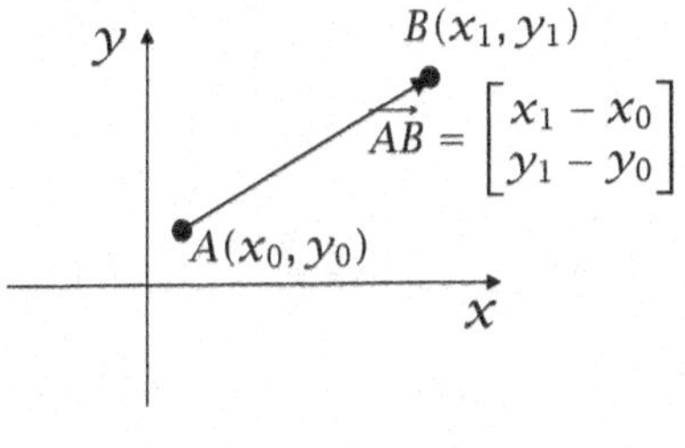

1.1 | The arrow from (x_0, y_0) to (x_1, y_1) is a picture of the vector $\begin{bmatrix}x_1 - x_0\\y_1 - y_0\end{bmatrix}$.

For example, if $A = (2, 3)$ and $B = (7, 4)$, the vector $\overrightarrow{AB}$ is $\begin{bmatrix}7-2\\4-3\end{bmatrix} = \begin{bmatrix}5\\1\end{bmatrix}$. This vector can be pictured by an arrow, in fact by many arrows, since we can start at any point (x_0, y_0). Each of the arrows in Figure 1 is a picture of the vector $\begin{bmatrix}1\\2\end{bmatrix}$.

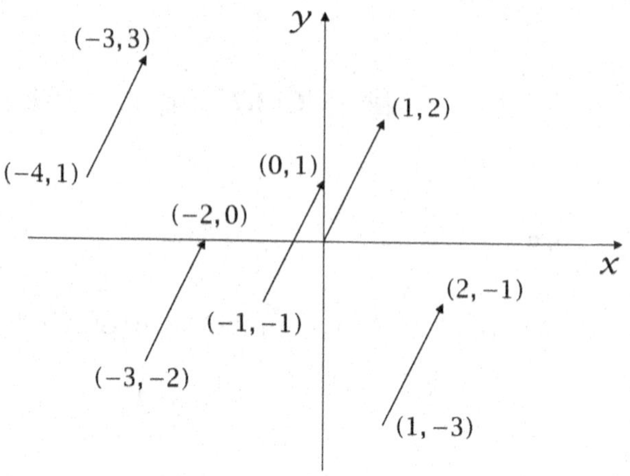

Figure 1: Five arrows, each a picture of the vector $\begin{bmatrix}1\\2\end{bmatrix}$.

In general, lots of different arrows picture the same vector, just by varying the tail A. How do we know if two arrows are pictures of the same vector?

1.2 Two arrows picture the same vector if and only if they have the same length and the same direction.

Scalar Multiplication of Vectors

We multiply vectors by numbers in the obvious way, "componentwise." As examples,

$$4\begin{bmatrix}-1\\3\end{bmatrix} = \begin{bmatrix}-4\\12\end{bmatrix}; \quad -2\begin{bmatrix}2\\-3\end{bmatrix} = \begin{bmatrix}-4\\6\end{bmatrix}; \quad 0\begin{bmatrix}4\\\sqrt{3}\end{bmatrix} = \begin{bmatrix}0\\0\end{bmatrix};$$

$$-\begin{bmatrix}-3\\8\end{bmatrix} = (-1)\begin{bmatrix}-3\\8\end{bmatrix} = \begin{bmatrix}3\\-8\end{bmatrix}.$$

The last example illustrates the agreement that $-\mathsf{v}$ means $(-1)\mathsf{v}$.

This multiplication of vectors by numbers is called *scalar multiplication. Scalar* just means "real number," so *scalar multiplication* means multiplication by a real number. By the term *scalar multiple* of a vector, we mean a vector like $2\mathsf{v}$ or $-\frac{1}{3}\mathsf{v}$.

Do you see the connection between an arrow for v and an arrow for $2\mathsf{v}$ or $-\mathsf{v}$? As illustrated in Figure 2, for a general scalar c, the vector $c\mathsf{v}$ has

- the same direction as v if $c > 0$,
- direction opposite to v if $c < 0$, and

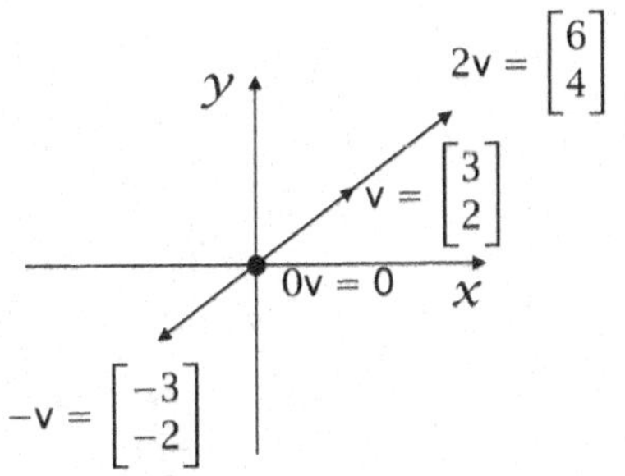

Figure 2: The vector 2v has the same direction as v but is twice as long; −v has direction opposite to v but the same length; 0v is a vector of zero length and no direction, which we picture as a single point.

- length $|c|$ times the length of v.

Thus 3v is three times the length of v while the length of $-\frac{1}{2}$v is $|-\frac{1}{2}| = +\frac{1}{2}$ the length of v.

The vector $0v = \begin{bmatrix}0\\0\end{bmatrix}$ is called the *zero vector* and denoted 0 (or $\underline{0}$ in handwriting).

Two-dimensional vectors are said to be *parallel* if arrows that picture them are parallel. Here is a definition that does not depend on pictures.

1.3 Definitions. Vectors u and v are *parallel* if one is a scalar multiple of the other, that is, if $u = cv$ or $v = cu$ for some scalar c. Parallel vectors v and cv have the *same direction* if $c > 0$ and *opposite direction* if $c < 0$.

By definition, the zero vector is parallel to any vector v since $0 = 0v$ is a scalar multiple of v. However, we do not talk about the zero vector having direction same or opposite to v since the zero vector has **no direction.**

Notice that the definition does **not** say that vectors are parallel if they are multiples of each other, $v = \begin{bmatrix}1\\2\end{bmatrix}$ and 0 are parallel because $0 = 0v$—0 is a multiple of v—but they are not multiples of each other because v is not a multiple of 0.

Thus $u = \begin{bmatrix}2\\-4\end{bmatrix}$ and $v = \begin{bmatrix}-\frac{1}{2}\\1\end{bmatrix}$ are parallel but have opposite direction because $u = -4v$ and $-4 < 0$. On the other hand, u and $w = \begin{bmatrix}1\\-2\end{bmatrix}$ are parallel and have the same direction.

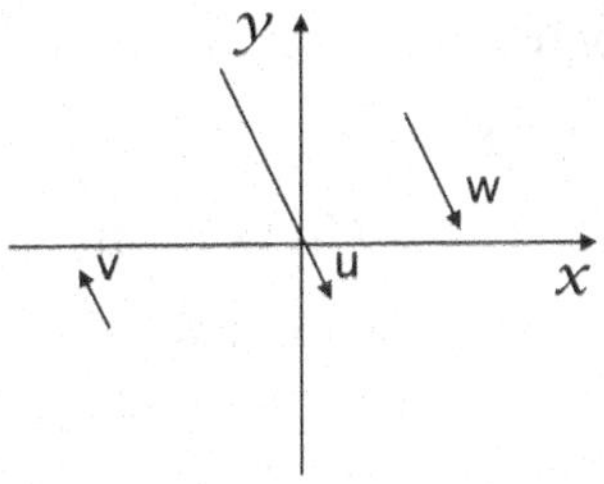

Vector Addition

We add vectors *componentwise*, the obvious way:

$$\begin{bmatrix}0\\-3\end{bmatrix} + \begin{bmatrix}2\\5\end{bmatrix} = \begin{bmatrix}2\\2\end{bmatrix}, \quad \begin{bmatrix}1\\5\end{bmatrix} + \begin{bmatrix}3\\-2\end{bmatrix} = \begin{bmatrix}4\\3\end{bmatrix}, \quad \begin{bmatrix}2\\3\end{bmatrix} + \begin{bmatrix}0\\0\end{bmatrix} = \begin{bmatrix}2\\3\end{bmatrix}.$$

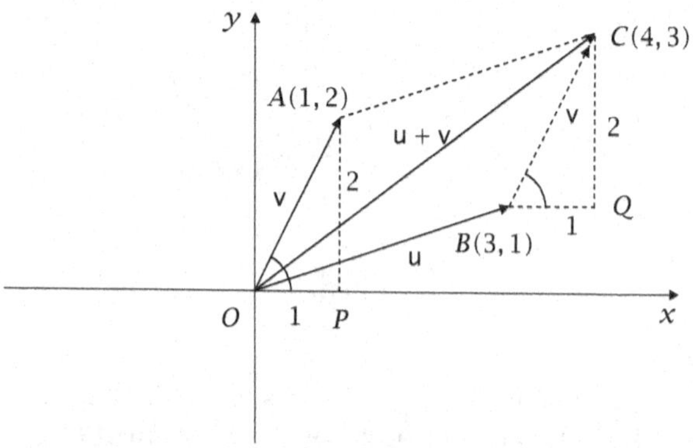

Figure 3: u + v can be pictured by the arrow that is the diagonal of the parallelogram with sides u and v.

There is a nice connection between arrows for vectors u and v and an arrow for the sum u + v.

1.4 Parallelogram Rule: If arrows for u and v are drawn with tails at the same point A, then an arrow for u + v is the diagonal of the parallelogram whose sides are u and v with tail at A.

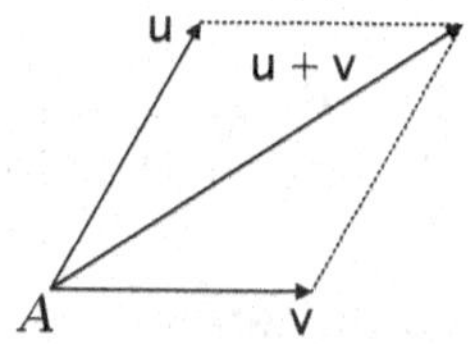

We illustrate this principle in Figure 3 for the vectors $\mathsf{u} = \begin{bmatrix} 3 \\ 1 \end{bmatrix}$ and $\mathsf{v} = \begin{bmatrix} 1 \\ 2 \end{bmatrix}$. The sum of these vectors is

$$\mathsf{u} + \mathsf{v} = \begin{bmatrix} 3 \\ 1 \end{bmatrix} + \begin{bmatrix} 1 \\ 2 \end{bmatrix} = \begin{bmatrix} 4 \\ 3 \end{bmatrix},$$

a vector that can be pictured by the arrow from the origin to $C(4, 3)$. The parallelogram rule says that the vector $\overrightarrow{OC} = \begin{bmatrix} 4 \\ 3 \end{bmatrix}$ is pictured by the arrow which is the diagonal of parallelogram $OACB$ pointing from 0 to C.

Let's verify the parallelogram rule by looking at parallelogram $OACB$. (See Figure 3.) Since OA and BC are parallel and have the same length, the right-angled triangles BQC and OPA are congruent, so BQ has length 1, QC has length 2 and the coordinates of C are $(4, 3)$. Thus $\overrightarrow{OC} = \begin{bmatrix} 4 \\ 3 \end{bmatrix}$ is pictured by the diagonal of the parallelogram, just as we said.

Here's something else. The fact that OA and BC have the same length and direction means $\overrightarrow{OA} = \overrightarrow{BC}$, so $\overrightarrow{OC} = \mathsf{u} + \mathsf{v} = \overrightarrow{OB} + \overrightarrow{BC}$: The vector $\overrightarrow{OC}$ is the third side of triangle OBC. This gives a second way to picture the sum of two vectors.

1.5 Triangle Rule: If an arrow for u goes from A to B and an arrow for v from B to C, then an arrow for $\mathsf{u}+\mathsf{v}$ runs from A to C: $\overrightarrow{AB}+\overrightarrow{BC}=\overrightarrow{AC}$.

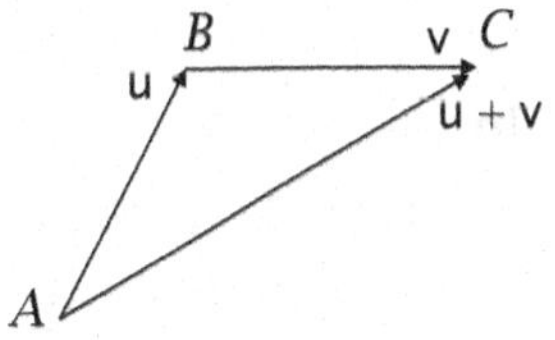

Subtracting Vectors

Just as $5-2$ means $5+(-2)$, we subtract vectors using the convention that

1.6
$$\boxed{\mathsf{u}-\mathsf{v}=\mathsf{u}+(-\mathsf{v}).}$$

1.7 Examples.

- $\begin{bmatrix}7\\2\end{bmatrix}-\begin{bmatrix}5\\3\end{bmatrix}=\begin{bmatrix}7\\2\end{bmatrix}+\left(-\begin{bmatrix}5\\3\end{bmatrix}\right)=\begin{bmatrix}7\\2\end{bmatrix}+\begin{bmatrix}-5\\-3\end{bmatrix}=\begin{bmatrix}2\\-1\end{bmatrix}$
- $\begin{bmatrix}-1\\3\end{bmatrix}-\begin{bmatrix}-2\\2\end{bmatrix}=\begin{bmatrix}1\\1\end{bmatrix}$
- $\begin{bmatrix}2\\0\end{bmatrix}-\begin{bmatrix}0\\-3\end{bmatrix}=\begin{bmatrix}2\\3\end{bmatrix}$

What's the connection between arrows for u and v, and an arrow for $\mathsf{u}-\mathsf{v}$? Since $\mathsf{v}+(\mathsf{u}-\mathsf{v})=\mathsf{u}$, it follows from the triangle rule that if we represent $\mathsf{u}-\mathsf{v}$ by an arrow with tail at the head of v, then an arrow for u follows the third side of the triangle, from the head of v to the head of u: $\overrightarrow{AB}-\overrightarrow{AC}=\overrightarrow{CB}$.

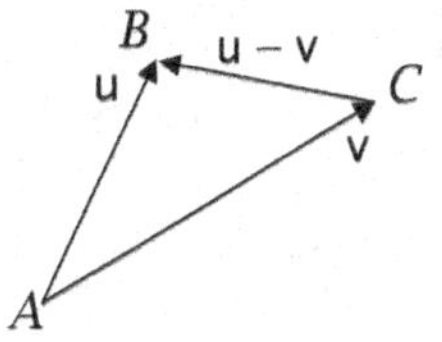

1.8

If vectors u and v are pictured by arrows with the same tail, then $\mathsf{u}-\mathsf{v}$ can be pictured by the arrow that goes from the head of v to the head of u.

Our definitions of vector addition and subtraction imply that the diagonals of the parallelogram with sides u and v are just $\mathsf{u}+\mathsf{v}$ and $\mathsf{u}-\mathsf{v}$.

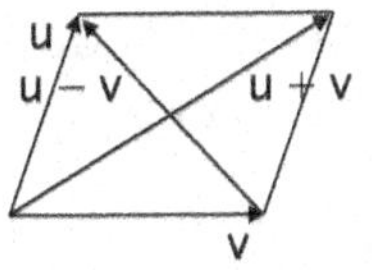

We have seen that vectors can be pictured by arrows and any two arrows with the same length and direction describe the same vector. These facts can be used to prove some familiar propositions from Euclidean geometry.

1.9 Problem. Show that the diagonals of a parallelogram bisect each other.

Solution. Let A, B, C, D be the vertices of the parallelogram and draw the diagonal AC. Let X be the midpoint of AC. Thus $\overrightarrow{AX} = \overrightarrow{XC}$. We wish to show that X is the midpoint of BD. With vectors, this is the equation $\overrightarrow{BX} = \overrightarrow{XD}$. Now

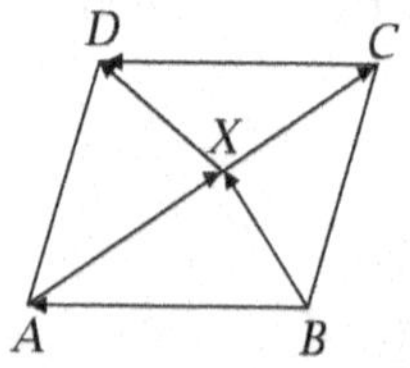

$$\overrightarrow{BX} = \overrightarrow{BA} + \overrightarrow{AX} \quad \text{and} \quad \overrightarrow{XD} = \overrightarrow{XC} + \overrightarrow{CD}.$$

Since $\overrightarrow{BA}$ and $\overrightarrow{CD}$ have the same length and direction, $\overrightarrow{BA} = \overrightarrow{CD}$. Since $\overrightarrow{AX} = \overrightarrow{XC}$ as well, we have the desired result.

We now list some properties of addition and scalar multiplication which, I admit, look rather boring. The names are the most important thing to remember because names make it easy to refer to different properties. Labels like "commutativity" and "associativity," perhaps even "distributivity" should be familiar.

1.10 Properties of Vector Addition and Scalar Multiplication

Let u, v and w be vectors, and let a and b denote scalars.

1. (Closure under addition) $\mathsf{u} + \mathsf{v}$ is a vector.
2. (Commutativity of addition) $\mathsf{u} + \mathsf{v} = \mathsf{v} + \mathsf{u}$.
3. (Associativity of addition) $(\mathsf{u} + \mathsf{v}) + \mathsf{w} = \mathsf{u} + (\mathsf{v} + \mathsf{w})$.
4. (Zero) $\mathsf{u} + 0 = 0 + \mathsf{u} = \mathsf{u}$.
5. (Additive inverse) There is a vector called the additive inverse of u, denoted $-\mathsf{u}$ and called "minus u", with the property that $\mathsf{u} + (-\mathsf{u}) = (-\mathsf{u}) + \mathsf{u} = 0$.
6. (Closure under scalar multiplication) $c\mathsf{u}$ is a vector.
7. (Scalar associativity) $a(b\mathsf{u}) = (ab)\mathsf{u}$.
8. (One) $1\mathsf{u} = \mathsf{u}$.
9. (Distributivity) $a(\mathsf{u} + \mathsf{v}) = a\mathsf{u} + a\mathsf{v}$ and $(a + b)\mathsf{u} = a\mathsf{u} + b\mathsf{u}$.

We describe the distributive property $a(\mathsf{u}+\mathsf{v}) = a\mathsf{u}+a\mathsf{v}$ by saying that scalar multiplication **distributes over** addition. "Scalar associativity" should remind you of ordinary associativity: $(xy)z = x(yz)$.

1.11 Examples. · $4(\mathsf{u} - 3\mathsf{v}) = 4\mathsf{u} + 4(-3\mathsf{v}) = 4\mathsf{u} + [4(-3)]\mathsf{v} = 4\mathsf{u} + (-12)\mathsf{v} = 4\mathsf{u} - 12\mathsf{v}$.

Here we showed all our steps. In the the remaining examples, we will not!

- $4(\mathsf{u} - 3\mathsf{v}) + 6(-2\mathsf{u} + 3\mathsf{v}) = -8\mathsf{u} + 6\mathsf{v}$.

- If $2u + 3v = 6x + 4u$, then $6x = -2u + 3v$, so $x = -\frac{1}{3}u + \frac{1}{2}v$.
- If $x = 3u - 2v$ and $y = u + v$, then u and v can be expressed in terms of x and y like this:

$$\begin{array}{rcl} x & = & 3u - 2v \\ 2y & = & 2u + 2v \\ \hline x + 2y & = & 5u \end{array}$$

so $u = \frac{1}{5}(x + 2y)$ and $v = y - u = -\frac{1}{5}x + \frac{3}{5}y$. ☺

Higher Dimensions

1.12 Definitions. A 3-*dimensional vector* is a triple of numbers written in a column and surrounded by brackets. The *components* of $\begin{bmatrix} a \\ b \\ c \end{bmatrix}$ are the three numbers a, b, and c. For any $n \geq 1$, an n-*dimensional vector* is a column $\begin{bmatrix} x_1 \\ x_2 \\ \vdots \\ x_n \end{bmatrix}$ of n numbers enclosed in brackets. The numbers $x_1, x_2, \ldots, x_n$ are called the *components* of the vector.

1.13 Examples.

- $\begin{bmatrix} 1 \\ 2 \\ 3 \end{bmatrix}$ is a 3-dimensional vector with components 1, 2, and 3;
- $\begin{bmatrix} 1 \\ 2 \\ 3 \\ 4 \end{bmatrix}$ is a 4-dimensional vector with components 1, 2, 3, 4;
- $\begin{bmatrix} -1 \\ 0 \\ 0 \\ 1 \\ 2 \end{bmatrix}$ is a 5-dimensional vector with components -1, 0, 0, 1, 2;
- the 3-dimensional *zero vector* is the vector $\begin{bmatrix} 0 \\ 0 \\ 0 \end{bmatrix}$ all of whose components are 0. ☺

1.14 Definition. *Euclidean n-space* is the set of all n-dimensional vectors. It is denoted R^n.

$$R^n = \left\{ \begin{bmatrix} x_1 \\ x_2 \\ \vdots \\ x_n \end{bmatrix} \mid x_1, x_2, \ldots, x_n \in R \right\}.$$

Euclidean 2-space, $\mathsf{R}^2 = \left\{ \begin{bmatrix} x \\ y \end{bmatrix} \mid x, y \in \mathsf{R} \right\}$ is the set of all 2-dimensional vectors, a set commonly called the *Euclidean plane* and we often refer to Euclidean 3-space,

$$\mathsf{R}^3 = \left\{ \begin{bmatrix} x \\ y \\ z \end{bmatrix} \mid x, y, z \in \mathsf{R} \right\},$$

as simply 3-*space.*

The general definition of vector *equality* is just like the definition in two dimensions.

1.15 Vectors $\mathsf{x} = \begin{bmatrix} x_1 \\ x_2 \\ \vdots \\ x_n \end{bmatrix}$ and $\mathsf{y} = \begin{bmatrix} y_1 \\ y_2 \\ \vdots \\ y_m \end{bmatrix}$ are *equal* if and only if $n = m$ and corresponding components are equal, that is, $x_1 = y_1, x_2 = y_2, \ldots, x_n = y_n$.

Just as with 2-dimensional vectors, addition and scalar multiplication of n-dimensional vectors are defined componentwise:

$$\begin{bmatrix} x_1 \\ x_2 \\ \vdots \\ x_n \end{bmatrix} + \begin{bmatrix} y_1 \\ y_2 \\ \vdots \\ y_n \end{bmatrix} = \begin{bmatrix} x_1 + y_1 \\ x_2 + y_2 \\ \vdots \\ x_n + y_n \end{bmatrix} \quad \text{and} \quad c \begin{bmatrix} x_1 \\ x_2 \\ \vdots \\ x_n \end{bmatrix} = \begin{bmatrix} cx_1 \\ cx_2 \\ \vdots \\ cx_n \end{bmatrix},$$

and all the properties of vector addition and scalar multiplication cited before for 2-dimensional vectors—see **1.10**—hold for n-dimensional vectors too. The n-dimensional *zero vector* is $\mathsf{0} = \begin{bmatrix} 0 \\ 0 \\ \vdots \\ 0 \end{bmatrix}$. This is the vector with the property that $\mathsf{x} + \mathsf{0} = \mathsf{x}$ for any n-dimensional vector x. For any vector x, the vector $-\mathsf{x} = (-1)\mathsf{x}$ has the property that $\mathsf{x} + (-\mathsf{x}) = (\mathsf{x}) + \mathsf{x} = \mathsf{0}$.

Linear Combinations

A vector like $-8\mathsf{u} + 6\mathsf{v}$, which is the sum of scalar multiples of vectors, has a special name.

1.16 Definition. A *linear combination* of vectors u and v is a vector of the form $a\mathsf{u} + b\mathsf{v}$, where a and b are scalars. More generally, a *linear combination* of k vectors $\mathsf{v}_1, \mathsf{v}_2, \ldots, \mathsf{v}_k$ is a vector of the form $c_1\mathsf{v}_1 + c_2\mathsf{v}_2 + \cdots + c_k\mathsf{v}_k$, where $c_1, \ldots, c_k$ are scalars.

1.17 Examples. • $\begin{bmatrix} -5 \\ 9 \end{bmatrix}$ is a linear combination of $\mathsf{u} = \begin{bmatrix} -2 \\ 3 \end{bmatrix}$ and $\mathsf{v} = \begin{bmatrix} -1 \\ 1 \end{bmatrix}$ since $\begin{bmatrix} -5 \\ 9 \end{bmatrix} = 4\begin{bmatrix} -2 \\ 3 \end{bmatrix} - 3\begin{bmatrix} -1 \\ 1 \end{bmatrix} = 4\mathsf{u} - 3\mathsf{v}$.

- $\begin{bmatrix} 2 \\ -6 \end{bmatrix}$ is a linear combination of $\mathsf{v}_1 = \begin{bmatrix} -2 \\ 3 \end{bmatrix}$, $\mathsf{v}_2 = \begin{bmatrix} 6 \\ -5 \end{bmatrix}$ and $\mathsf{v}_3 = \begin{bmatrix} 4 \\ 5 \end{bmatrix}$ since $\begin{bmatrix} 2 \\ -6 \end{bmatrix} = 3\begin{bmatrix} -2 \\ 3 \end{bmatrix} + 2\begin{bmatrix} 6 \\ -5 \end{bmatrix} - \begin{bmatrix} 4 \\ 5 \end{bmatrix} = 3\mathsf{v}_1 + 2\mathsf{v}_2 + (-1)\mathsf{v}_3$.

- The equation $\begin{bmatrix} -11 \\ 7 \\ -4 \\ 2 \end{bmatrix} = 2\begin{bmatrix} -2 \\ 0 \\ 5 \\ 1 \end{bmatrix} - 7\begin{bmatrix} 1 \\ -1 \\ 2 \\ 0 \end{bmatrix}$

  says that $\begin{bmatrix} -11 \\ 7 \\ -4 \\ 2 \end{bmatrix}$ is a linear combination of $\begin{bmatrix} -2 \\ 0 \\ 5 \\ 1 \end{bmatrix}$ and $\begin{bmatrix} 1 \\ -1 \\ 2 \\ 0 \end{bmatrix}$. ☺

1.18 Example. The vector $\begin{bmatrix} 1 \\ 5 \\ 2 \\ 0 \\ -1 \end{bmatrix}$ is not a linear combination of $\begin{bmatrix} 1 \\ 2 \\ -1 \\ 3 \\ 4 \end{bmatrix}$ and $\begin{bmatrix} 3 \\ 5 \\ 0 \\ 1 \\ 2 \end{bmatrix}$ because

$$\begin{bmatrix} 1 \\ 5 \\ 2 \\ 0 \\ -1 \end{bmatrix} = a\begin{bmatrix} 1 \\ 2 \\ -1 \\ 3 \\ 4 \end{bmatrix} + b\begin{bmatrix} 3 \\ 5 \\ 0 \\ 1 \\ 2 \end{bmatrix}$$

leads to the system of equations

$$\begin{aligned} a + 3b &= 1 \\ 2a + 5b &= 5 \\ -a &= 2 \\ 3a + b &= 0 \\ 4a + 2b &= -1, \end{aligned}$$

which has no solution. The third equation says $a = -2$. Substituting in the first equation, we get $b = 1$, but the pair $a = -2, b = 1$ does not satisfy the second equation. ☺

1.19 Example. Is $\mathsf{x} = \begin{bmatrix} -1 \\ -2 \\ 2 \end{bmatrix}$ a linear combination of $\mathsf{u} = \begin{bmatrix} 0 \\ 1 \\ 4 \end{bmatrix}$, $\mathsf{v} = \begin{bmatrix} -1 \\ 1 \\ 2 \end{bmatrix}$, and $\mathsf{w} = \begin{bmatrix} 3 \\ 1 \\ -2 \end{bmatrix}$?

The question asks whether or not there exist scalars a, b, c so that

$$\mathsf{x} = a\mathsf{u} + b\mathsf{v} + c\mathsf{w};$$

that is, such that

$$\begin{bmatrix} -1 \\ -2 \\ 2 \end{bmatrix} = a\begin{bmatrix} 0 \\ 1 \\ 4 \end{bmatrix} + b\begin{bmatrix} -1 \\ 1 \\ 2 \end{bmatrix} + c\begin{bmatrix} 3 \\ 1 \\ -2 \end{bmatrix}.$$

The vector on the right is $\begin{bmatrix} -b+3c \\ a+b+c \\ 4a+2b-2c \end{bmatrix}$, so the question is, are there numbers a, b, c such that

$$\begin{aligned} -\ \ b+3c &= -1 \\ a+\ b+\ c &= -2 \\ 4a+2b-2c &= 2 \end{aligned} \quad ?$$

Whether or not you can find a, b, and c yourself at this point isn't important. We'll talk about how to solve systems of equations such as the one here later in the course. Here, without explanation, we simply note that $a = 1$, $b = -2$, $c = -1$ is a solution. Thus $x = u - 2v - w$ is a linear combination of u, v, and w.

1.20 **Problem.** Show that $\begin{bmatrix} 3 \\ 0 \end{bmatrix}$ is a linear combination of $\begin{bmatrix} -1 \\ 2 \end{bmatrix}$ and $\begin{bmatrix} -3 \\ 3 \end{bmatrix}$.

Solution. We must produce scalars a and b so that $\begin{bmatrix} 3 \\ 0 \end{bmatrix} = a\begin{bmatrix} -1 \\ 2 \end{bmatrix} + b\begin{bmatrix} -3 \\ 3 \end{bmatrix}$. Now $a\begin{bmatrix} -1 \\ 2 \end{bmatrix} + b\begin{bmatrix} -3 \\ 3 \end{bmatrix} = \begin{bmatrix} -a-3b \\ 2a+3b \end{bmatrix}$. Remembering that vectors are equal if and only if they have the same corresponding components,

$$\begin{bmatrix} 3 \\ 0 \end{bmatrix} = \begin{bmatrix} -a-3b \\ 2a+3b \end{bmatrix}$$

means $3 = -a-3b$ and $0 = 2a+3b$. Adding these equations gives $a = 3$, so $-3b = 3+a = 6$ and $b = -2$.

Actually, this example has nothing to do with the vector $\begin{bmatrix} 3 \\ 0 \end{bmatrix}$: Any 2-dimensional vector is a linear combination of $\begin{bmatrix} -1 \\ 2 \end{bmatrix}$ and $\begin{bmatrix} -3 \\ 3 \end{bmatrix}$!

1.21

> If 2-dimensional vectors u and v not parallel, then any vector is a linear combination of u and v; that is, the set of linear combinations of u and v is the entire xy-plane.

In Figure 4, we show two vectors u and v that are not parallel and an arbitrary vector w. It is easy to make a parallelogram whose diagonal is w and whose sides are multiples of u and v. Thus the sides have the form au and bv and w is their sum. To repeat, we have the following:

1.22

> If 2-dimensional vectors u and v are not parallel, then the set of linear combinations of u and v is the entire xy-plane.

We can in fact deduce a more general fact from Figure 4. By lifting your book from your desk and moving it around so that the vectors u and v lie in the plane of your choice, we

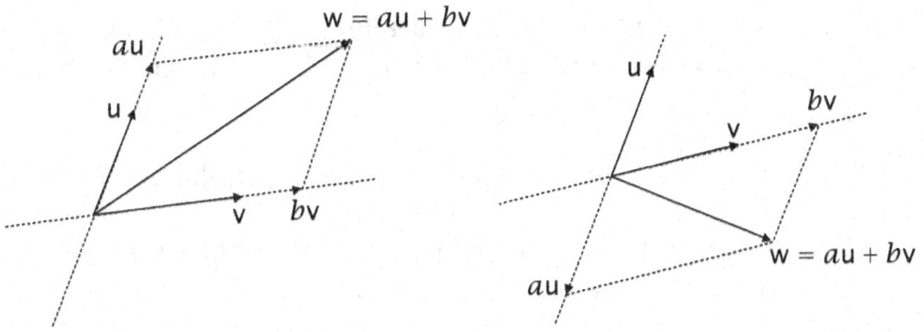

Figure 4: Any vector w is a linear combination of u and v.

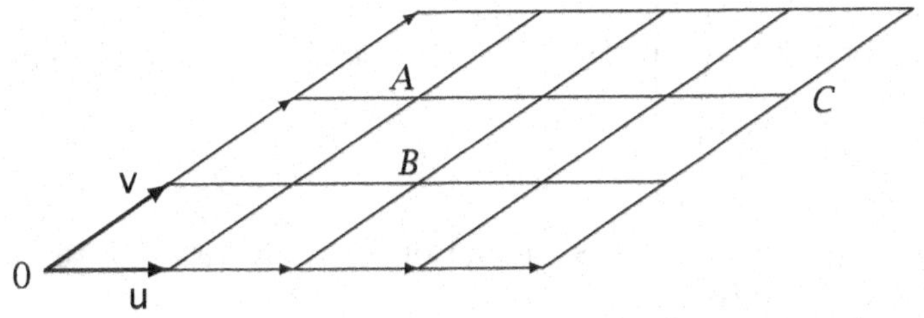

Figure 5: The set of all linear combinations of u and v is a plane.

see that

1.23

> If 3-dimensional vectors u and v are not parallel, then the set of all linear combinations of u and v is a plane. This is called the plane *spanned by* u *and* v.

Perhaps this important idea is reinforced by Figure 5. The arrow from 0 to A is a picture of u + 2v and the arrow from 0 to B is a picture of 2u + v. Every vector in the plane shown in the figure is a linear combination of u and v.

The Span of Vectors

1.24 Definition. The *span* of vectors $v_1, v_2, \ldots, v_k$ is the set of all linear combinations of these vectors.

For example, we have seen that the span of two nonparallel vectors in R^2 is R^2 and more generally that the span of two nonparallel vectors in R^3 is a plane. We extend this concept to higher dimensions.

1.25 Definitions. A *plane* in R^n, for any $n \geq 1$, is the span of two nonparallel vectors u and v.

Notice how we simply transfer concepts that are true in R^3 to higher dimensions simply by making a definition. We have a concept of "parallel," for instance. In R^2 or R^3, arrows that picture two vectors are parallel if and only if one of the vectors is a scalar multiple of the other. In R^n, for any n, therefore, we **define** two vectors to be parallel if one is a scalar multiple of the other. In R^3, two nonparallel vectors span a plane. In R^n, therefore, we **define** a plane as the span of two nonparallel vectors.

The word "span" is used both as both a noun—the span of vectors—and as a verb. We say that the vectors $\mathsf{v}_1, \mathsf{v}_2, \ldots, \mathsf{v}_k$ *span* (verb) something or other. If $k = 2$, we say that v_1 and v_2 span a plane (if they are not parallel).

1.26 Problem. Let $\mathsf{u} = \begin{bmatrix} -1 \\ 0 \\ 2 \\ 1 \end{bmatrix}$ and $\mathsf{v} = \begin{bmatrix} 2 \\ 3 \\ -4 \\ 1 \end{bmatrix}$. Describe the plane spanned by u and v. Does $\begin{bmatrix} 1 \\ 6 \\ -1 \\ 0 \end{bmatrix}$ belong to this plane?

Solution. The plane spanned by u and v is the set of linear combinations of u and v. So the plane consists of vectors of the form

$$a\mathsf{u} + b\mathsf{v} = a\begin{bmatrix} -1 \\ 0 \\ 2 \\ 1 \end{bmatrix} + b\begin{bmatrix} 2 \\ 3 \\ -4 \\ 1 \end{bmatrix} = \begin{bmatrix} -a + 2b \\ 3b \\ 2a - 4b \\ a + b \end{bmatrix}.$$

The vector $\begin{bmatrix} 1 \\ 6 \\ -1 \\ 0 \end{bmatrix}$ is in the plane if and only if there are numbers a and b so that

$$\begin{aligned} -a + 2b &= 1 \\ 3b &= 6 \\ 2a - 4b &= -1 \\ a + b &= 0. \end{aligned}$$

So we would need $b = 2$ and $a = -b = -2$ (equations two and four), but then the first equation is not satisfied because $-a + 2b = 2 + 4 = 6 \neq 1$. No such a and b exist, so the vector is **not** in the plane. ☝

The Standard Basis Vectors

It is easy to check **1.21** if the vectors are particularly nice. For example, if $\mathsf{i} = \begin{bmatrix} 1 \\ 0 \end{bmatrix}$ and $\mathsf{j} = \begin{bmatrix} 0 \\ 1 \end{bmatrix}$, the vector $\begin{bmatrix} 3 \\ 4 \end{bmatrix}$, for instance, is certainly a linear combination of i and j: $\begin{bmatrix} 3 \\ 4 \end{bmatrix} =$

$3\begin{bmatrix}1\\0\end{bmatrix}+4\begin{bmatrix}0\\1\end{bmatrix}=3\mathsf{i}+4\mathsf{j}$. This has nothing to do with the vector $\begin{bmatrix}3\\4\end{bmatrix}$.

1.27 Every 2-dimensional vector is a linear combination of i and j: $\begin{bmatrix}a\\b\end{bmatrix}=a\mathsf{i}+b\mathsf{j}$.

Similarly, every 3-dimensional vector is a linear combination of $\mathsf{i}=\begin{bmatrix}1\\0\\0\end{bmatrix}$, $\mathsf{j}=\begin{bmatrix}0\\1\\0\end{bmatrix}$ and $\mathsf{k}=\begin{bmatrix}0\\0\\1\end{bmatrix}$. These vectors have a special name.

1.28 Definition. The *2-dimensional standard basis vectors* are $\mathsf{i}=\begin{bmatrix}1\\0\end{bmatrix}$ and $\mathsf{j}=\begin{bmatrix}0\\1\end{bmatrix}$. The *3-dimensional standard basis vectors* are $\mathsf{i}=\begin{bmatrix}1\\0\\0\end{bmatrix}$, $\mathsf{j}=\begin{bmatrix}0\\1\\0\end{bmatrix}$ and $\mathsf{k}=\begin{bmatrix}0\\0\\1\end{bmatrix}$.

These vectors have obvious analogues in R^n where they are typically labelled $\mathsf{e}_1,\mathsf{e}_2,\dots,\mathsf{e}_n$.

1.29 Definition. The *standard basis vectors* in R^n are the vectors $\mathsf{e}_1,\mathsf{e}_2,\dots,\mathsf{e}_n$ where e_i has ith component 1 and all other components 0. Thus

$$\mathsf{e}_1=\begin{bmatrix}1\\0\\0\\0\\\vdots\\0\end{bmatrix},\quad \mathsf{e}_2=\begin{bmatrix}0\\1\\0\\0\\\vdots\\0\end{bmatrix},\quad \mathsf{e}_3=\begin{bmatrix}0\\0\\1\\0\\\vdots\\0\end{bmatrix},\quad \dots,\quad \mathsf{e}_n=\begin{bmatrix}0\\0\\0\\\vdots\\0\\1\end{bmatrix}.$$

Thus, in R^2, $\mathsf{e}_1=\begin{bmatrix}1\\0\end{bmatrix}$ and $\mathsf{e}_2=\begin{bmatrix}0\\1\end{bmatrix}$ are the vectors we labelled i and j just above, and in R^3, $\mathsf{e}_1=\begin{bmatrix}1\\0\\0\end{bmatrix}$, $\mathsf{e}_2=\begin{bmatrix}0\\1\\0\end{bmatrix}$ and $\mathsf{e}_3=\begin{bmatrix}0\\0\\1\end{bmatrix}$ are also denoted i, j, k, respectively.

Just as every 2-dimensional vector is a linear combination of i and j,

1.30 Every vector in R^n is a linear combination of the standard basis vectors: $\begin{bmatrix}x_1\\x_2\\\vdots\\x_n\end{bmatrix}=x_1\mathsf{e}_1+x_2\mathsf{e}_2+\cdots+x_n\mathsf{e}_n$.

And the argument for R^n is the same as with R^2. If $x_1,x_2,\dots,x_n$ are any scalars,

$$x_1\mathsf{e}_1=\begin{bmatrix}x_1\\0\\0\\\vdots\\0\end{bmatrix},\quad x_2\mathsf{e}_2=\begin{bmatrix}0\\x_2\\0\\\vdots\\0\end{bmatrix},\quad \dots,\quad x_n\mathsf{e}_n=\begin{bmatrix}0\\0\\\vdots\\0\\x_n\end{bmatrix},$$

so if we add, $x_1\mathsf{e}_1 + x_2\mathsf{e}_2 + \cdots + x_n\mathsf{e}_n = \begin{bmatrix} x_1 \\ x_2 \\ \vdots \\ x_n \end{bmatrix}$.

1.31 Examples.

- $\begin{bmatrix} 3 \\ -4 \\ 5 \end{bmatrix} = 3\begin{bmatrix} 1 \\ 0 \\ 0 \end{bmatrix} - 4\begin{bmatrix} 0 \\ 1 \\ 0 \end{bmatrix} + 5\begin{bmatrix} 0 \\ 0 \\ 1 \end{bmatrix} = 3\mathsf{i} - 4\mathsf{j} + 5\mathsf{k}$;

- $\begin{bmatrix} -2 \\ 1 \\ 5 \\ -9 \end{bmatrix} = -2\begin{bmatrix} 1 \\ 0 \\ 0 \\ 0 \end{bmatrix} + \begin{bmatrix} 0 \\ 1 \\ 0 \\ 0 \end{bmatrix} + 5\begin{bmatrix} 0 \\ 0 \\ 1 \\ 0 \end{bmatrix} - 9\begin{bmatrix} 0 \\ 0 \\ 0 \\ 1 \end{bmatrix} = -2\mathsf{e}_1 + \mathsf{e}_2 + 5\mathsf{e}_3 - 9\mathsf{e}_4$.

True/False Questions for Week 1

Decide, with as little calculation as possible, whether each of the following statements is true or false and explain your answer whenever you say "false."

1. There exist numbers a and b so that $\begin{bmatrix} a+1 \\ 2b \\ a-1 \end{bmatrix} = \begin{bmatrix} 1 \\ 2 \\ 3 \end{bmatrix}$.
2. If $A = (1,2)$ and $B = (-3,5)$, the vector $\overrightarrow{AB} = \begin{bmatrix} 4 \\ -3 \end{bmatrix}$.
3. A vector is an arrow.
4. $\mathsf{u} = \mathsf{v}$
5. $\mathsf{u} = \mathsf{w}$

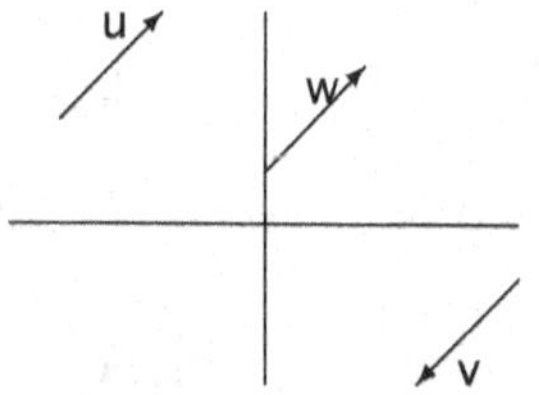

6. The zero vector is a scalar multiple of $\begin{bmatrix} 1 \\ 2 \\ 3 \end{bmatrix}$.
7. Vectors u and v are parallel if $\mathsf{u} = c\mathsf{v}$ for some scalar c.
8. The vectors $\begin{bmatrix} 3 \\ -2 \end{bmatrix}$ and $\begin{bmatrix} 1 \\ -\frac{2}{3} \end{bmatrix}$ are parallel and have the same direction.
9. Associativity of vector addition is the statement $\mathsf{u} + \mathsf{v} = \mathsf{v} + \mathsf{u}$.
10. If vectors u and v are pictured by arrows with the same tail, then $\mathsf{u} - \mathsf{v}$ can be pictured by the arrow from the head of u to the head of v.
11. $\begin{bmatrix} 1 \\ 2 \\ 3 \end{bmatrix}$ is a linear combination of $\begin{bmatrix} 1 \\ 0 \\ 1 \end{bmatrix}$ and $\begin{bmatrix} 2 \\ 0 \\ -1 \end{bmatrix}$.
12. $4\mathsf{u} - 9\mathsf{v}$ is a linear combination of $2\mathsf{u}$ and $7\mathsf{v}$.

13. If a vector u is the sum of vectors v and w, then w is a linear combination of u and v.

14. Every 2-dimensional vector is a linear combination of $\mathsf{u} = \begin{bmatrix} -1 \\ 2 \end{bmatrix}$ and $\mathsf{v} = \begin{bmatrix} 3 \\ 4 \end{bmatrix}$.

15. If c is a scalar and v is a vector, the length of $c\mathsf{v}$ is c times the length of v.

16. If u and v are vectors that are not parallel, then any vector w is a linear combination of u and v.

17. Every vector in R^3 is a linear combination of i, j, and k.

Week 1 Test Yourself

Here are a few problems with short answers that you can use to test your understanding of the concepts you have met this week.

1. What is the name given to the numbers 1, 2, 3 in the vector $\begin{bmatrix} 1 \\ 2 \\ 3 \end{bmatrix}$?

2. Given $A = (-2, 1)$ and $B = (1, 4)$, find the vector $\overrightarrow{AB}$ and illustrate with a picture.

3. Find B, given $A = (1, 4)$ and $\overrightarrow{AB} = \begin{bmatrix} -1 \\ 2 \end{bmatrix}$.

4. If possible, express $\mathsf{x} = \begin{bmatrix} -2 \\ 7 \\ 4 \end{bmatrix}$ as a scalar multiple of $\mathsf{u} = \begin{bmatrix} 8 \\ -28 \\ -16 \end{bmatrix}$, as a scalar multiple of $\mathsf{v} = \begin{bmatrix} 0 \\ 0 \\ 0 \end{bmatrix}$, and as a scalar multiple of $\mathsf{w} = \begin{bmatrix} \frac{2}{7} \\ -1 \\ -\frac{4}{7} \end{bmatrix}$. Justify your answers.

5. What is the meaning of *parallel vectors*?

6. What does it mean to say the u and v have *opposite direction*?

7. What does vector *closure under addition* mean?

8. Find $3\begin{bmatrix} 2 \\ 1 \\ 3 \end{bmatrix} - 2\begin{bmatrix} 1 \\ 0 \\ -5 \end{bmatrix} - 4\begin{bmatrix} 0 \\ -1 \\ 2 \end{bmatrix}$.

9. Express $\begin{bmatrix} 7 \\ 7 \end{bmatrix}$ as a linear combination of $\begin{bmatrix} 2 \\ 6 \end{bmatrix}$ and $\begin{bmatrix} 14 \\ -8 \end{bmatrix}$.

10. Let $A = (1, 2)$, $B = (4, -3)$ and $C = (-1, -2)$ be three points in the plane. Find a fourth point D such that A, B, C and D are the vertices of a parallelogram and justify your answer. Is the answer unique?

11. Show that any linear combination of $\begin{bmatrix} 1 \\ \frac{3}{2} \\ 0 \end{bmatrix}$ and $\begin{bmatrix} 0 \\ 3 \\ 6 \end{bmatrix}$ is also a linear combination of $\begin{bmatrix} 2 \\ 3 \\ 0 \end{bmatrix}$ and $\begin{bmatrix} 0 \\ 1 \\ 2 \end{bmatrix}$.

12. What are the standard basis vectors in R^3?

13. Write the vector $\begin{bmatrix} 1 \\ 3 \\ 2 \\ 4 \end{bmatrix}$ as a linear combination of the standard basis vectors in R^4.

Week 2

Length and Angles

How long is an arrow that represents the two-dimensional vector $\mathsf{v} = \begin{bmatrix} a \\ b \end{bmatrix}$? If we represent v as the arrow from $(0,0)$ to (a,b), the Theorem of Pythagoras gives this length to be $\sqrt{a^2+b^2}$. See Figure 6. We use the notation $\|\mathsf{v}\|$ for the length of v.

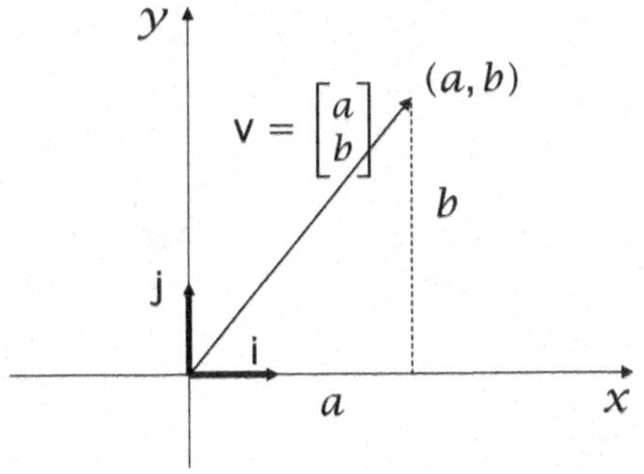

Figure 6: The length of v is $\|\mathsf{v}\| = \sqrt{a^2+b^2}$.

2.1 Definition. The *length* of the vector $\mathsf{v} = \begin{bmatrix} a \\ b \end{bmatrix}$ is the number $\|\mathsf{v}\| = \sqrt{a^2+b^2}$. The *length* of the vector $\mathsf{v} = \begin{bmatrix} a \\ b \\ c \end{bmatrix}$ in R^3 is $\sqrt{a^2+b^2+c^2}$. Most generally, for any $n \geq 1$, the *length* of the vector $\mathsf{x} = \begin{bmatrix} x_1 \\ x_2 \\ \vdots \\ x_n \end{bmatrix}$ is the number $\|\mathsf{x}\| = \sqrt{x_1^2 + x_2^2 + \cdots + x_n^2}$.

Notice again how we extend "reality" in R^2 or R^3 to higher dimensions with a mere definition.

2.2 Examples. · If $\mathsf{u} = \begin{bmatrix} 3 \\ -4 \end{bmatrix}$, then $\|\mathsf{u}\| = \sqrt{3^2 + (-4)^2} = 5$.

· $\left\| \begin{bmatrix} -1 \\ 2 \\ 1 \end{bmatrix} \right\| = \sqrt{6}$

· If $\mathsf{u} = \begin{bmatrix} 1 \\ 2 \\ 3 \\ -4 \end{bmatrix}$, then $\|\mathsf{u}\| = \sqrt{1^2 + 2^2 + 3^2 + (-4)^2} = \sqrt{30}$. ⌣̈

Suppose $\mathsf{x} = \begin{bmatrix} x_1 \\ x_2 \\ \vdots \\ x_n \end{bmatrix}$ and k is a scalar. Then $k\mathsf{x} = \begin{bmatrix} kx_1 \\ kx_2 \\ \vdots \\ kx_n \end{bmatrix}$, so

$$\begin{aligned}\|k\mathsf{x}\| &= \sqrt{(kx_1)^2 + (kx_2)^2 + \cdots + (kx_n)^2} \\ &= \sqrt{k^2(x_1^2 + x_2^2 + \cdots + x_n^2)} = \sqrt{k^2}\sqrt{x_1^2 + x_2^2 + \cdots + x_n^2} = |k|\,\|\mathsf{x}\|.\end{aligned}$$

2.3 $$\boxed{\|k\mathsf{x}\| = |k|\,\|\mathsf{x}\|.}$$

Equation **2.3** is just an algebraic way to express what we have already observed pictorially in R^2 or R^3—see Figure 2. The length of $2\mathsf{x}$ is twice the length of x; the length of $-\frac{1}{2}\mathsf{v}$ is one half the length of x. The length of $k\mathsf{x}$ is $|k|$ times the length of x.

2.4 Definition. A *unit vector* is a vector of length 1.

2.5 Examples. · Each of the standard basis vectors in R^n has length 1. For example, in R^3, $\mathsf{i} = \begin{bmatrix} 1 \\ 0 \\ 0 \end{bmatrix}$, so $\|\mathsf{i}\| = \sqrt{1^2 + 0^2 + 0^2} = \sqrt{1} = 1$.

· The vector $\begin{bmatrix} \frac{1}{\sqrt{2}} \\ \frac{1}{\sqrt{2}} \end{bmatrix}$ is a unit vector in the plane and $\begin{bmatrix} \frac{1}{\sqrt{6}} \\ -\frac{2}{\sqrt{6}} \\ \frac{1}{\sqrt{6}} \end{bmatrix}$ is a unit vector in 3-space.

:)

Manufacturing Unit Vectors

We can use **2.3** to make unit vectors because, if x is any vector (other than 0) and we put $k = \frac{1}{\|\mathsf{x}\|}$, then

$$\|k\mathsf{x}\| = |k|\,\|\mathsf{x}\| = \frac{1}{\|\mathsf{x}\|}\,\|\mathsf{x}\| = 1.$$

Thus $\frac{1}{\|\mathsf{x}\|}\mathsf{x}$ is a unit vector, and it has the same direction as x since it is a **positive** scalar multiple of x.

2.6 For any x, the vector $\frac{1}{\|\mathsf{x}\|}\mathsf{x}$ is a unit vector in the direction of x.

2.7 Problem. Find a vector of length 2 with the same direction as $\mathsf{x} = \begin{bmatrix} 2 \\ -1 \\ 1 \\ 3 \end{bmatrix}$ and a vector of length 3 with direction opposite x.

Solution. The length of x is $\sqrt{4+1+1+9} = \sqrt{14}$, so $\frac{1}{\sqrt{14}}\mathsf{x}$ is a unit vector with the direction of x. Multiplying by 2, we have $\frac{2}{\sqrt{14}}\mathsf{x}$ a vector of length 2 with the direction of x. Multiplying by 3 gives a vector three times as long, and changing sign changes direction. Thus $-\frac{3}{\sqrt{14}}\mathsf{x}$ is a vector of length 3 with direction opposite x.

Dot Product and the Angle Between Vectors

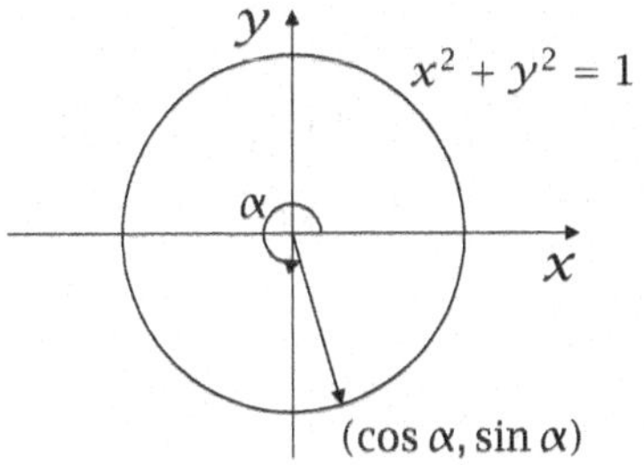

Let u be a unit vector. If u is pictured by an arrow with tail at the origin $(0,0)$, then the head of u lies on the unit circle. Since any point on the unit circle has coordinates of the form $(\cos\alpha, \sin\alpha)$ it follows that $\mathsf{u} = \begin{bmatrix} \cos\alpha \\ \sin\alpha \end{bmatrix}$ for some angle α.

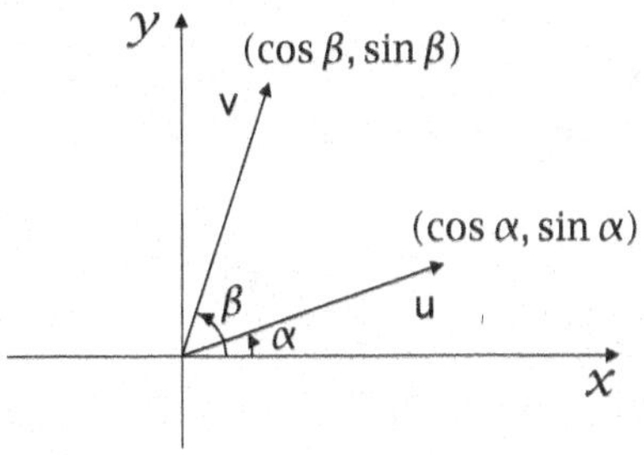

Suppose we have two unit vectors, say $\mathsf{u} = \begin{bmatrix} \cos\alpha \\ \sin\alpha \end{bmatrix}$ and $\mathsf{v} = \begin{bmatrix} \cos\beta \\ \sin\beta \end{bmatrix}$, and suppose too that $0 < \alpha < \beta < \frac{\pi}{2}$. The angle between u and v is $\beta - \alpha$. Now I hope you remember one of the so-called **addition formulas** in trigonometry:
$\cos(\beta - \alpha) = \cos\beta\cos\alpha + \sin\beta\sin\alpha$.

The right hand side of this formula is just the product of the first coordinates of u and v added to the product of the second coordinates of u and v. Letting $\theta = \beta - \alpha$ be the angle between $\mathsf{u} = \begin{bmatrix} a_1 \\ a_2 \end{bmatrix}$ and $\mathsf{v} = \begin{bmatrix} b_1 \\ b_2 \end{bmatrix}$, we have shown that

$$\cos\theta = a_1 b_1 + a_2 b_2$$

(assuming $\|\mathsf{u}\| = 1$ and $\|\mathsf{v}\| = 1$).
The expression on the right has a special name: it's called the **dot product** of u and v.

2.8 Definition. The *dot product* of $\mathsf{u} = \begin{bmatrix} a_1 \\ a_2 \end{bmatrix}$ and $\mathsf{v} = \begin{bmatrix} b_1 \\ b_2 \end{bmatrix}$ is the number $a_1 b_1 + a_2 b_2$. It is denoted $\mathsf{u} \cdot \mathsf{v}$. Thus $\mathsf{u} \cdot \mathsf{v} = a_1 b_1 + a_2 b_2$. More generally, for any $n \geq 1$, the *dot product* of

$\mathsf{u} = \begin{bmatrix} a_1 \\ a_2 \\ \vdots \\ a_n \end{bmatrix}$ and $\mathsf{v} = \begin{bmatrix} b_1 \\ b_2 \\ \vdots \\ b_n \end{bmatrix}$ is the number

$$\boxed{\mathsf{u}\cdot\mathsf{v} = a_1b_1 + a_2b_2 + \cdots + a_nb_n.}$$

2.9 Examples. · If $\mathsf{u} = \begin{bmatrix} -1 \\ 2 \end{bmatrix}$ and $\mathsf{v} = \begin{bmatrix} 4 \\ -3 \end{bmatrix}$, then $\mathsf{u}\cdot\mathsf{v} = -1(4) + 2(-3) = -10$.

· If $\mathsf{u} = \begin{bmatrix} -1 \\ 0 \\ 2 \end{bmatrix}$ and $\mathsf{v} = \begin{bmatrix} 1 \\ 2 \\ 3 \end{bmatrix}$, then $\mathsf{u}\cdot\mathsf{v} = -1(1) + 0(2) + 2(3) = 5$. :)

The dot product is useful in a number of ways. First note that if $\mathsf{u} = \begin{bmatrix} a_1 \\ a_2 \\ \vdots \\ a_n \end{bmatrix}$, then $\mathsf{u}\cdot\mathsf{u} = \begin{bmatrix} a_1 \\ a_2 \\ \vdots \\ a_n \end{bmatrix} \cdot \begin{bmatrix} a_1 \\ a_2 \\ \vdots \\ a_n \end{bmatrix} = a_1^2 + a_2^2 + \cdots + a_n^2$ is precisely square of the length of u.

2.10

$$\boxed{\|\mathsf{u}\|^2 = \mathsf{u}\cdot\mathsf{u}}$$

Secondly, if u and v are unit vectors in R^2 (or indeed, in R^3), then $\mathsf{u}\cdot\mathsf{v}$ is the cosine of the angle between u and v.

The vectors $\mathsf{u} = \begin{bmatrix} \frac{1}{\sqrt{2}} \\ \frac{1}{\sqrt{2}} \end{bmatrix}$ and $\mathsf{v} = \begin{bmatrix} -\frac{3}{5} \\ -\frac{4}{5} \end{bmatrix}$ are unit vectors, so, denoting by θ the angle between them, we have $\cos\theta = \mathsf{u}\cdot\mathsf{v} = -\frac{3}{5\sqrt{2}} - \frac{4}{5\sqrt{2}} = -\frac{6}{5\sqrt{2}} \approx -0.849$, so $\theta \approx 2.584$ rads $\approx 148.05°$.

What if u and v are not unit vectors? What is the angle between u and v in this case? Here we make a useful observation. The angle between u and v has nothing to do with their lengths: the angle between them is the same as the angle between unit vectors that have the same directions as u and v. Remember that $\frac{1}{\|\mathsf{u}\|}\mathsf{u}$ and $\frac{1}{\|\mathsf{v}\|}\mathsf{v}$ are just such vectors—see **2.6**. Thus

$$\cos\theta = \frac{1}{\|\mathsf{u}\|}\mathsf{u}\cdot\frac{1}{\|\mathsf{v}\|}\mathsf{v} = \frac{1}{\|\mathsf{u}\|\|\mathsf{v}\|}\mathsf{u}\cdot\mathsf{v}. \tag{1}$$

This is worth repeating!

2.11

$$\boxed{\cos\theta = \frac{\mathsf{u}\cdot\mathsf{v}}{\|\mathsf{u}\|\,\|\mathsf{v}\|}}$$

I think this is an amazing fact and certainly not obvious.

Remark. By the way, we have been referring to the "angle between vectors." This is somewhat ambiguous as shown in the figure. Is the angle between the vectors θ or is it $2\pi - \theta$? To avoid this difficulty, by general agreement, we mean the smaller angle. The angle between vectors is always an angle θ with $0 \le \theta \le \pi$.

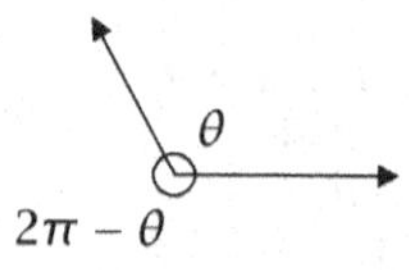

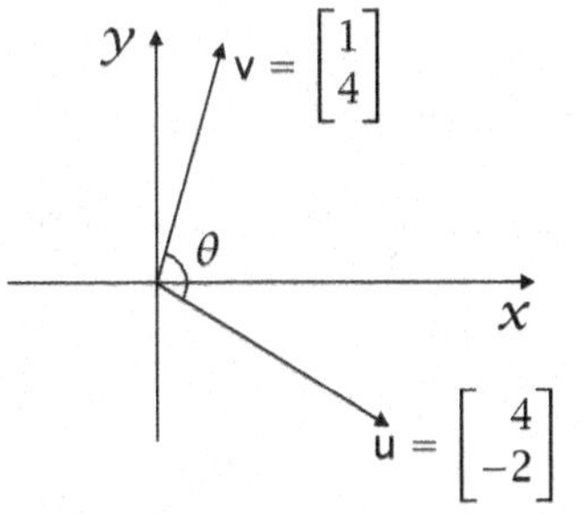

Suppose $u = \begin{bmatrix} 4 \\ -2 \end{bmatrix}$ and $v = \begin{bmatrix} 1 \\ 4 \end{bmatrix}$ and, as a student in Grade 11, you are asked to find the angle between (arrows representing) these vectors. Can you think of some geometrical construction that might help?

But now we have a formula! The cosine of the angle between these vectors is

$$\cos\theta = \frac{u \cdot v}{\|u\|\,\|v\|} = \frac{-4}{\sqrt{20}\sqrt{17}} = \frac{-4}{\sqrt{340}} \approx -0.2169.$$

"$\cos^{-1}$" is common notation for the **inverse** of the cosine (**arccos** is also used) which is defined by the assertion that

$$\cos^{-1} x = y \quad \text{means the same as} \quad x = \cos y.$$

Most calculators have a $\cos^{-1}$ key. Thus, $\cos\theta \approx -0.2169$ means $\theta \approx \cos^{-1}(-.2169) \approx 1.7895$ radians $= 1.7895\left(\frac{180}{\pi}\right)^\circ \approx 103^\circ$. Isn't this cool?

While we have only suggested a proof of the formula for $\cos\theta$ in R^2, the same holds true in R^3.

2.12 Example. Suppose $u = \begin{bmatrix} 1 \\ 2 \\ 3 \end{bmatrix}$ and $v = \begin{bmatrix} -2 \\ 1 \\ -1 \end{bmatrix}$. Then

$$\|u\| = \sqrt{1^2 + 2^2 + 3^2} = \sqrt{14}, \qquad \|v\| = \sqrt{(-2)^2 + 1^2 + (-1)^2} = \sqrt{6}$$

and $u \cdot v = 1(-2) + 2(1) + 3(-1) = -3$. So $-3 = \sqrt{14}\sqrt{6}\cos\theta = \sqrt{84}\cos\theta$ and hence $\cos\theta = \frac{-3}{\sqrt{84}} \approx -0.327$ and $\theta \approx \cos^{-1}(-.327) = 1.904$ radians $\approx 109^\circ$. ☺

Vectors u and v are perpendicular (in linear algebra, "orthogonal" is commonly used) if the angle between them is $\theta = \frac{\pi}{2}$ radians. Look carefully at the formula for $\cos\theta$ in **2.11** and remember that a fraction can only be 0 if the numerator is 0. This leads to a definition.

2.13 Definition. In any Euclidean space, vectors u and v are said to be *orthogonal* if and only if $u \cdot v = 0$.

This is another case of "reality versus definition." In R^2, where we have a geometric concept of "orthogonal," arrows representing u and v really are perpendicular if and only if $u \cdot v = 0$. In higher dimensions, we simply **define** orthogonality using the dot product, as in 2.13: $u \cdot v = 0$.

2.14 u and v are orthogonal if and only if $u \cdot v = 0$.

2.15 Examples. · The vectors $\mathsf{i} = \begin{bmatrix} 1 \\ 0 \end{bmatrix}$ and $\mathsf{j} = \begin{bmatrix} 0 \\ 1 \end{bmatrix}$ are orthogonal because $\mathsf{i} \cdot \mathsf{j} = 1(0) + 0(1) = 0$.

· The vectors $\mathsf{u} = \begin{bmatrix} 3 \\ 4 \end{bmatrix}$ and $\mathsf{v} = \begin{bmatrix} -4 \\ 3 \end{bmatrix}$ are orthogonal because $\mathsf{u} \cdot \mathsf{v} = 3(-4) + 4(3) = 0$.

· The vectors $\mathsf{u} = \begin{bmatrix} -1 \\ 3 \\ 5 \\ 2 \end{bmatrix}$ and $\mathsf{v} = \begin{bmatrix} 2 \\ -1 \\ 3 \\ -5 \end{bmatrix}$ are orthogonal because $\mathsf{u} \cdot \mathsf{v} = (-1)2 + 3(-1) + 5(3) + 2(-5) = 0$.

· The zero vector in R^n is orthogonal to any vector $\mathsf{x} = \begin{bmatrix} x_1 \\ x_2 \\ \vdots \\ x_n \end{bmatrix}$ because $\mathsf{x} \cdot \begin{bmatrix} 0 \\ 0 \\ \vdots \\ 0 \end{bmatrix} = x_1(0) + x_2(0) + \cdots + x_n(0) = 0$.

We summarize the basic properties of the dot product, again urging you to pay particular attention to names. We describe the distributive property $\mathsf{u} \cdot (\mathsf{v} + \mathsf{w})$ by saying that the dot product **distributes over** addition. "Scalar associativity" should remind you of ordinary associativity of numbers: $(xy)z = x(yz)$.

Properties of the Dot Product

Let u, v, and w be vectors. Let c be a scalar.

1. (Commutativity) $\mathsf{u} \cdot \mathsf{v} = \mathsf{v} \cdot \mathsf{u}$.
2. (Distributivity) $\mathsf{u} \cdot (\mathsf{v} + \mathsf{w}) = \mathsf{u} \cdot \mathsf{v} + \mathsf{u} \cdot \mathsf{w}$ and $(\mathsf{u} + \mathsf{v}) \cdot \mathsf{w} = \mathsf{u} \cdot \mathsf{w} + \mathsf{v} \cdot \mathsf{w}$.
3. (Scalar associativity) $(c\mathsf{u}) \cdot \mathsf{v} = c(\mathsf{u} \cdot \mathsf{v}) = \mathsf{u} \cdot (c\mathsf{v})$.

We illustrate all these properties with a single example.

2.16 Example. $(2\mathsf{u} + 3\mathsf{v}) \cdot (\mathsf{u} + 4\mathsf{v})$

$$
\begin{aligned}
&= (2\mathsf{u} + 3\mathsf{v}) \cdot \mathsf{u} + (2\mathsf{u} + 3\mathsf{v}) \cdot (4\mathsf{v}) && \text{distributivity} \\
&= (2\mathsf{u}) \cdot \mathsf{u} + (3\mathsf{v}) \cdot \mathsf{u} + (2\mathsf{u}) \cdot (4\mathsf{v}) + (3\mathsf{v}) \cdot (4\mathsf{v}) && \text{distributivity again} \\
&= 2(\mathsf{u} \cdot \mathsf{u}) + 3(\mathsf{v} \cdot \mathsf{u}) + 2(4)(\mathsf{u} \cdot \mathsf{v}) + 3(4)(\mathsf{v} \cdot \mathsf{v}) && \text{scalar associativity} \\
&= 2(\mathsf{u} \cdot \mathsf{u}) + 3(\mathsf{u} \cdot \mathsf{v}) + 8(\mathsf{u} \cdot \mathsf{v}) + 12(\mathsf{v} \cdot \mathsf{v}) && \text{commutativity} \\
&= 2\,\|\mathsf{u}\|^2 + 11\mathsf{u} \cdot \mathsf{v} + 12\,\|\mathsf{v}\|^2 && \mathsf{x} \cdot \mathsf{x} = \|\mathsf{x}\|^2.
\end{aligned}
$$

Notice how this computation resembles the calculation of the product $(2x + 3y)(x + 4y) = 2x^2 + 11xy + 12y^2$.

Some Inequalities

I have tried to convince you that if u and v are vectors in the plane, then the cosine of the angle between arrows for these vectors is $\frac{\mathbf{u}\cdot\mathbf{v}}{\|\mathbf{u}\|\|\mathbf{v}\|}$:

$$\cos\theta = \frac{\mathbf{u}\cdot\mathbf{v}}{\|\mathbf{u}\|\,\|\mathbf{v}\|}.$$

I would like to use this very formula to extend the definition of angle from the plane to $\mathbf{R}^n$, for any n. In order to do this, we need to know that the fraction on the right is always between -1 and 1 because if it could ever be $\frac{4}{3}$, say, then it sure isn't the cosine of any angle I know!

2.17 Example. Suppose $\mathbf{u} = \begin{bmatrix}-1\\0\\2\\1\end{bmatrix}$ and $\mathbf{v} = \begin{bmatrix}2\\-1\\1\\-4\end{bmatrix}$. Then $\|\mathbf{u}\| = \sqrt{1+0+4+1} = \sqrt{6}$, $\|\mathbf{v}\| = \sqrt{4+1+1+16} = \sqrt{22}$ and $\mathbf{u}\cdot\mathbf{v} = -2+0+2-4 = -4$. Notice that $\frac{\mathbf{u}\cdot\mathbf{v}}{\|\mathbf{u}\|\|\mathbf{v}\|} = \frac{-4}{\sqrt{6}\sqrt{22}} = \frac{-4}{\sqrt{132}} \approx -0.348$. This is certainly between -1 and 1 and it isn't luck! ☺

A famous inequality called the *Cauchy-Schwarz Inequality* usually appears in the form

2.18 $$\boxed{|\mathbf{u}\cdot\mathbf{v}| \leq \|\mathbf{u}\|\,\|\mathbf{v}\|}$$ **Cauchy-Schwarz Inequality.**

The proof is in my book if you are interested. What is important, however, is that it follows from this important inequality that

$$-1 \leq \frac{\mathbf{u}\cdot\mathbf{v}}{\|\mathbf{u}\|\,\|\mathbf{v}\|} \leq 1$$

so that the fraction in the middle is indeed a cosine.

It's "reality versus definition" again. In $\mathbf{R}^2$ or $\mathbf{R}^3$ where we know what angle means, the cosine of the angle between arrows for vectors u and v is $\frac{\mathbf{u}\cdot\mathbf{v}}{\|\mathbf{u}\|\,\|\mathbf{v}\|}$. In higher dimensions, we take this simply as the definition of angle, a definition which is justified because of the Cauchy-Schwarz Inequality. We summarize

2.19 $$\boxed{\cos\theta = \frac{\mathbf{u}\cdot\mathbf{v}}{\|\mathbf{u}\|\,\|\mathbf{v}\|}}$$

A formula for all dimensions!

2.20 Problem. Verify the Cauchy-Schwarz inequality for the vectors $\mathbf{u} = \begin{bmatrix}-1\\1\\2\\1\end{bmatrix}$ and $\mathbf{v} = \begin{bmatrix}2\\0\\1\\-3\end{bmatrix}$.

What's the angle between these two vectors?

Solution. We have $\|\mathsf{u}\| = \sqrt{1+1+4+1} = \sqrt{7}$, $\|\mathsf{v}\| = \sqrt{4+0+1+9} = \sqrt{14}$ and $\mathsf{u}\cdot\mathsf{v} = -2+0+2-3 = -3$. The Cauchy-Schwarz inequality says that $|\mathsf{u}\cdot\mathsf{v}| \le \|\mathsf{u}\|\,\|\mathsf{v}\|$; that is, $|-3| \le \sqrt{7}\sqrt{14}$, which is certainly true: $3 \le \sqrt{98}$. The cosine of the angle between u and v is $\cos\theta = \dfrac{\mathsf{u}\cdot\mathsf{v}}{\|\mathsf{u}\|\,\|\mathsf{v}\|} = \dfrac{-3}{\sqrt{98}} \approx -0.303$, so $\theta = \cos^{-1}(-0.303) \approx 1.263$ rads $\approx 72.4°$.

The Cauchy-Schwarz inequality implies a number of other inequalities.

2.21 Example. Given real numbers a, b, c, d, let $\mathsf{u} = \begin{bmatrix} a \\ c \end{bmatrix}$ and $\mathsf{v} = \begin{bmatrix} b \\ d \end{bmatrix}$. Then $\mathsf{u}\cdot\mathsf{v} = ab + cd$, $\|\mathsf{u}\| = \sqrt{a^2+c^2}$ and $\|\mathsf{v}\| = \sqrt{b^2+d^2}$. The Cauchy-Schwarz inequality gives $|ab+cd| \le \sqrt{a^2+c^2}\sqrt{b^2+d^2}$. For example, $|2(3)+4(5)| \le sqrt4+16\sqrt{9+25}$, that is, $26 \le \sqrt{26.8}$. Close, eh?

Suppose a and b are nonnegative real numbers and we take $\mathsf{u} = \begin{bmatrix} \sqrt{a} \\ \sqrt{b} \end{bmatrix}$ and $\mathsf{v} = \begin{bmatrix} \sqrt{b} \\ \sqrt{a} \end{bmatrix}$. Then $\mathsf{u}\cdot\mathsf{v} = \sqrt{a}\sqrt{b} + \sqrt{a}\sqrt{b} = 2\sqrt{ab}$ while $\|\mathsf{u}\| = \|\mathsf{v}\| = \sqrt{a+b}$. The Cauchy-Schwarz inequality says that

$$|2\sqrt{ab}| \le \sqrt{a+b}\sqrt{a+b} = a+b,$$

which produces another useful inequality (with a name).

2.22 Arithmetic Mean–Geometric Mean Inequality

$$\boxed{\sqrt{ab} \le \frac{a+b}{2} \text{ for any } a, b \ge 0.}$$

The expression $\frac{a+b}{2}$ on the right-hand side is the average or **arithmetic mean** of a and b. The expression on the left, $\sqrt{ab}$, is their **geometric mean**. This explains the name for this particular inequality. For example, with $a = 4$, $b = 9$, it confirms that $6 = \sqrt{4(9)} \le \frac{4+9}{2} = 6\frac{1}{2}$. It says that $\sqrt{2(3)} \le \frac{2+3}{2} = 2.5$, and so on.

A glance at the figure to the right shows how the next inequality got its name. The length of any side of a triangle ($\|\mathsf{u}+\mathsf{v}\|$) is never more than the sum of the lengths of the other two sides ($\|\mathsf{u}\| + \|\mathsf{v}\|$).

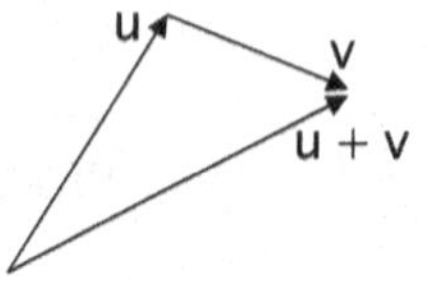

2.23 The Triangle Inequality

$$\boxed{\text{For any vectors } \mathsf{u} \text{ and } \mathsf{v} \text{ in } \mathsf{R}^n,\ \|\mathsf{u}+\mathsf{v}\| \le \|\mathsf{u}\| + \|\mathsf{v}\|.}$$

The proof just uses $\|x\|^2 = x \cdot x$ and basic properties of the dot product.

$$\begin{aligned}\|u+v\|^2 &= (u+v)\cdot(u+v) = u\cdot u + 2u\cdot v + v\cdot v \\ &= \|u\|^2 + 2u\cdot v + \|v\|^2 \qquad (2)\\ &\le \|u\|^2 + 2|u\cdot v| + \|v\|^2\end{aligned}$$

because $a \le |a|$ for any real number a. Now we apply the Cauchy-Schwarz inequality to $|u \cdot v|$ and get

$$\|u\|^2 + 2|u\cdot v| + \|v\|^2 \le \|u\|^2 + 2\,\|u\|\,\|v\| + \|v\|^2.$$

Since the expression on the right is $(\|u\| + \|v\|)^2$, we obtain finally

$$\|u+v\|^2 \le (\|u\| + \|v\|)^2,$$

and the triangle inequality follows by taking square roots. (Each number a that is squared is nonnegative and $\sqrt{a^2} = a$).

At the step we labelled (2), we had $\|u+v\|^2 = \|u\|^2 + 2u\cdot v + \|v\|^2$. Since vectors u and v are orthogonal if and only if $u \cdot v = 0$, the Theorem of Pythagoras is true in any Euclidean space, not just in 3-space. Moreover, so it is converse: If the square of one side of a triangle is the sum of the squares of the other two sides, then that triangle is right-angled.

2.24 Theorem. (Pythagoras) *Vectors* u *and* v *in* R^n *are orthogonal if and only if* $\|u+v\|^2 = \|u\|^2 + \|v\|^2$.

True/False Questions for Week 2

Decide, with as little calculation as possible, whether each of the following statements is true or false and explain your answer whenever you say "false."

1. $\sqrt{4} = \pm 2$.
2. $\sqrt{k^2} = k$.
3. The solution to $x^2 = 4$ is $x = \sqrt{4}$.
4. The solution to $x^2 = 5$ is $x = \pm\sqrt{5}$.
5. The length of $\begin{bmatrix}-1\\2\\2\end{bmatrix}$ is 3.
6. $\begin{bmatrix}1\\1\\1\end{bmatrix}$ is a unit vector.
7. The vector $\begin{bmatrix}\frac{1}{\sqrt{6}}\\-\frac{2}{\sqrt{6}}\\\frac{1}{\sqrt{6}}\end{bmatrix}$ is a unit vector.

8. If v is a nonzero vector, a unit vector with the same direction as v is $\frac{1}{\|\mathsf{v}\|}\mathsf{v}$.

9. The cosine of the angle between vectors u and v is $\mathsf{u}\cdot\mathsf{v}$.

10. If u is a vector and $\mathsf{u}\cdot\mathsf{u}=0$, then $\mathsf{u}=0$.

11. There exist unit vectors u and v with $\mathsf{u}\cdot\mathsf{v}=2$.

12. The vectors $\begin{bmatrix}1\\1\\1\\1\\1\end{bmatrix}$ and $\begin{bmatrix}1\\0\\-1\\-1\\1\end{bmatrix}$ are orthogonal.

13. For vectors u and v, $\|\mathsf{u}+\mathsf{v}\| = \|\mathsf{u}\|+\|\mathsf{v}\|$.

14. For any real numbers a,b,c,d, $|ab+cd| \le \sqrt{a^2+b^2}\sqrt{c^2+d^2}$.

15. For any real numbers a and b, $a+b \le 2\sqrt{ab}$.

16. For any vectors u and v, $\mathsf{u}\cdot\mathsf{v} \le \|\mathsf{u}\|\,\|\mathsf{v}\|$.

Week 2 Test Yourself

Here are a few problems with short answers that you can use to test your understanding of the concepts you have met this week.

1. Show that $\begin{bmatrix}1\\0\end{bmatrix}$ is in the plane spanned by $\mathsf{u}=\begin{bmatrix}1\\1\end{bmatrix}$ and $\mathsf{v}=\begin{bmatrix}-2\\1\end{bmatrix}$.

2. Determine whether $\mathsf{w}=\begin{bmatrix}-18\\5\\9\end{bmatrix}$ is in the plane spanned by $\mathsf{u}=\begin{bmatrix}-1\\0\\3\end{bmatrix}$ and $\mathsf{v}=\begin{bmatrix}3\\-1\\0\end{bmatrix}$.

3. What does it mean to say that the dot product is *commutative*?

4. Let $c=2$, $\mathsf{u}=\begin{bmatrix}1\\2\\3\end{bmatrix}$ and $\mathsf{v}=\begin{bmatrix}-2\\1\\1\end{bmatrix}$. Write down $c\mathsf{u}$ and $(c\mathsf{u})\cdot\mathsf{v}$ and check that this is the same as $c(\mathsf{u}\cdot\mathsf{v})$. What is the name of this property of the dot product?

5. Let $\mathsf{u}=\begin{bmatrix}0\\3\\4\end{bmatrix}$ and $\mathsf{v}=\begin{bmatrix}1\\2\\2\end{bmatrix}$.

   (a) Find $\|\mathsf{u}\|$ and the length of v.

   (b) Find a unit vector in the direction of u and a unit vector in the direction opposite to v.

6. Susie is asked to find the length of the vector $\mathsf{u}=\frac{4}{7}\begin{bmatrix}-2\\1\\5\end{bmatrix}$. She proceeds like this.

$$\mathsf{u}=\begin{bmatrix}-\frac{8}{7}\\ \frac{4}{7}\\ \frac{20}{7}\end{bmatrix},\ \text{so } \|\mathsf{u}\| = \sqrt{(\tfrac{8}{7})^2+(\sqrt{\tfrac{4}{7}})^2+(\tfrac{20}{7})^2} = \sqrt{480/7}.$$

This is correct, but isn't there an easier method?

7. In each case, find $\mathsf{u} \cdot \mathsf{v}$ and the angle between u and v in radians and in degrees.

(a) $\mathsf{u} = \begin{bmatrix} 3 \\ 4 \end{bmatrix}$, $\mathsf{v} = \begin{bmatrix} -1 \\ 1 \end{bmatrix}$ (b) $\mathsf{u} = \begin{bmatrix} 2 \\ -1 \\ 1 \end{bmatrix}$ and $\mathsf{v} = \begin{bmatrix} 1 \\ 1 \\ 2 \end{bmatrix}$

8. Let $\mathsf{u} = \begin{bmatrix} 3 \\ 4 \\ 0 \end{bmatrix}$, $\mathsf{v} = \begin{bmatrix} 2 \\ 1 \\ 2 \end{bmatrix}$, and $\mathsf{w} = \begin{bmatrix} 1 \\ 1 \\ 1 \end{bmatrix}$. Find $\|\mathsf{u}\|$, $\|\mathsf{u} + \mathsf{v}\|$ and $\left\| \frac{\mathsf{w}}{\|\mathsf{w}\|} \right\|$.

9. Let u, v, and w be vectors of lengths 3, 2, and 6, respectively. Suppose $\mathsf{u} \cdot \mathsf{v} = 3$, $\mathsf{u} \cdot \mathsf{w} = 5$ and $\mathsf{v} \cdot \mathsf{w} = 2$. Find $(\mathsf{u} - 2\mathsf{v}) \cdot (3\mathsf{w} + 2\mathsf{u})$.

10. Determine whether or not u and v are orthogonal in each case.

(a) $\mathsf{u} = \begin{bmatrix} -2 \\ 3 \\ 1 \end{bmatrix}$, $\mathsf{v} = \begin{bmatrix} 1 \\ 2 \\ 3 \end{bmatrix}$ (b) $\mathsf{u} = \begin{bmatrix} -1 \\ 0 \\ 2 \\ 5 \end{bmatrix}$, $\mathsf{v} = \begin{bmatrix} 3 \\ 4 \\ -1 \\ 1 \end{bmatrix}$

11. Find the angle between the 4-dimensional vectors $\mathsf{u} = \begin{bmatrix} -4 \\ 0 \\ 2 \\ -2 \end{bmatrix}$ and $\mathsf{v} = \begin{bmatrix} 2 \\ 0 \\ -1 \\ 1 \end{bmatrix}$.

12. Let $\mathsf{u} = \begin{bmatrix} 1 \\ 1 \\ 0 \\ -2 \end{bmatrix}$ and $\mathsf{v} = \begin{bmatrix} -3 \\ 0 \\ 0 \\ 1 \end{bmatrix}$. Find all scalars c with the property that

(a) $\|c\mathsf{u}\| = 8$; (b) u is orthogonal to $\mathsf{u} + c\mathsf{v}$.

13. Let $\mathsf{u} = \begin{bmatrix} 1 \\ 1 \\ 0 \end{bmatrix}$ and $\mathsf{v} = \begin{bmatrix} x \\ 1 \\ 1 \end{bmatrix}$. Find all numbers x (if any) such that the angle between u and v is 60°.

14. Verify the Cauchy-Schwarz and triangle inequalities for each of the following pairs of vectors:

(a) $\mathsf{u} \begin{bmatrix} 1 \\ 0 \\ -2 \end{bmatrix}$, $\mathsf{v} = \begin{bmatrix} -3 \\ 1 \\ 0 \end{bmatrix}$ (b) $\mathsf{u} = \begin{bmatrix} -1 \\ 0 \\ 1 \\ 2 \end{bmatrix}$, $\mathsf{v} = \begin{bmatrix} 3 \\ -1 \\ 1 \\ 0 \end{bmatrix}$

15. Use the Cauchy-Schwarz inequality to prove that $(a \cos\theta + b \sin\theta)^2 \le a^2 + b^2$ for all $a, b \in \mathbf{R}$.

16. Does there exist a triangle with sides of lengths 1, 2, 4? Find a "proof without words."

Week 3

The Equation of a Plane

In this section, we derive the equation of a plane in 3-space. First, let's be sure we know what we mean by the equation of a curve or surface.

3.1

The equation of a blank is an equation satisfied by all points lying on blank, and only those points.

3.2 Examples. · A point (x, y) is on the x-axis in the Euclidean plane if and only if its y-coordinate is 0. Thus $y = 0$ is the equation of the x-axis.

- A point (x, y) in the Euclidean plane is on the circle with centre $(0, 0)$ and radius 1 if and only if the distance from (x, y) to $(0, 0)$ is 1. Since this distance is $\sqrt{x^2 + y^2}$, the point (x, y) is on the circle if and only if $\sqrt{x^2 + y^2} = 1$, which is equivalent to $x^2 + y^2 = 1$. Thus the equation of the unit circle centred at the origin is $x^2 + y^2 = 1$.

☺

In these examples, it is important to note that we were in the Euclidean plane. What is described by the equation $x^2 + y^2 = 1$, for example, if we are in 3-space, where points have three coordinates? Notice that all the points (x, y, z) listed below satisfy $x^2 + y^2 = 1$—

$$(1, 0, 0), (1, 0, 1), (1, 0, 2), (1, 0, 3), (1, 0, 4), \ldots, (1, 0, -1), (1, 0, -2)$$

so they all lie on the graph of this equation (in 3-space). The point $(1, 0)$ lies on the unit circle in the Euclidean plane with equation $x^2 + y^2 = 1$. In 3-space, however, any point directly above or below $(1, 0, 0)$ satisfies the same equation. Are you ready to guess what the graph of this equation in R^3 is? Did you guess **cylinder**? In R^3, the equation $x^2 + y^2 = 1$ is satisfied by precisely those points on the cylinder of radius 1 and central axis the z-axis. It cuts through the xy-plane in the familiar unit circle with centre the origin. See the picture on the left in Figure 7.

What object in 3-space is described by the equation $y = 0$? In 3-space, where points have three coordinates, not only do the points on the x-axis satisfy $y = 0$, but so do all points directly above and below the x-axis. In 3-space, the graph of $y = 0$ is a **plane**, the plane perpendicular to the xy-plane which slices through this plane along the x-axis. See the picture on the right of Figure 7.

We want to discover the nature of the equation of a plane in 3-space, and remember—**3.1**—this means an equation satisfied by the coordinates of any point (x, y, z) on the plane, and only such points.

A key idea in discovering the equation is the concept of **normal**.

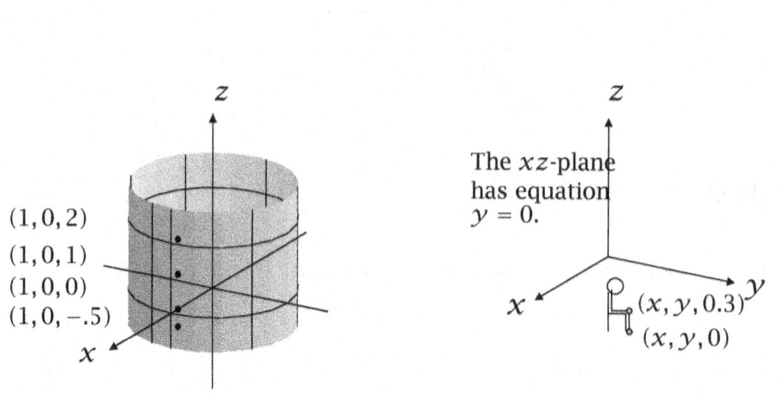

Figure 7: In 3-space, the graph of $x^2 + y^2 = 1$ is a cylinder. The xy-plane has equation $z = 0$. The xz-plane has equation $y = 0$.

3.3 Definition. A *normal* to a plane is a vector orthogonal to a plane.

For example, the vector k is a normal to the xy-plane. It is also normal to any plane parallel to the xy-plane. Specifying a normal to a plane specifies a parallel class of planes and we can identify one of these planes as soon as we specify a point on the plane. For instance, of all the planes with normal k, the xy-plane is the only one containing the origin.

3.4

A plane is uniquely determined by a normal and a point on it.

This idea is the key to finding the equation a plane.

Let $\mathsf{n} = \begin{bmatrix} a \\ b \\ c \end{bmatrix}$ be a normal to a plane π and suppose $P_0(x_0, y_0, z_0)$ is a point on π. The general point $P(x, y, z)$ lies on π if and only if the arrow from P_0 to P is orthogonal to n; that is, if and only if $\overrightarrow{P_0P}$ and n are orthogonal vectors.

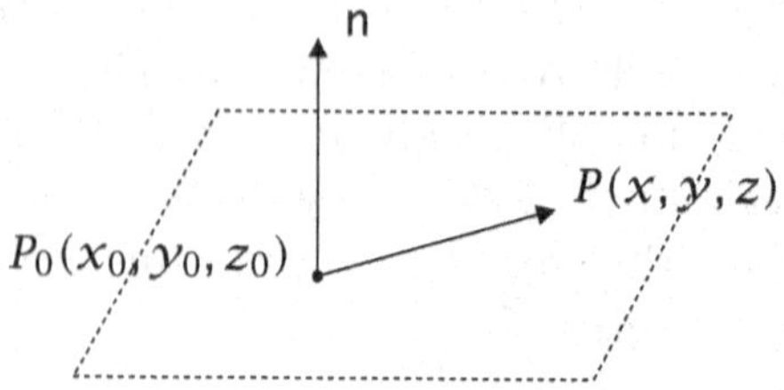

Remembering that vectors are orthogonal if and only if their dot product is 0, we see that (x, y, z) is in the plane if and only if $\begin{bmatrix} x - x_0 \\ y - y_0 \\ z - z_0 \end{bmatrix} \cdot \begin{bmatrix} a \\ b \\ c \end{bmatrix} = 0$; that is, if and only if

$$a(x - x_0) + b(y - y_0) + c(z - z_0) = 0,$$

which is the same as

$$ax + by + cz = ax_0 + by_0 + cz_0.$$

Since x_0, y_0 and z_0 are given numbers (as are a, b and c), the right side of this equation is just a number—call it d. So we have shown that planes in R^3 have equations of the form

$$ax + by + cz = d, \tag{3}$$

the coefficients a, b, c being the components of a normal.

3.5

> In 3-space, planes have equations of the form $ax + by + cz = d$.
> The vector $\begin{bmatrix} a \\ b \\ c \end{bmatrix}$ is a normal to the plane with this equation.

3.6 Examples. · The equation of the xy-plane in R^3 is $z = 0$ because $\mathsf{k} = \begin{bmatrix} 0 \\ 0 \\ 1 \end{bmatrix}$ is a normal (so the equation has the form $0x + 0y + 1z = d$ and $(0, 0, 0)$ is a point on the plane (so $x = 0$, $y = 0$, $z = 0$ satisfy the equation).

· The equation $2x - 3y + 4z = 10$ describes the plane with normal $\begin{bmatrix} 2 \\ -3 \\ 4 \end{bmatrix}$ passing through $(5, 0, 0)$ and $(1, -1, 0)$ and $(\frac{1}{2}, 0, 1)$, and so on. There are many planes with this normal, all parallel. Each has an equation of the form $2x - 3y + 4z = d$. Different values of d correspond to different planes. If the plane passes through $(1, 1, 1)$, for instance, then $x = 1$, $y = 1$, $z = 1$ must satisfy the equation. So $2 - 3 + 4 = d$ and $d = 3$. The equation of the plane with normal $\begin{bmatrix} 2 \\ -3 \\ 4 \end{bmatrix}$ and passing through $(1, 1, 1)$ is $2x - 3y + 4z = 3$. ☺

3.7 Remark. (Vectors in a Plane) In this section, we have talked about **points** lying in planes. It is convenient to be able also to talk about **vectors** lying in planes.

As with points, when we say that a vector $\begin{bmatrix} x \\ y \\ z \end{bmatrix}$ lies in a plane π, we mean that the components of the vector satisfy the equation of the plane, but also, by convention, that **the plane passes through the origin,** so that the arrow from $(0, 0, 0)$ to $P(x, y, z)$ lies in the plane.

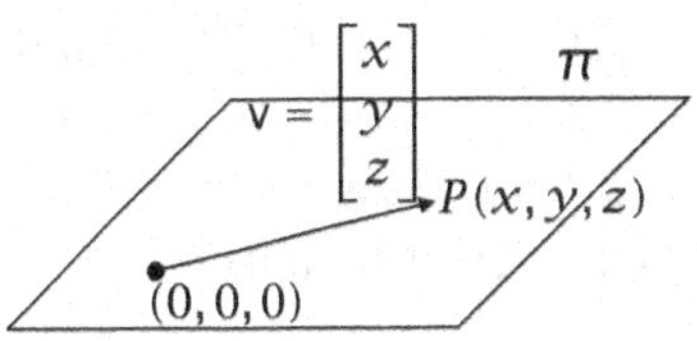

3.8 Example. The vector $v = \begin{bmatrix} -1 \\ 1 \\ -1 \end{bmatrix}$ is in the plane π with equation $x - 4y - 5z = 0$ because both $(0,0,0)$ and $(-1,1,-1)$ are in π: the coordinates of each point satisfy the equation. :)

Remember that the set of all linear combinations of two nonparallel vectors u and v is a plane called the plane "spanned" by u and v and, conversely, a plane in R^3 is spanned by any two nonparallel vectors it contains–see **1.23**.

3.9 Example. Let π be the plane with equation $x - 3y + 5z = 0$. This is the span of $\begin{bmatrix} -1 \\ 3 \\ 2 \end{bmatrix}$ and $\begin{bmatrix} 3 \\ 1 \\ 0 \end{bmatrix}$ because these are nonparallel vectors and each is in π (the plane contains $(0,0,0)$ and the components of each vector satisfy the equation). On p. 11, we gave a geometrical explanation. Here's an algebraic argument that shows that π is the span of two particular nonparallel vectors. Rewrite the equation of π in the form $x = 3y - 5z$. Then $w = \begin{bmatrix} x \\ y \\ z \end{bmatrix}$ is in the plane if and only if $w = \begin{bmatrix} 3y - 5z \\ y \\ z \end{bmatrix} = \begin{bmatrix} 3y \\ y \\ 0 \end{bmatrix} + \begin{bmatrix} -5z \\ 0 \\ z \end{bmatrix} = y \begin{bmatrix} 3 \\ 1 \\ 0 \end{bmatrix} + z \begin{bmatrix} -5 \\ 0 \\ 1 \end{bmatrix}$. So we see that π consists precisely of all linear combinations of (the nonparallel vectors) $\begin{bmatrix} 3 \\ 1 \\ 0 \end{bmatrix}$ and $\begin{bmatrix} -5 \\ 0 \\ 1 \end{bmatrix}$. :)

The Cross Product

We have seen (and just repeated the idea) that the set of all linear combinations of two nonparallel vectors—the span of two nonparallel vectors—is a plane. But what's the equation of such a plane? Say $u = \begin{bmatrix} 1 \\ 3 \\ 2 \end{bmatrix}$ and $v = \begin{bmatrix} 0 \\ 2 \\ -1 \end{bmatrix}$. What's the equation of the plane that they span? Since the plane passes through the origin, we seek an equation of the form $ax + by + cz = 0$. The coefficients a, b, and c are the components of a vector normal to the plane, so the question becomes "How do we find a normal?" The answer is something called "the cross product" which, in order to define, requires the notion of 2×2 (read "two by two") **determinant**.

3.10 Definition. If a, b, c, and d are four numbers, the 2×2 *determinant* $\begin{vmatrix} a & b \\ c & d \end{vmatrix}$ is the number $ad - bc$:

$$\boxed{\begin{vmatrix} a & b \\ c & d \end{vmatrix} = ad - bc.}$$

For example, $\begin{vmatrix} 2 & 1 \\ -4 & 5 \end{vmatrix} = 2(5) - 1(-4) = 14.$

3.11 Definition. The *cross product* of vectors $\mathsf{u} = \begin{bmatrix} u_1 \\ u_2 \\ u_3 \end{bmatrix}$ and $\mathsf{v} = \begin{bmatrix} v_1 \\ v_2 \\ v_3 \end{bmatrix}$ is the vector

$$\mathsf{u} \times \mathsf{v} = \begin{vmatrix} u_2 & u_3 \\ v_2 & v_3 \end{vmatrix} \mathsf{i} \underset{\uparrow}{-} \begin{vmatrix} u_1 & u_3 \\ v_1 & v_3 \end{vmatrix} \mathsf{j} + \begin{vmatrix} u_1 & u_2 \\ v_1 & v_2 \end{vmatrix} \mathsf{k}.$$

As indicated, the cross product of u and v is denoted u × v, read "u cross v." Expanding the three determinants, we obtain

$$\mathsf{u} \times \mathsf{v} = (u_2v_3 - u_3v_2)\mathsf{i} - (u_1v_3 - u_3v_1)\mathsf{j} + (u_1v_2 - u_2v_1)\mathsf{k} = \begin{bmatrix} u_2v_3 - u_3v_2 \\ -(u_1v_3 - u_3v_1) \\ u_1v_2 - u_2v_1 \end{bmatrix}.$$

This is a horrible looking formula which few people remember. Instead, most of us simply remember the following scheme:

3.12

$$\boxed{\mathsf{u} \times \mathsf{v} = \begin{vmatrix} \mathsf{i} & \mathsf{j} & \mathsf{k} \\ u_1 & u_2 & u_3 \\ v_1 & v_2 & v_3 \end{vmatrix}.}$$

Here's how this works.

To find the coefficient of i in a cross product, look at the array $\begin{vmatrix} \mathsf{i} & \mathsf{j} & \mathsf{k} \\ u_1 & u_2 & u_3 \\ v_1 & v_2 & v_3 \end{vmatrix}$, mentally remove the row and column in which i appears (the first row and the first column), and compute the determinant of the four numbers that remain,

$$\begin{vmatrix} u_2 & u_3 \\ v_2 & v_3 \end{vmatrix} = u_2v_3 - u_3v_2.$$

The coefficient of k is obtained by removing the row and column in which k appears (the first row and third column) and computing the determinant of the four numbers that remain,

$$\begin{vmatrix} u_1 & u_2 \\ v_1 & v_2 \end{vmatrix} = u_1v_2 - u_2v_1.$$

The coefficient of j is obtained in a similar way, except for a sign change. Notice the minus sign highlighted by the arrow in 3.11.

Why are we worried about cross products? Remember that in order to find an equation for the plane containing we need a vector $\mathsf{u} = \begin{bmatrix} 1 \\ 3 \\ 2 \end{bmatrix}$ and $\mathsf{v} = \begin{bmatrix} 0 \\ 2 \\ -1 \end{bmatrix}$, we need a vector orthogonal to both u and v. And now we know how to find such a vector. It's the cross product! Let's check. Since

$$(\mathsf{u} \times \mathsf{v}) \cdot \mathsf{u} = \begin{bmatrix} -7 \\ 1 \\ 2 \end{bmatrix} \cdot \begin{bmatrix} 1 \\ 3 \\ 2 \end{bmatrix} = -7 + 3 + 4 = 0$$

and

$$(\mathsf{u} \times \mathsf{v}) \cdot \mathsf{v} = \begin{bmatrix} -7 \\ 1 \\ 2 \end{bmatrix} \cdot \begin{bmatrix} 0 \\ -2 \\ 1 \end{bmatrix} = -2 + 2 = 0,$$

$\begin{bmatrix} -7 \\ 1 \\ 2 \end{bmatrix}$ is orthogonal to u and to v, so, as we can easily prove (next theorem), it's a normal vector to the plane spanned by u and v. The equation of the plane is $7x - y - 2z = 0$.

The cross product of vectors u and v is orthogonal to u and to v, but why should it be orthogonal to the entire plane spanned by u and v? Read on!

3.13 Theorem. *Suppose a plane π is spanned by vectors* u *and* v. *If a vector* n *is orthogonal to* u *and* v, *then* n *is orthogonal to π, that is, orthogonal to every vector in π.*

Proof. Let w be a vector in π. We wish to show that $\mathsf{n} \cdot \mathsf{w} = 0$. We know that $\mathsf{w} = a\mathsf{u} + b\mathsf{v}$ is a linear combination of u and v, so $\mathsf{n} \cdot \mathsf{w} = \mathsf{n} \cdot (a\mathsf{u} + b\mathsf{v}) = a(\mathsf{n} \cdot \mathsf{u}) + b(\mathsf{n} \cdot \mathsf{v}) = 0 + 0 = 0$ because $\mathsf{n} \cdot \mathsf{u} = 0$ and $\mathsf{n} \cdot \mathsf{v} = 0$. ■

The symbol ■ is often used in mathematics to denote the end of a proof.

The cross product was introduced because the equation of a plane requires a normal vector, so it useful to be able to find a vector perpendicular to two given vectors. Theorem 3.13 says that such a vector is normal to the entire span of those vectors.

3.14 Example. To find the equation of the plane through the points $A(1, 3, -2)$, $B(1, 1, 5)$ and $C(2, -2, 3)$. we first find a normal vector to the plane. This will be a vector orthogonal to both $\overrightarrow{AB} = \begin{bmatrix} 0 \\ -2 \\ 7 \end{bmatrix}$ and $\overrightarrow{AC} = \begin{bmatrix} 1 \\ -5 \\ 5 \end{bmatrix}$. The cross product of $\overrightarrow{AB}$ and $\overrightarrow{AC}$ is

$$\overrightarrow{AB} \times \overrightarrow{AC} = \begin{vmatrix} \mathsf{i} & \mathsf{j} & \mathsf{k} \\ 0 & -2 & 7 \\ 1 & -5 & 5 \end{vmatrix} = 25\mathsf{i} - (-7)\mathsf{j} + (-(-2)\mathsf{k} = 25\mathsf{i} + 7\mathsf{j} + 2\mathsf{k} = \begin{bmatrix} 25 \\ 7 \\ 2 \end{bmatrix}.$$

This is a normal to the plane, so the plane has equation $25x + 7y + 2z = d$, where d is determined by the fact that $A(1, 3, -2)$ (or B or C) is on the plane. Thus $25(1) + 7(3) + 2(-2) = d$. So $d = 42$ and the equation of the plane is $25x + 7y + 2z = 42$. ☺

Properties of the Cross Product

If u, v, w are vectors in 3-space and c is a scalar,

1. (Anticommutativity) $\mathsf{v} \times \mathsf{u} = -(\mathsf{u} \times \mathsf{v})$.
2. (Distributivity) $\mathsf{u} \times (\mathsf{v} + \mathsf{w}) = (\mathsf{u} \times \mathsf{v}) + (\mathsf{u} \times \mathsf{w})$.
3. (Scalar associativity) $(c\mathsf{u}) \times \mathsf{v} = c(\mathsf{u} \times \mathsf{v}) = \mathsf{u} \times (c\mathsf{v})$.
4. $\mathsf{u} \times \mathsf{0} = \mathsf{0} = \mathsf{0} \times \mathsf{u}$.
5. $\mathsf{u} \times \mathsf{v}$ is orthogonal to both u and v.
6. $\mathsf{u} \times \mathsf{u} = \mathsf{0}$.
7. $\mathsf{u} \times \mathsf{v} = \mathsf{0}$ if and only if u is a scalar multiple of v or v is a scalar multiple of u, that is, if and only if u and v are parallel.
8. $\|\mathsf{u} \times \mathsf{v}\| = \|\mathsf{u}\|\,\|\mathsf{v}\| \sin\theta$, where θ is the angle between u and v. (Remember that we always assume $0 \le \theta \le \pi$.)
9. $\|\mathsf{u} \times \mathsf{v}\|$ is the area of the parallelogram having u and v as adjacent sides.
10. (No associativity) In general, $(\mathsf{u} \times \mathsf{v}) \times \mathsf{w} \neq \mathsf{u} \times (\mathsf{v} \times \mathsf{w})$

We discuss these properties in order.

First, that really is a minus sign in Property 1. Remember that an operation $\star$ (like vector addition) is *commutative* if $a \star b = b \star a$ for any a and b. Many operations are commutative, such as ordinary addition and multiplication of real numbers and vector addition. Some are not, like subtraction and division on the natural numbers $1, 2, 3, \ldots$. The cross product also is not commutative, but it does have a redeeming quality. Instead of being commutative, the cross product is *anticommutative*: $\mathsf{u} \times \mathsf{v} = -(\mathsf{v} \times \mathsf{u})$. For example, with $\mathsf{u} = \begin{bmatrix} 1 \\ 3 \\ 2 \end{bmatrix}$ and $\mathsf{v} = \begin{bmatrix} 0 \\ 2 \\ -1 \end{bmatrix}$,

$$\mathsf{u} \times \mathsf{v} = \begin{vmatrix} \mathsf{i} & \mathsf{j} & \mathsf{k} \\ 1 & 3 & 2 \\ 0 & 2 & -1 \end{vmatrix} = -7\mathsf{i} - (-1)\mathsf{j} + 2\mathsf{k} = \begin{bmatrix} -7 \\ 1 \\ 2 \end{bmatrix}$$

whereas

$$\mathsf{v} \times \mathsf{u} = \begin{vmatrix} \mathsf{i} & \mathsf{j} & \mathsf{k} \\ 0 & 2 & -1 \\ 1 & 3 & 2 \end{vmatrix} = 7\mathsf{i} - (1)\mathsf{j} + (-2)\mathsf{k} = \begin{bmatrix} 7 \\ -1 \\ -2 \end{bmatrix} = -(\mathsf{u} \times \mathsf{v}).$$

Distributivity we have met before. We say the **distributes over** addition. Scalar associativity should be easy to remember because it reminds us of associativity of ordinary multiplication: $(xy)z = x(yz)$. Both properties are easy to check, straight from the definition in 3.11. The fact that $\mathsf{u} \times \mathsf{0} = \mathsf{0}$ and $\mathsf{0} \times \mathsf{u} = \mathsf{0}$ should be obvious; just imagine a row of 0s in the array 3.12. Property 5, which asserts that $\mathsf{u} \times \mathsf{v}$ is orthogonal to u and to v, involves some calculation, but it's not hard. Let $\mathsf{u} = \begin{bmatrix} u_1 \\ u_2 \\ u_3 \end{bmatrix}$ and $\mathsf{v} = \begin{bmatrix} v_1 \\ v_2 \\ v_3 \end{bmatrix}$. Then $\mathsf{u} \times \mathsf{v} = \begin{bmatrix} u_2v_3 - u_3v_2 \\ u_1v_3 - u_3v_1 \\ u_1v_2 - u_2v_1 \end{bmatrix}$. Now simply check that the dot products of $\mathsf{u} \times \mathsf{v}$ with both u and v are 0.

Here's another example.

3.15 Example. If $\mathsf{u} = \begin{bmatrix} 1 \\ -4 \\ 1 \end{bmatrix}$ and $\mathsf{v} = \begin{bmatrix} 2 \\ 3 \\ 0 \end{bmatrix}$, then $\mathsf{u} \times \mathsf{v} = \begin{vmatrix} \mathsf{i} & \mathsf{j} & \mathsf{k} \\ 1 & -4 & 1 \\ 2 & 3 & 0 \end{vmatrix} = -3\mathsf{i} - (-2)\mathsf{j} + 11\mathsf{k} = \begin{bmatrix} -3 \\ 2 \\ 11 \end{bmatrix}$. We compute $\begin{bmatrix} -3 \\ 2 \\ 11 \end{bmatrix} \cdot \mathsf{u} = -3(1) + 2(-4) + 11(1) = 0$ and $\begin{bmatrix} -3 \\ 2 \\ 11 \end{bmatrix} \cdot \mathsf{v} = -3(2) + 2(3) + 11(0) = 0$ so that, indeed, $\mathsf{u} \times \mathsf{v}$ is orthogonal to both u and v. ☺

That $\mathsf{u} \times \mathsf{u} = \mathsf{0}$ for any u follows immediately from Definition 3.11—look at what you get when each $v_i = u_i$. From scalar associativity, it follows in fact that $(c\mathsf{u}) \times \mathsf{v} = \mathsf{0}$ for any scalar c. To get the converse, that **if** $\mathsf{u} \times \mathsf{v} = \mathsf{0}$, **then** either u or v is a scalar multiple of the other is rather tricky and I'll leave things at that.

Property 8 is curious (especially when you consider that it implies Property 9). Its proof follows quickly from an identity you should check

$$\|\mathsf{u} \times \mathsf{v}\|^2 + (\mathsf{u} \cdot \mathsf{v})^2 = \|\mathsf{u}\|^2 \, \|\mathsf{v}\|^2 .$$

and is a favourite homework question of mine. Here, I'll just illustrate with an example.

3.16 Example. Let $\mathsf{u} = \begin{bmatrix} 1 \\ 2 \\ 3 \end{bmatrix}$ and $\mathsf{v} = \begin{bmatrix} -2 \\ 1 \\ 4 \end{bmatrix}$. Then

$$\mathsf{u} \times \mathsf{v} = \begin{vmatrix} \mathsf{i} & \mathsf{j} & \mathsf{k} \\ 1 & 2 & 3 \\ -2 & 1 & 4 \end{vmatrix} = 5\mathsf{i} - 10\mathsf{j} + 5\mathsf{k} = \begin{bmatrix} 5 \\ -10 \\ 5 \end{bmatrix} = 5 \begin{bmatrix} 1 \\ -2 \\ 1 \end{bmatrix},$$

so that $\|\mathsf{u} \times \mathsf{v}\| = 5 \left\| \begin{bmatrix} 1 \\ -2 \\ 1 \end{bmatrix} \right\| = 5\sqrt{6}$. Let's check that this is $\|\mathsf{u}\| \, \|\mathsf{v}\| \sin\theta$, with θ the angle between u and v. We have $\|\mathsf{u}\| = \sqrt{14}$ and $\|\mathsf{v}\| = \sqrt{21}$, so we need $\sin\theta$. Remember that $\cos\theta = \dfrac{\mathsf{u} \cdot \mathsf{v}}{\|\mathsf{u}\| \, \|\mathsf{v}\|}$. Here then, $\cos\theta = \dfrac{\mathsf{u} \cdot \mathsf{v}}{\|\mathsf{u}\| \, \|\mathsf{v}\|} = \dfrac{12}{7\sqrt{6}}$, so

$$\sin^2\theta = 1 - \cos^2\theta = 1 - \frac{144}{49(6)} = 1 - \frac{24}{49} = \frac{25}{49}.$$

Thus $\sin\theta = \pm\frac{5}{7}$ and, since $0 \le \theta \le \pi$, we must have $\sin\theta \ge 0$, so $\sin\theta = +\frac{5}{7}$. Finally then $\|\mathsf{u} \times \mathsf{v}\| = 5\sqrt{6}$, and

$$\|\mathsf{u}\|\,\|\mathsf{v}\|\sin\theta = \sqrt{14}\sqrt{21}(\frac{5}{7}) = \sqrt{2}\sqrt{7}\sqrt{3}\sqrt{7}(\frac{5}{7}) = 5\sqrt{2}\sqrt{3} = 5\sqrt{6}$$

too, so Property 8 indeed holds, at least in this particular case. :)

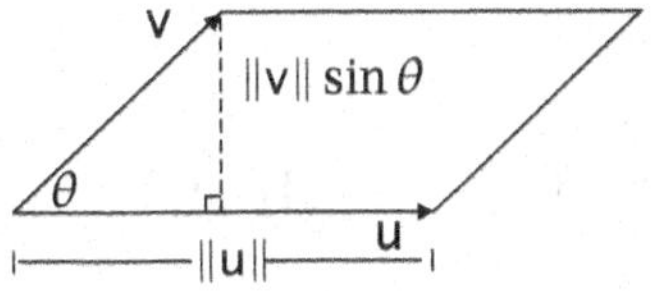

Figure 8: The area of the parallelogram is base times height: $\|\mathsf{u}\|\,(\|\mathsf{v}\|\sin\theta)$, and this is precisely $\|\mathsf{u} \times \mathsf{v}\|$.

Figure 8 explains Property 9 which I have always found amazing and certainly not intuitive! The area of a parallelogram is the product of the lengths of the base and the height, and the units are squared. Apparently, the numerical value of this number is the length (in linear units) of a certain vector orthogonal to the plane! Say the parallelogram has sides u and v. Then the base has length $\|\mathsf{u}\|$ and the height is $\|\mathsf{v}\|\sin\theta$, so the area is $\|\mathsf{u}\|\,\|\mathsf{v}\|\sin\theta$ which is precisely $\|\mathsf{u} \times \mathsf{v}\|$.

3.17 Example. Suppose we want the area of the parallelogram with adjacent sides $\mathsf{u} = \begin{bmatrix} -3 \\ 4 \\ 1 \end{bmatrix}$ and $\mathsf{v} = \begin{bmatrix} 1 \\ 0 \\ -2 \end{bmatrix}$. We compute $\mathsf{u} \times \mathsf{v} = \begin{vmatrix} \mathsf{i} & \mathsf{j} & \mathsf{k} \\ -3 & 4 & 1 \\ 1 & 0 & -2 \end{vmatrix} = -8\mathsf{i} - 5\mathsf{j} - 4\mathsf{k} = \begin{bmatrix} -8 \\ -5 \\ -4 \end{bmatrix}$ and conclude that the area of the parallelogram is $\|\mathsf{u} \times \mathsf{v}\| = \sqrt{64 + 25 + 16} = \sqrt{105}$. :)

Finally we come to the last property of the cross product. You may consider this a "non-property," but what's in a word?? An operation $\star$ is *associative* if $(a \star b) \star c = a \star (b \star c)$ for all a, b, and c. Ordinary addition and multiplication of real numbers, and vector addition, are examples of associative operations. The cross product of vectors, however, is **not** associative. In general,

3.18

$$(\mathsf{u} \times \mathsf{v}) \times \mathsf{w} \ne \mathsf{u} \times (\mathsf{v} \times \mathsf{w}).$$

3.19 Example. Here's an easy example showing that the cross product is, in general, not associative. Let i, j and k be the standard three-dimensional basis vectors. Then

$$\mathsf{i}\times\mathsf{i} = \begin{vmatrix} \mathsf{i} & \mathsf{j} & \mathsf{k} \\ 1 & 0 & 0 \\ 1 & 0 & 0 \end{vmatrix} = 0\mathsf{i} - 0\mathsf{j} + 0\mathsf{k} = \begin{bmatrix} 0 \\ 0 \\ 0 \end{bmatrix}$$

(this is another illustration of the fact that $\mathsf{u}\times\mathsf{u} = \mathsf{0}$ for any u), and so $(\mathsf{i}\times\mathsf{i})\times\mathsf{j} = \mathsf{0}\times\mathsf{j} = \mathsf{0}$. On the other hand,

$$\mathsf{i}\times\mathsf{j} = \begin{vmatrix} \mathsf{i} & \mathsf{j} & \mathsf{k} \\ 1 & 0 & 0 \\ 0 & 1 & 0 \end{vmatrix} = 0\mathsf{i} - 0\mathsf{j} + 1\mathsf{k} = \begin{bmatrix} 0 \\ 0 \\ 1 \end{bmatrix} = \mathsf{k} \tag{4}$$

so that $\mathsf{i}\times(\mathsf{i}\times\mathsf{j}) = \mathsf{i}\times\mathsf{k} = \begin{vmatrix} \mathsf{i} & \mathsf{j} & \mathsf{k} \\ 1 & 0 & 0 \\ 0 & 0 & 1 \end{vmatrix} = 0\mathsf{i} - 1\mathsf{j} + 0\mathsf{k} = -\mathsf{j}$. Most certainly, $(\mathsf{i}\times\mathsf{i})\times\mathsf{j} \neq \mathsf{i}\times(\mathsf{i}\times\mathsf{j})$.

:-)

In (4), we showed that $\mathsf{i}\times\mathsf{j} = \mathsf{k}$. The cross products of any two of i, j, k are quite easy to remember. The cross product of any of these with itself is $\mathsf{0}$ ($\mathsf{u}\times\mathsf{u} = \mathsf{0}$ for any u) and thinking of i, j, k arranged clockwise in a circle, the cross product of any two adjacent vectors is the next; that is,

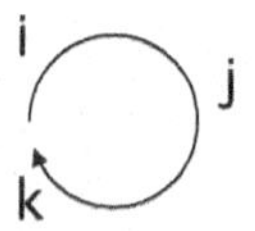

3.20 $$\boxed{\mathsf{i}\times\mathsf{j} = \mathsf{k}, \quad \mathsf{j}\times\mathsf{k} = \mathsf{i}, \quad \mathsf{k}\times\mathsf{i} = \mathsf{j}.}$$

Cross products in the reverse order are the negatives of these, by Property 1.

$$\mathsf{j}\times\mathsf{i} = -\mathsf{k}, \quad \mathsf{k}\times\mathsf{j} = -\mathsf{i}, \quad \mathsf{i}\times\mathsf{k} = -\mathsf{j}.$$

The Equation of a Line

We now turn our attention to the equation of a line in 3-space. While an equation like $y = 2x$ describes a line in the plane, we now know that it describes a plane in 3-space, so the equation of a line in 3-space is going to look different from what you might expect.

Remembering the connection between a geometric shape and its equation—see **3.1**—we seek an equation that is satisfied by x, y, z if and only if the point (x, y, z) lies on the line. With reference to Figure 9, we note that a line ℓ is specified by a direction vector $\mathsf{d} = \begin{bmatrix} a \\ b \\ c \end{bmatrix}$ and a point $P_0(x_0, y_0, z_0)$ on it. In fact, $P(x, y, z)$ is on ℓ if and only if the vector $\overrightarrow{P_0P}$ is a multiple of d; that is, if and only if

$$\begin{bmatrix} x - x_0 \\ y - y_0 \\ z - z_0 \end{bmatrix} = t\mathsf{d} = \begin{bmatrix} ta \\ tb \\ tc \end{bmatrix}$$

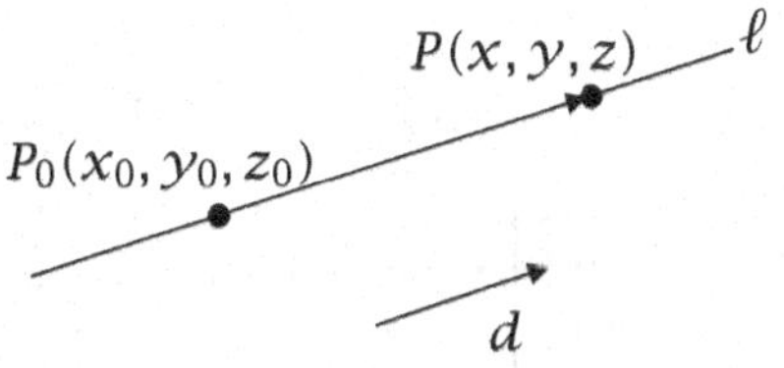

Figure 9: A line is determined by a direction d and a point P_0 on the line.

for some scalar t, and this occurs if and only if, for some t,

3.21 $$\boxed{\begin{bmatrix} x \\ y \\ z \end{bmatrix} = \begin{bmatrix} x_0 \\ y_0 \\ z_0 \end{bmatrix} + t \begin{bmatrix} a \\ b \\ c \end{bmatrix}}$$ **Vector equation of a line.**

3.22 Example. The line through $P(-1,2,5)$ in the direction of $\mathsf{d} = \begin{bmatrix} 2 \\ -3 \\ 4 \end{bmatrix}$ has vector equation $\begin{bmatrix} x \\ y \\ z \end{bmatrix} = \begin{bmatrix} -1 \\ 2 \\ 5 \end{bmatrix} + t \begin{bmatrix} 2 \\ -3 \\ 4 \end{bmatrix}$. Points on the line correspond to values of the "parameter" t. For instance

$$\begin{aligned} t &= 0 \quad (-1,2,5) \\ t &= 1 \quad (1,-1,9) \\ t &= -2 \quad (-5,8,-3). \end{aligned}$$

Does $(1,2,3)$ lie on the line? If so, there is a value of t so that

$$\begin{aligned} 1 &= -1 + 2t \\ 2 &= 2 - 3t \\ 3 &= 5 + 4t. \end{aligned}$$

The first equation gives $2t = 2$, so $t = 1$, but this is not a solution to either of the last two equations. No t exists; the point $(1,2,3)$ does not lie on the line.

Does the line pass through the point $P(7,-10,21)$? If so, there is a t so that

$$\begin{aligned} 7 &= -1 + 2t \\ -10 &= 2 - 3t \\ 21 &= 5 + 4t. \end{aligned}$$

We find that $t = 4$ satisfies all three equations so, yes, the line does pass through $(7,-10,21)$. ⌣̈

3.23 Example. The line through $P(2,0,1)$ and $Q(4,-1,1)$ has direction $\mathsf{d} = \overrightarrow{PQ} = \begin{bmatrix} 2 \\ -1 \\ 0 \end{bmatrix}$. So the line has vector equation $\begin{bmatrix} x \\ y \\ z \end{bmatrix} = \begin{bmatrix} 2 \\ 0 \\ 1 \end{bmatrix} + t \begin{bmatrix} 2 \\ -1 \\ 0 \end{bmatrix}$. Another correct answer is $\begin{bmatrix} x \\ y \\ z \end{bmatrix} = \begin{bmatrix} 4 \\ -1 \\ 1 \end{bmatrix} + t \begin{bmatrix} 2 \\ -1 \\ 0 \end{bmatrix}$, and another is $\begin{bmatrix} x \\ y \\ z \end{bmatrix} = \begin{bmatrix} -2 \\ 2 \\ 1 \end{bmatrix} + t \begin{bmatrix} -6 \\ 3 \\ 0 \end{bmatrix}$. (Can you see why?) Lots of vector equations describe the same line. ☺

3.24 Example. The line ℓ with equation $\begin{bmatrix} x \\ y \\ z \end{bmatrix} = \begin{bmatrix} -3 \\ 1 \\ 6 \end{bmatrix} + t \begin{bmatrix} 4 \\ -1 \\ -7 \end{bmatrix}$ has direction $\mathsf{d} = \begin{bmatrix} 4 \\ -1 \\ -7 \end{bmatrix}$, so any line parallel to ℓ has direction d too. For instance, the line through $P(3,-1,2)$ that is parallel to ℓ has (vector) equation $\begin{bmatrix} x \\ y \\ z \end{bmatrix} = \begin{bmatrix} 3 \\ -1 \\ 2 \end{bmatrix} + t \begin{bmatrix} 4 \\ -1 \\ -7 \end{bmatrix}$. ☺

3.25 Example. Are the lines with equations

$$\begin{bmatrix} x \\ y \\ z \end{bmatrix} = \begin{bmatrix} 1 \\ 2 \\ 1 \end{bmatrix} + t \begin{bmatrix} -3 \\ 5 \\ 1 \end{bmatrix} \quad \text{and} \quad \begin{bmatrix} x \\ y \\ z \end{bmatrix} = \begin{bmatrix} -1 \\ 3 \\ 1 \end{bmatrix} + t \begin{bmatrix} 1 \\ -4 \\ -1 \end{bmatrix},$$

parallel? Their direction vectors are $\begin{bmatrix} -3 \\ 5 \\ 1 \end{bmatrix}$ and $\begin{bmatrix} 1 \\ -4 \\ -1 \end{bmatrix}$. Neither is a scalar multiple of the other, so the lines are not parallel. Do they intersect? Do I hear people screaming "Yes! Yes!"??? Well, let's see if they do. If they intersect, say at the point (x, y, z), then the coordinates of this point must satisfy each equation. So there is a t such that

$$\begin{bmatrix} x \\ y \\ z \end{bmatrix} = \begin{bmatrix} 1 \\ 2 \\ 1 \end{bmatrix} + t \begin{bmatrix} -3 \\ 5 \\ 1 \end{bmatrix}$$

and an s (**not necessarily the same as** t) such that

$$\begin{bmatrix} x \\ y \\ z \end{bmatrix} = \begin{bmatrix} -1 \\ 3 \\ 1 \end{bmatrix} + s \begin{bmatrix} 1 \\ -4 \\ -1 \end{bmatrix}.$$

The question then is whether there exist parameters t and s so that

$$\begin{bmatrix} 1 \\ 2 \\ 1 \end{bmatrix} + t \begin{bmatrix} -3 \\ 5 \\ 1 \end{bmatrix} = \begin{bmatrix} -1 \\ 3 \\ 1 \end{bmatrix} + s \begin{bmatrix} 1 \\ -4 \\ -1 \end{bmatrix} ?$$

We try to solve

$$\begin{aligned} 1-3t &= -1+s \\ 2+5t &= 3-4s \; ; \\ 1+t &= 1-s \end{aligned} \qquad \text{that is,} \qquad \begin{aligned} s+3t &= 2 \\ 4s+5t &= 1\ , \\ s+t &= 0 \end{aligned}$$

and find that $s = -1$, $t = 1$ is a solution. The lines indeed intersect, where

$$\begin{bmatrix} x \\ y \\ z \end{bmatrix} = \begin{bmatrix} 1 \\ 2 \\ 1 \end{bmatrix} + t\begin{bmatrix} -3 \\ 5 \\ 1 \end{bmatrix} = \begin{bmatrix} 1 \\ 2 \\ 1 \end{bmatrix} + 1\begin{bmatrix} -3 \\ 5 \\ 1 \end{bmatrix} = \begin{bmatrix} -2 \\ 7 \\ 2 \end{bmatrix};$$

that is, at the point $(-2, 7, 2)$. So our attention is drawn to the fact that in R^3, it is possible (quite likely, actually) that two lines are not parallel and they do not intersect. Such lines are called *skew*. :)

3.26 Example. Consider the line with equation $\begin{bmatrix} x \\ y \\ z \end{bmatrix} = \begin{bmatrix} -1 \\ 7 \\ 0 \end{bmatrix} + t\begin{bmatrix} 1 \\ 2 \\ -1 \end{bmatrix}$ and the plane with equation $2x - 3y + 5z = 4$. Do these intersect?

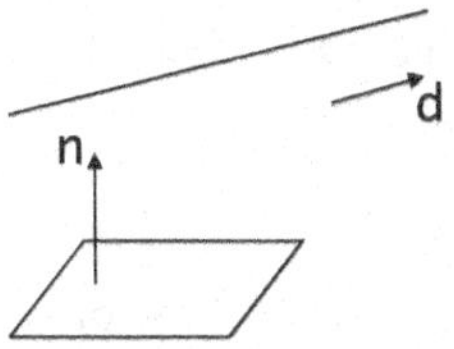

The direction of the line is $\mathsf{d} = \begin{bmatrix} 1 \\ 2 \\ -1 \end{bmatrix}$. A normal to the plane is $\mathsf{n} = \begin{bmatrix} 2 \\ -3 \\ 5 \end{bmatrix}$. Since $\mathsf{d} \cdot \mathsf{n} \neq 0$, d and n are not orthogonal, so the line is not parallel to the plane. The plane and the line must intersect. To find the point of intersection, substitute $x = -1 + t$, $y = 7 + 2t$, $z = -t$ in the equation for the plane as follows:

$$\begin{aligned} 2(-1+t) - 3(7+2t) + 5(-t) &= 4 \\ -23 - 9t &= 4 \\ -9t &= 27 \\ t &= -3. \end{aligned}$$

So the point of intersection occurs where $x = -1 + t = -4$, $y = 7 + 2t = 1$, $z = -t = 3$; that is, at $(-4, 1, 3)$. :)

True/False Questions for Week 3

Decide, with as little calculation as possible, whether each of the following statements is true or false and explain your answer whenever you say "false."

1. The point $(1, 1, 1)$ lies on the plane with equation $2x - 3y + 5z = 4$.
2. The arrow from $(0, 0, 0)$ to $(1, -2, 2)$ lies in the plane with equation $2x - 3y - 4z = 0$.
3. The vector $\begin{bmatrix} -1 \\ 2 \\ 0 \end{bmatrix}$ is in the plane with equation $2x + y - 5z = 1$.

4. The vector $\begin{bmatrix} -1 \\ 2 \\ 0 \end{bmatrix}$ is in the plane with equation $2x + y - 5z = 0$.

5. The planes with equations $-2x + 3y - z = 2$ and $4x - 6y + 2z = 0$ are parallel.

6. The distance from $(-2, 3, 1)$ to the plane with equation $x - y + 5z = 0$ is 0.

7. $\begin{vmatrix} 2 & -3 \\ 5 & 4 \end{vmatrix} = -7.$

8. Subtraction is an anticommutative operation.

9. The cross product is an associative operation.

10. If θ is the angle between vectors u and v, then $|\mathsf{u} \times \mathsf{v}| = \|\mathsf{u}\|\,\|\mathsf{v}\| \sin\theta$.

11. If θ is the angle between vectors u and v, then $|\mathsf{u} \cdot \mathsf{v}| = \|\mathsf{u}\|\,\|\mathsf{v}\| \cos\theta$.

12. If $\|\mathsf{u} \times \mathsf{v}\| = \|\mathsf{u}\|\,\|\mathsf{v}\|$, then u and v are orthogonal.

13. There exist vectors u and v such that $\|\mathsf{u}\| = 4$, $\|\mathsf{v}\| = 5$ and $\|\mathsf{u} \times \mathsf{v}\| = -10$.

14. $\mathsf{i} \times (\mathsf{i} \times \mathsf{j}) = -\mathsf{j}$.

15. The line with vector equation $\begin{bmatrix} x \\ y \\ z \end{bmatrix} = \begin{bmatrix} 4 \\ -3 \\ 5 \end{bmatrix} + t \begin{bmatrix} -1 \\ 0 \\ 1 \end{bmatrix}$ passes through the point $(4, -3, 5)$.

16. The line with vector equation $\begin{bmatrix} x \\ y \\ z \end{bmatrix} = \begin{bmatrix} 4 \\ -3 \\ 5 \end{bmatrix} + t \begin{bmatrix} -1 \\ 0 \\ 1 \end{bmatrix}$ passes through the point $(-1, 0, 1)$.

17. The line with vector equation $\begin{bmatrix} x \\ y \\ z \end{bmatrix} = \begin{bmatrix} 4 \\ -3 \\ 5 \end{bmatrix} + t \begin{bmatrix} -1 \\ 0 \\ 1 \end{bmatrix}$ passes through the point $(3, -3, 6)$.

18. The line whose vector equation is $\begin{bmatrix} x \\ y \\ z \end{bmatrix} = \begin{bmatrix} 0 \\ -3 \\ 5 \end{bmatrix} + t \begin{bmatrix} 2 \\ -4 \\ 1 \end{bmatrix}$ is parallel to the plane whose equation is $2x - 4y + z = 0$.

19. The line with vector equation $\begin{bmatrix} x \\ y \\ z \end{bmatrix} = \begin{bmatrix} -1 \\ 0 \\ 2 \end{bmatrix} + t \begin{bmatrix} 1 \\ 2 \\ 3 \end{bmatrix}$ is parallel to the plane with equation $x + 2y + 3z = 0$.

Week 3 Test Yourself

Here are a few problems with short answers that you can use to test your understanding of the concepts you have met this week.

1. Determine whether or not the vector $\mathsf{v} = \begin{bmatrix} 2 \\ -1 \\ 3 \end{bmatrix}$ is in each of the following planes:

   (a) with equation $x - y - z = 0$;
   (b) with equation $x - y - z = 1$;

(c) with equation $5x + y - 3z = 0$.

2. Find a normal to the plane with equation $-x + 2y + z = 4$. Also, find a point in this plane.

3. Express the plane with equation $2x - 3y + 4z = 0$ as the set of all linear combinations of two nonparallel vectors. (There are many possibilities.)

4. Verify that $\mathsf{u} \times \mathsf{v} = -(\mathsf{v} \times \mathsf{u})$ given $\mathsf{u} = \begin{bmatrix} 4 \\ -2 \\ -1 \end{bmatrix}$ and $\mathsf{v} = \begin{bmatrix} 5 \\ 6 \\ -3 \end{bmatrix}$.

5. Verify that $\|\mathsf{u} \times \mathsf{v}\| = \|\mathsf{u}\| \, \|\mathsf{v}\| \sin\theta$ given $\mathsf{u} = \begin{bmatrix} -2 \\ -4 \\ 1 \end{bmatrix}$ and $\mathsf{v} = \begin{bmatrix} 1 \\ 5 \\ -2 \end{bmatrix}$.

6. Find the equation of the plane parallel to the plane with equation $18x + 6y - 5z = 0$ and passing through the point $(-1, 1, 7)$.

7. Find the equation of the plane passing through $A(-1, 2, 1)$, $B(0, 1, 1)$, $C(7, -3, 0)$.

8. Let c_1 and c_2 denote arbitrary scalars, let $\mathsf{u} = \begin{bmatrix} 1 \\ -3 \\ 0 \end{bmatrix}$ and $\mathsf{v} = \begin{bmatrix} 1 \\ 2 \\ 6 \end{bmatrix}$. Find $c_1\mathsf{u} + c_2\mathsf{v}$. Then find the equation of the plane that consists of all linear combinations of u and v.

9. Find a unit vector that is perpendicular to both $\begin{bmatrix} 4 \\ 3 \\ -1 \end{bmatrix}$ and $\begin{bmatrix} 2 \\ 3 \\ 0 \end{bmatrix}$.

10. With $\mathsf{i}, \mathsf{j}, \mathsf{k}$ the standard basis vectors in R^3, write down $\mathsf{j} \times \mathsf{i}$, $\mathsf{j} \times \mathsf{k}$ and $\mathsf{j} \times \mathsf{j}$.

11. What is a vector equation of the line through $(1, 2, 3)$ with direction $\begin{bmatrix} -2 \\ 1 \\ 7 \end{bmatrix}$?

12. Determine whether or not the points $(1, 2, 3)$ and $(7, -2, 11)$ lie on the line that has vector equation $\begin{bmatrix} x \\ y \\ z \end{bmatrix} = \begin{bmatrix} 3 \\ -4 \\ 1 \end{bmatrix} + t \begin{bmatrix} 2 \\ 1 \\ 5 \end{bmatrix}$.

13. (a) How do you know that the line ℓ with equation $\begin{bmatrix} x \\ y \\ z \end{bmatrix} = \begin{bmatrix} 1 \\ 1 \\ 2 \end{bmatrix} + t \begin{bmatrix} -1 \\ 1 \\ 2 \end{bmatrix}$ intersects the plane π with equation $2x + y - z = 7$?

    (b) Find the point of intersection of π and ℓ.

14. Determine whether or not the line ℓ in Exercise 13(a) intersects the line with vector equation $\begin{bmatrix} x \\ y \\ z \end{bmatrix} = \begin{bmatrix} 3 \\ -1 \\ -2 \end{bmatrix} + t \begin{bmatrix} 1 \\ 0 \\ 1 \end{bmatrix}$.

15. Determine, with a simple reason, whether the line with equation $\begin{bmatrix} x \\ y \\ z \end{bmatrix} = \begin{bmatrix} -1 \\ 1 \\ 2 \end{bmatrix} + t \begin{bmatrix} 2 \\ 1 \\ -4 \end{bmatrix}$ and the plane with equation $3x - 10y - z = 8$ intersect. Find any point of intersection.

Week 4

Projections

In this section, we discuss the notion of the **projection** of a vector onto a vector or a plane. These are important concepts with many applications, as we shall soon see.

Projections onto Vectors

In Figure 10, we show the projection of a vector u on a line ℓ with direction vector v in two different situations according to whether the angle between u and v is acute or obtuse.

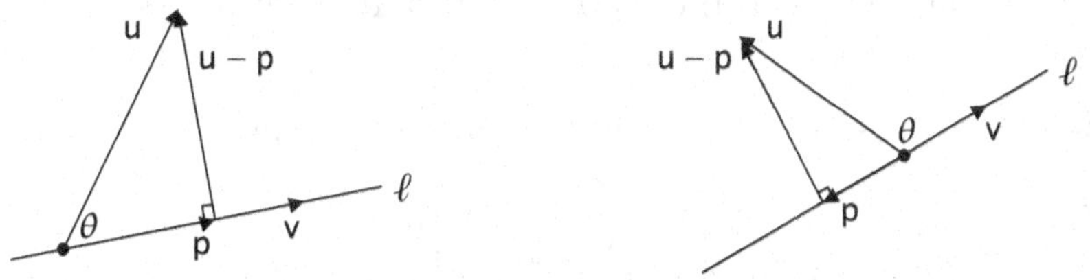

Figure 10: In both cases, $\mathsf{p} = \operatorname{proj}_\ell \mathsf{u}$ is the projection of u on ℓ.

4.1 **Definition.** The *projection* of a vector u on a line ℓ with direction v is that multiple $c\mathsf{v}$ of v with the property that $\mathsf{u} - \mathsf{p}$ is orthogonal to v. We use the notation $\operatorname{proj}_\ell \mathsf{u}$ or $\operatorname{proj}_{\mathsf{v}} \mathsf{u}$.

The alternative notation $\operatorname{proj}_{\mathsf{v}} \mathsf{u}$ reflects the fact that we talk about the projection of a vector u onto a **vector** as well as onto the line with that vector as direction.

The definition says that $\mathsf{p} = \operatorname{proj}_\ell \mathsf{u}$ is a multiple $c\mathsf{v}$ for a certain scalar c. It is not hard to find c. We want $\mathsf{p} = c\mathsf{v}$ with $\mathsf{u} - \mathsf{p}$ orthogonal to v, so

$$0 = (\mathsf{u} - \mathsf{p}) \cdot \mathsf{v} = \mathsf{u} \cdot \mathsf{v} - \mathsf{p} \cdot \mathsf{v} = \mathsf{u} \cdot \mathsf{v} - c(\mathsf{v} \cdot \mathsf{v}).$$

This gives $c(\mathsf{v} \cdot \mathsf{v}) = \mathsf{u} \cdot \mathsf{v}$. Assuming $\mathsf{v} \neq 0$, we know $\|\mathsf{v}\| \neq 0$, so $\mathsf{v} \cdot \mathsf{v} = \|\mathsf{v}\|^2 \neq 0$ and $c = \frac{\mathsf{u} \cdot \mathsf{v}}{\mathsf{v} \cdot \mathsf{v}}$.

4.2 The projection of a vector u on a line ℓ with direction the (nonzero) vector v is the vector $\operatorname{proj}_\ell \mathsf{u} = \frac{\mathsf{u} \cdot \mathsf{v}}{\mathsf{v} \cdot \mathsf{v}} \mathsf{v}$.

Although our pictures and examples so far have been in the plane, everything we have done carries over to 3-space.

4.3 Example. With $\mathsf{u} = \begin{bmatrix} 1 \\ -3 \\ 2 \end{bmatrix}$ and ℓ with direction $\mathsf{v} = \begin{bmatrix} -2 \\ 1 \\ 1 \end{bmatrix}$, we have $\frac{\mathsf{u}\cdot\mathsf{v}}{\mathsf{v}\cdot\mathsf{v}} = \frac{-3}{6} = -\frac{1}{2}$, so the projection of u on ℓ is $\text{proj}_{\mathsf{v}}\,\mathsf{u} = -\frac{1}{2}\mathsf{v} = \begin{bmatrix} 1 \\ -\frac{1}{2} \\ -\frac{1}{2} \end{bmatrix}$. ☺

The next example illustrates a principle I hope you will remember.

4.4 Example. It is frequently the case that we want to find two orthogonal vectors e and f in a plane. To do this, we first find **any** nonparallel vectors u and v in the plane. Then we find the projection p of u on v and take e = v and f = u − p as our orthogonal vectors. See Figure 10. For instance, say we have the plane with equation $3x + y - z = 0$. Since π goes through the origin, $\begin{bmatrix} x \\ y \\ z \end{bmatrix}$ is in the plane if and only if its components satisfy the equation of the plane. Two obvious nonparallel vectors in the plane (just experiment) are $\mathsf{u} = \begin{bmatrix} 1 \\ -3 \\ 0 \end{bmatrix}$ and $\mathsf{v} = \begin{bmatrix} 0 \\ 1 \\ 1 \end{bmatrix}$. Then $\mathsf{p} = \text{proj}_{\mathsf{v}}\,\mathsf{u} = \frac{\mathsf{u}\cdot\mathsf{v}}{\mathsf{v}\cdot\mathsf{v}}\,\mathsf{v} = \frac{-3}{2}\begin{bmatrix} 0 \\ 1 \\ 1 \end{bmatrix}$, $\mathsf{u} - \mathsf{p} = \begin{bmatrix} 1 \\ -3 \\ 0 \end{bmatrix} + \frac{3}{2}\begin{bmatrix} 0 \\ 1 \\ 1 \end{bmatrix} = \begin{bmatrix} 1 \\ -\frac{3}{2} \\ \frac{3}{2} \end{bmatrix}$, so take $\mathsf{e} = \begin{bmatrix} 0 \\ 1 \\ 1 \end{bmatrix}$ and $\mathsf{f} = \frac{1}{2}\begin{bmatrix} 2 \\ -3 \\ 3 \end{bmatrix}$. If we wanted to do something else with these vectors, we would probably multiply f by 2 (this does not affect direction) and use $\mathsf{e} = \begin{bmatrix} 0 \\ 1 \\ 1 \end{bmatrix}$ and $\mathsf{f} = \begin{bmatrix} 2 \\ -3 \\ 3 \end{bmatrix}$.

We encourage you to check that this answer is correct: The vectors are orthogonal and each lie in the plane with with equation $3x + y - z = 0$. ☺

The Distance from a Point to a Plane

Suppose we want the distance of a point P from a plane π. This time, we let Q be any point on the plane and note that the distance we want is precisely the length of the projection of $\overrightarrow{PQ}$ on n, a normal to π. (You might want to refer to the picture on the right of Figure 10 which shows the projection of one vector on another in the case that the angle between these vectors is more than 90°.)

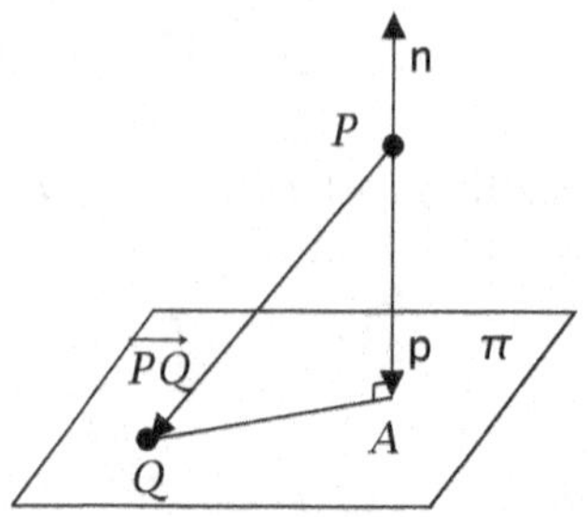

4.5 Example. To find the distance from $P(-1, 2, 1)$ to the plane with equation $5x + y + 3z = 7$, we begin with a point on the plane, say $Q(1,-1,1)$. A normal to the plane is $\mathsf{n} = \begin{bmatrix} 5 \\ 1 \\ 3 \end{bmatrix}$ and $\overrightarrow{PQ} = \begin{bmatrix} 2 \\ -3 \\ 0 \end{bmatrix}$. The projection of $\overrightarrow{PQ}$ on n is $\operatorname{proj}_{\mathsf{n}} \overrightarrow{PQ} = \frac{\overrightarrow{PQ}\cdot\mathsf{n}}{\mathsf{n}\cdot\mathsf{n}}\,\mathsf{n} = \frac{7}{35}\,\mathsf{n} = \frac{1}{5}\mathsf{n}$ and the distance from P to π is $\left\|\frac{1}{5}\,\mathsf{n}\right\| = \frac{1}{5}\,\|\mathsf{n}\| = \frac{1}{5}\sqrt{35}$. ☺

The Distance from a Point to a Line

Suppose we want to find the distance from a point P to a line ℓ whose direction vector is d. Here's what to do. Find some point Q on ℓ, any point at all, then project $\overrightarrow{QP}$ onto d. Call the projection p and observe that the required distance is $\left\|\overrightarrow{QP} - \mathsf{p}\right\|$.

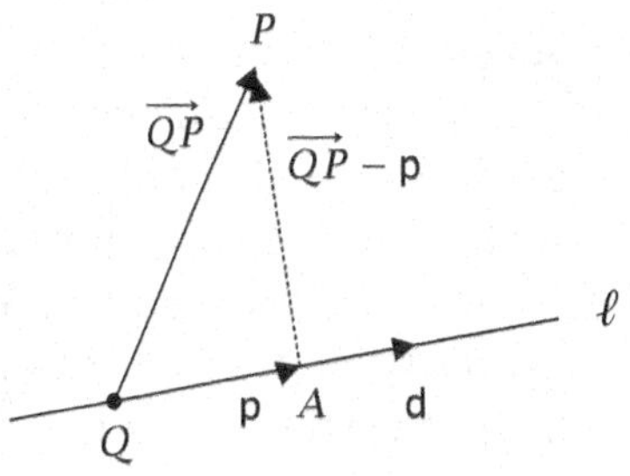

4.6 Example. To find the distance from $P(1,-1,2)$ to the line with (vector) equation $\begin{bmatrix} x \\ y \\ z \end{bmatrix} = \begin{bmatrix} -1 \\ 0 \\ 2 \end{bmatrix} + t\begin{bmatrix} -2 \\ 3 \\ 1 \end{bmatrix}$, we first find a point on the line, say $Q(-1,0,2)$ and compute $\overrightarrow{QP} = \begin{bmatrix} 2 \\ -1 \\ 0 \end{bmatrix}$. The line has direction $\mathsf{d} = \begin{bmatrix} -2 \\ 3 \\ 1 \end{bmatrix}$ and the distance we want is the length of $\overrightarrow{QP} - \mathsf{p}$, $\mathsf{p} = \operatorname{proj}_{\mathsf{d}} \overrightarrow{QP}$ the projection of $\overrightarrow{QP}$ on d. We have $\mathsf{p} = \operatorname{proj}_{\mathsf{d}} \overrightarrow{QP} = \frac{\overrightarrow{QP}\cdot\mathsf{d}}{\mathsf{d}\cdot\mathsf{d}}\,\mathsf{d} = \frac{-7}{14}\begin{bmatrix} -2 \\ 3 \\ 1 \end{bmatrix} = -\frac{1}{2}\begin{bmatrix} -2 \\ 3 \\ 1 \end{bmatrix}$, so

$$\overrightarrow{QP} - \mathsf{p} = \begin{bmatrix} 2 \\ -1 \\ 0 \end{bmatrix} + \frac{1}{2}\begin{bmatrix} -2 \\ 3 \\ 1 \end{bmatrix} = \begin{bmatrix} 1 \\ \frac{1}{2} \\ \frac{1}{2} \end{bmatrix} = \frac{1}{2}\begin{bmatrix} 2 \\ 1 \\ 1 \end{bmatrix}.$$ The required distance is $\left\|\overrightarrow{QP} - \mathsf{p}\right\| = \frac{1}{2}\sqrt{6}$.

If you wanted also the point A on ℓ closest to P, let A have coordinates (x,y,z) and use the fact that $\overrightarrow{QA} = \mathsf{p}$; that is, $\begin{bmatrix} x+1 \\ y \\ z-2 \end{bmatrix} = -\frac{1}{2}\begin{bmatrix} -2 \\ 3 \\ 1 \end{bmatrix}$. So $x+1 = 1$, $y = -\frac{3}{2}$, $z-2 = -\frac{1}{2}$ and A is $(0, -\frac{3}{2}, \frac{3}{2})$. ☺

Projections onto Planes through the Origin

In this section, we will regularly be talking about vectors in planes, so we assume that all planes pass through the origin. See Remark 3.7.

4.7 **Definition.** The *projection of a vector* w *on a plane* π (containing $(0, 0, 0)$) is a vector p in π with the property that w − p is orthogonal to every vector in π. We denote this vector $\text{proj}_\pi \mathsf{w}$.

This concept is illustrated in Figure 11. In the discussion that follows, we make use of the important idea which was stated as Theorem 3.13: if a vector is orthogonal to two nonparallel vectors, then it is orthogonal to the entire plane spanned by those two vectors.

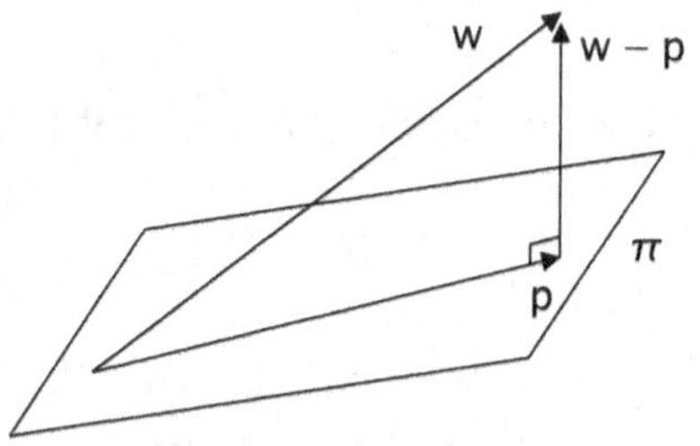

Figure 11: Vector $\mathsf{p} = \text{proj}_\pi \mathsf{w}$ is the projection of w on the plane π.

Suppose a plane π is spanned by **orthogonal** vectors e and f. (Remember that we know how to find such e and f—see Example 4.4.) This projection p is in π, so it is a linear combination of e and f, $\mathsf{p} = a\mathsf{e} + b\mathsf{f}$, for scalars a and b. Moreover, p must have the property that w − p is perpendicular to every vector in π. By Theorem 3.13, it is sufficient to have w − p perpendicular to just e and f, so we require $(\mathsf{w} - \mathsf{p}) \cdot \mathsf{e} = 0$ and $(\mathsf{w} - \mathsf{p}) \cdot \mathsf{f} = 0$. Equivalently, we want $\mathsf{w} \cdot \mathsf{e} = \mathsf{p} \cdot \mathsf{e}$ and $\mathsf{w} \cdot \mathsf{f} = \mathsf{p} \cdot \mathsf{f}$.

Here is where the orthogonality of e and f is useful:

$$\mathsf{w} \cdot \mathsf{e} = (a\mathsf{e} + b\mathsf{f}) \cdot \mathsf{e} = a(\mathsf{e} \cdot \mathsf{e}) \qquad (*)$$

because $\mathsf{f} \cdot \mathsf{e} = 0$. Assuming $\mathsf{e} \neq 0$ and hence $\mathsf{e} \cdot \mathsf{e} = \|\mathsf{e}\|^2 \neq 0$, equation (*) gives $a = \frac{\mathsf{w} \cdot \mathsf{e}}{\mathsf{e} \cdot \mathsf{e}}$. Similarly, $b = \frac{\mathsf{w} \cdot \mathsf{f}}{\mathsf{f} \cdot \mathsf{f}}$, so

$$\mathsf{p} = a\mathsf{e} + b\mathsf{f} = \frac{\mathsf{w} \cdot \mathsf{e}}{\mathsf{e} \cdot \mathsf{e}}\mathsf{e} + \frac{\mathsf{w} \cdot \mathsf{f}}{\mathsf{f} \cdot \mathsf{f}}\mathsf{f}.$$

4.8 **Theorem.** *The projection of a vector* w *on the plane spanned by* ***orthogonal*** *vectors* e *and* f *is*

$$\boxed{\text{proj}_\pi \mathsf{w} = \frac{\mathsf{w} \cdot \mathsf{e}}{\mathsf{e} \cdot \mathsf{e}}\mathsf{e} + \frac{\mathsf{w} \cdot \mathsf{f}}{\mathsf{f} \cdot \mathsf{f}}\mathsf{f}.}$$

Note how similar this formula is to the one given in **4.2** for the projection of w on a single vector e: $\text{proj}_\mathsf{e} \mathsf{w} = \frac{\mathsf{w} \cdot \mathsf{e}}{\mathsf{e} \cdot \mathsf{e}}\mathsf{e}$.

4.9 **Example.** Suppose we want the projection of $\mathsf{w} = \begin{bmatrix} 3 \\ 1 \\ 1 \end{bmatrix}$ on the plane π with equation $3x + y - z = 0$. In Example 4.4, we found orthogonal vectors $\mathsf{e} = \begin{bmatrix} 0 \\ 1 \\ 1 \end{bmatrix}$ and $\mathsf{f} = \begin{bmatrix} 2 \\ -3 \\ 3 \end{bmatrix}$ in π, so

$$\text{proj}_\pi \mathsf{w} = \frac{\mathsf{w}\cdot\mathsf{e}}{\mathsf{e}\cdot\mathsf{e}}\mathsf{e} + \frac{\mathsf{w}\cdot\mathsf{f}}{\mathsf{f}\cdot\mathsf{f}}\mathsf{f} = \frac{2}{2}\begin{bmatrix}0\\1\\1\end{bmatrix} + \frac{6}{22}\begin{bmatrix}2\\-3\\3\end{bmatrix} = \begin{bmatrix}0\\1\\1\end{bmatrix} + \frac{3}{11}\begin{bmatrix}2\\-3\\3\end{bmatrix} = \begin{bmatrix}\frac{6}{11}\\\frac{2}{11}\\\frac{20}{11}\end{bmatrix}.$$

:)

Linear Independence and Dependence

We conclude the first unit of this course with a short section devoted to one of the most important concepts in linear algebra, **linear independence**, and its negation, **linear dependence.**

Suppose we have some vectors $\mathsf{v}_1, \mathsf{v}_2, \ldots, \mathsf{v}_n$. It is easy to find a linear combination of these that equals the zero vector:

$$0\mathsf{v}_1 + 0\mathsf{v}_2 + \cdots + 0\mathsf{v}_n = \mathsf{0}.$$

This linear combination, with all coefficients 0, is called the *trivial* one. Vectors are linearly independent if this is the **only** way to get $\mathsf{0}$.

4.10 **Definition.** Vectors $\mathsf{v}_1, \mathsf{v}_2, \ldots, \mathsf{v}_n$ are *linearly independent* if

$$c_1\mathsf{v}_1 + c_2\mathsf{v}_2 + \cdots + c_n\mathsf{v}_n = \mathsf{0} \quad \text{implies} \quad c_1 = 0, c_2 = 0, \ldots, c_n = 0.$$

Vectors are *linearly dependent* if they are not linearly independent; that is, there is an equation of the form $c_1\mathsf{v}_1 + c_2\mathsf{v}_2 + \cdots + c_n\mathsf{v}_n = \mathsf{0}$ with **at least one** coefficient not 0.

4.11 **Example.** Are the vectors $\mathsf{v}_1 = \begin{bmatrix}1\\2\end{bmatrix}$ and $\mathsf{v}_2 = \begin{bmatrix}-1\\0\end{bmatrix}$ linearly independent? Certainly

$$0\begin{bmatrix}1\\2\end{bmatrix} + 0\begin{bmatrix}-1\\0\end{bmatrix} = \begin{bmatrix}0\\0\end{bmatrix},$$

but this is not the point! Here's the question: Is $0\begin{bmatrix}1\\2\end{bmatrix} + 0\begin{bmatrix}-1\\0\end{bmatrix}$ the **only** way for a linear combination of v_1 and v_2 to be $\mathsf{0}$?

Suppose $c_1\mathsf{v}_1 + c_2\mathsf{v}_2 = \mathsf{0}$. This is

$$c_1\begin{bmatrix}1\\2\end{bmatrix} + c_2\begin{bmatrix}-1\\0\end{bmatrix} = \begin{bmatrix}0\\0\end{bmatrix}, \quad \text{which is} \quad \begin{bmatrix}c_1 - c_2\\2c_1\end{bmatrix} = \begin{bmatrix}0\\0\end{bmatrix}.$$

So we must solve the two equations

$$\begin{aligned} c_1 - c_2 &= 0 \\ 2c_1 \quad\; &= 0 \end{aligned}$$

for the unknowns c_1 and c_2. The second equation gives $c_1 = 0$. Then putting $c_1 = 0$ in the first equation gives $c_2 = 0$. All the coefficients are forced to be 0. The vectors v_1 and v_2 are linearly independent. ☺

4.12 Problem. Are the vectors $\mathrm{v}_1 = \begin{bmatrix} 2 \\ -4 \end{bmatrix}$ and $\mathrm{v}_2 = \begin{bmatrix} -3 \\ 6 \end{bmatrix}$ linearly dependent or linearly independent?

Solution. We must determine whether $c_1 = c_2 = 0$ is the only way to get $c_1\mathrm{v}_1 + c_2\mathrm{v}_2 = 0$. This equation is

$$c_1\begin{bmatrix} 2 \\ -4 \end{bmatrix} + c_2\begin{bmatrix} -3 \\ 6 \end{bmatrix} = \begin{bmatrix} 0 \\ 0 \end{bmatrix}, \quad \text{which is} \quad \begin{bmatrix} 2c_1 - 3c_2 \\ -4c_1 + 6c_2 \end{bmatrix} = \begin{bmatrix} 0 \\ 0 \end{bmatrix}.$$

There are many **nontrivial** solutions, such as $c_1 = 3$, $c_2 = 2$. Therefore, the vectors v_1 and v_2 are linearly dependent. ♧

4.13 Example. Let $\mathrm{v}_1 = \begin{bmatrix} 1 \\ -2 \\ 5 \end{bmatrix}$, $\mathrm{v}_2 = \begin{bmatrix} 0 \\ 5 \\ -7 \end{bmatrix}$ and $\mathrm{v}_3 = \begin{bmatrix} 2 \\ 1 \\ 3 \end{bmatrix}$. To determine whether these vectors are linearly independent or linearly dependent, we suppose $c_1\mathrm{v}_1 + c_2\mathrm{v}_2 + c_3\mathrm{v}_3 = 0$. This equation leads to the three equations

$$\begin{aligned} c_1 \qquad\quad + 2c_3 &= 0 \\ -2c_1 + 5c_2 + \;\; c_3 &= 0 \\ 5c_1 - 7c_2 + 3c_3 &= 0 \end{aligned}$$

The first equation gives $c_1 = -2c_3$ and substituting this into the second and third equations gives

$$\begin{aligned} 5c_2 + 5c_3 &= 0 \\ -7c_2 - 7c_3 &= 0 \end{aligned}$$

So $c_3 = -c_2$, but there seems no way to get all the coefficients 0. In fact, $c_2 = 1$, $c_3 = -c_2 = -1$ and $c_1 = -2c_3 = 2$ gives a solution that is not all 0s: $2\mathrm{v}_1 + \mathrm{v}_2 - \mathrm{v}_3 = 0$. We conclude that the vectors are linearly dependent. ☺

4.14 Problem. Let $\mathrm{v}_1, \mathrm{v}_2, \mathrm{v}_3$ be as above and let $\mathrm{v}_4 = \begin{bmatrix} -3 \\ 1 \\ 5 \end{bmatrix}$. Are $\mathrm{v}_1, \mathrm{v}_2, \mathrm{v}_3, \mathrm{v}_4$ linearly independent or linearly dependent?

Solution. The four vectors are linearly dependent. We just noted that $2\mathsf{v}_1 + \mathsf{v}_2 - \mathsf{v}_3 = \mathbf{0}$, so $2\mathsf{v}_1 + \mathsf{v}_2 - \mathsf{v}_3 + 0\mathsf{v}_4 = \mathbf{0}$. Some linear combination is the zero vector without **all** coefficients 0.

4.15 Problem. Are $\mathsf{v}_1 = \begin{bmatrix} 1 \\ 2 \\ 5 \\ -1 \end{bmatrix}$, $\mathsf{v}_2 = \begin{bmatrix} 1 \\ 0 \\ 0 \\ 3 \end{bmatrix}$, and $\mathsf{v}_3 = \begin{bmatrix} -2 \\ 1 \\ 4 \\ 8 \end{bmatrix}$ linearly independent or linearly dependent?

Solution. Suppose $c_1\mathsf{v}_1 + c_2\mathsf{v}_2 + c_3\mathsf{v}_3 = \mathbf{0}$. Then

$$\begin{bmatrix} c_1 + c_2 - 2c_3 \\ 2c_1 + c_3 \\ 5c_1 + 4c_3 \\ -c_1 + 3c_2 + 8c_3 \end{bmatrix} = \begin{bmatrix} 0 \\ 0 \\ 0 \\ 0 \end{bmatrix}, \text{ which is } \begin{aligned} c_1 + c_2 - 2c_3 &= 0 \\ 2c_1 + c_3 &= 0 \\ 5c_1 + 4c_3 &= 0 \\ -c_1 + 3c_2 + 8c_3 &= 0. \end{aligned}$$

The second equation says $c_3 = -2c_1$. Substituting in the third, we get $5c_1 - 8c_1 = 0$. So $-3c_1 = 0$ and $c_1 = 0$. Then $c_3 = -2c_1 = 0$ and the first equation gives $c_2 = 0$ too. The only linear combination of v_1, v_2, v_3 which gives $\mathbf{0}$ is the trivial one: all coefficients 0. So the vectors are linearly independent.

4.16 Problem. Let $\mathsf{v}_1 = \begin{bmatrix} 3 \\ 2 \\ 5 \end{bmatrix}$, $\mathsf{v}_2 = \begin{bmatrix} 2 \\ 0 \\ 0 \end{bmatrix}$, $\mathsf{v}_3 = \begin{bmatrix} 0 \\ 1 \\ 2 \end{bmatrix}$, $\mathsf{v}_4 = \begin{bmatrix} 1 \\ 1 \\ 1 \end{bmatrix}$. Are the vectors $\mathsf{v}_1, \mathsf{v}_2, \mathsf{v}_3, \mathsf{v}_4$ linearly dependent or linearly independent?

Solution. The equation $c_1\mathsf{v}_1 + c_2\mathsf{v}_2 + c_3\mathsf{v}_3 + c_4\mathsf{v}_4 = \mathbf{0}$ gives

$$\begin{bmatrix} 3c_1 + 2c_2 + c_4 \\ 2c_1 + c_3 + c_4 \\ 5c_1 + 2c_3 + c_4 \end{bmatrix} = \begin{bmatrix} 0 \\ 0 \\ 0 \end{bmatrix}, \text{ which says } \begin{aligned} 3c_1 + 2c_2 + c_4 &= 0 \\ 2c_1 + c_3 + c_4 &= 0 \\ 5c_1 + 2c_3 + c_4 &= 0. \end{aligned}$$

Subtracting the last two equations gives $3c_1 + c_3 = 0$, so $c_3 = -3c_1$. Substituting in the second equation, $2c_1 - 3c_1 + c_4 = 0$, so $-c_1 + c_4 = 0$ and $c_4 = c_1$. Substituting in the first equation, $3c_1 + 2c_2 + c_1 = 0$, so $4c_1 + 2c_2 = 0$, $2c_2 = -4c_1$, $c_2 = -2c_1$. Letting $c_1 = 1$, $c_2 = -2c_1 = -2$, $c_3 = -3c_1 = -3$ and $c_4 = c_1 = 1$, we see that $\mathsf{v}_1 - 2\mathsf{v}_2 - 3\mathsf{v}_3 + \mathsf{v}_4 = \mathbf{0}$. The vectors are linearly dependent.

In Conclusion

Linear independence/dependence has to do with direction. Roughly speaking, vectors are linearly independent if and only if they have different directions. Suppose, for example, that two vectors u and v are linearly dependent. Then we have some nontrivial linear combination $a\mathsf{u} + b\mathsf{v} = \mathbf{0}$ with at least one of a or b not 0. Say $a \neq 0$. Then $\mathsf{u} = -\frac{b}{a}\mathsf{v}$, so u and v are parallel. In a similar way, if three vectors are linearly dependent, then they are either all parallel, or one lies in the plane spanned by the other two. In general

4.17 Theorem. *Vectors are linearly dependent if and only if one of them is a linear combination of the others.*

True/False Questions for Week 4

Decide, with as little calculation as possible, whether each of the following statements is true or false and explain your answer whenever you say "false."

1. The projection of a vector u on a nonzero vector v is $\frac{u \cdot v}{v \cdot v} u$.
2. If u and v are nonzero vectors and $\text{proj}_v u = 0$, then u and v are orthogonal.
3. If nonzero vectors u and v are orthogonal, you cannot find the projection of u on v.
4. The projection of a vector w on the plane spanned by vectors e and f is given by the formula $\text{proj}_\pi w = \frac{w \cdot e}{e \cdot e} e + \frac{w \cdot f}{f \cdot f} f$.
5. The projection of $\begin{bmatrix} 1 \\ 1 \\ 1 \end{bmatrix}$ on the xy-plane is $\begin{bmatrix} 1 \\ 1 \\ 0 \end{bmatrix}$.
6. $u = \begin{bmatrix} 0 \\ 0 \\ 0 \end{bmatrix}$, $v = \begin{bmatrix} 7 \\ 11 \\ 9 \end{bmatrix}$ and $w = \begin{bmatrix} -1 \\ 6 \\ 4 \end{bmatrix}$ are linearly dependent.
7. If u, v, w are linearly independent and $au + bv + cw = 0$, then we must have $a = 0$, $b = 0$ and $c = 0$.
8. If u, v, w are linearly dependent, then $au + bv + cw = 0$ implies at least one of the coefficients a, b, c is not zero.
9. Linear independence of vectors u and v means that $0u + 0v = 0$.
10. If u, v and w are linearly dependent, then one of these vectors is a scalar multiple of the others.
11. $u = \begin{bmatrix} -1 \\ 1 \end{bmatrix}$ and $v = \begin{bmatrix} 3 \\ -3 \end{bmatrix}$ are linearly dependent.

Week 4 Test Yourself

Here are a few problems with short answers that you can use to test your understanding of the concepts you have met this week.

1. Find the projection of u on v and the projection of v on u given $u = \begin{bmatrix} 7 \\ 0 \\ 5 \end{bmatrix}$ and $v = \begin{bmatrix} 3 \\ 1 \\ -4 \end{bmatrix}$.
2. Draw a picture that shows how to find the distance of a point P from a line ℓ.
3. Find the distance from point $P(4, -4, 17)$ to line ℓ with equation $\begin{bmatrix} x \\ y \\ z \end{bmatrix} = \begin{bmatrix} -1 \\ 2 \\ 0 \end{bmatrix} + t \begin{bmatrix} 4 \\ -1 \\ 8 \end{bmatrix}$. Also find the point on the line closest to P.

4. Find the distance from the point $P(-3, 1, 12)$ to the plane with equation $2x - y - 4z = 8$. Find also the point in the plane that is closest to P.

5. Find the projection of $\mathsf{w} = \begin{bmatrix} -2 \\ 3 \\ 0 \end{bmatrix}$ on the plane π with equation $x + y - 2z = 0$.

6. Find the distance from $P(2, 0, -1)$ to the plane with equation $x - y + 2z = 2$.

7. (a) Find two orthogonal vectors that span the plane π with equation $x + y + z = 0$.

   (b) Use your answer to part (a) to find the projection of $\mathsf{w} = \begin{bmatrix} -3 \\ 1 \\ 1 \end{bmatrix}$ on π.

8. Determine whether each of the given sets of vectors is linearly independent or linearly dependent.

   (a) $\mathsf{v}_1 = \begin{bmatrix} -1 \\ 1 \\ 2 \end{bmatrix}$, $\mathsf{v}_2 = \begin{bmatrix} 0 \\ 1 \\ 4 \end{bmatrix}$, $\mathsf{v}_3 = \begin{bmatrix} 3 \\ 1 \\ 2 \end{bmatrix}$, $\mathsf{v}_4 = \begin{bmatrix} 2 \\ -2 \\ -4 \end{bmatrix}$

   (b) $\mathsf{v}_1 = \begin{bmatrix} -1 \\ 1 \end{bmatrix}$, $\mathsf{v}_2 = \begin{bmatrix} 3 \\ 0 \end{bmatrix}$

9. If u is in the plane spanned by v and w, prove that u, v and w are linearly dependent.

Week 5

UNIT TWO: MATRICES AND LINEAR EQUATIONS

Matrices

Most of linear algebra is concerned with "systems of linear equations."
Here is an example.

$$\begin{aligned} 2x_1 + x_2 - x_3 &= 2 \\ 3x_1 - 2x_2 + 6x_3 &= -1. \end{aligned} \tag{5}$$

The idea is to find numbers x_1, x_2, and x_3 that make all the equations true at once. Notice that each row is a statement about the dot product of a vector with the vector $\mathsf{x} = \begin{bmatrix} x_1 \\ x_2 \\ x_3 \end{bmatrix}$ of unknowns. Row 1, for instance, says that the dot product of $\begin{bmatrix} 2 \\ 1 \\ -1 \end{bmatrix}$ and x is 2. Letting

$$\mathsf{r}_1 = \begin{bmatrix} 2 \\ 1 \\ -1 \end{bmatrix} \quad \text{and} \quad \mathsf{r}_2 = \begin{bmatrix} 3 \\ -2 \\ 6 \end{bmatrix}$$

(the coefficients of x_1, x_2, x_3 written as columns), the two given equations are just the assertions

$$\begin{aligned} \mathsf{r}_1 \cdot \mathsf{x} &= 2 \\ \mathsf{r}_2 \cdot \mathsf{x} &= -1. \end{aligned}$$

As we shall see, these equations can also be written

$$\begin{bmatrix} 2 & 1 & -1 \end{bmatrix} \begin{bmatrix} x_1 \\ x_2 \\ x_3 \end{bmatrix} = 2$$

$$\begin{bmatrix} 3 & -2 & 6 \end{bmatrix} \begin{bmatrix} x_1 \\ x_2 \\ x_3 \end{bmatrix} = -1$$

and summarized in the single **matrix equation**

$$\begin{bmatrix} 2 & 1 & -1 \\ 3 & -2 & 6 \end{bmatrix} \begin{bmatrix} x_1 \\ x_2 \\ x_3 \end{bmatrix} = \begin{bmatrix} 2 \\ -1 \end{bmatrix}.$$

This equation can be expressed simply as $A\mathsf{x} = \mathsf{b}$ where

$$A = \begin{bmatrix} 2 & 1 & -1 \\ 3 & -2 & 6 \end{bmatrix}, \quad \mathsf{x} = \begin{bmatrix} x_1 \\ x_2 \\ x_3 \end{bmatrix}, \quad \text{and} \quad \mathsf{b} = \begin{bmatrix} 2 \\ -1 \end{bmatrix}.$$

Look at the rows of A. They are the vectors r_1 and r_2 written as rows. In the form $A\mathsf{x} = \mathsf{b}$, the system doesn't look much different from the single equation $ax = b$ with a and b given real numbers and x the unknown.

5.1 Definition. A *matrix* is a rectangular arrangement of numbers, all surrounded by square brackets. If the rectangle has m rows and n columns, the matrix said to be $m \times n$ or to have *size* $m \times n$.

For example, $\begin{bmatrix} 2 & 1 & -1 \\ 3 & -2 & 6 \end{bmatrix}$ is a 2×3 ("two by three") matrix, $\begin{bmatrix} 1 & 2 \\ 3 & 4 \\ 5 & 6 \end{bmatrix}$ is a 3×2 matrix, and $\begin{bmatrix} -4 \\ 0 \\ 5 \end{bmatrix}$ is a 3×1 matrix. (When we say $m \times n$, the first number m is the number of rows and the second number n is the number of columns.) An n-dimensional vector $\begin{bmatrix} x_1 \\ x_2 \\ \vdots \\ x_n \end{bmatrix}$ is just an $n \times 1$ matrix, which we naturally call a *column matrix.* A $1 \times n$ matrix is called a *row matrix* because such a matrix looks like $\begin{bmatrix} x_1 & x_2 & \dots & x_n \end{bmatrix}$.

Ignoring the technicality that a vector, strictly speaking, should be enclosed in brackets, it is often convenient to think of an $m \times n$ matrix as n column vectors written side by side and enclosed in a single pair of brackets. We will write

$$A = \begin{bmatrix} \mathsf{a}_1 & \mathsf{a}_2 & \dots & \mathsf{a}_n \\ \downarrow & \downarrow & & \downarrow \\ & & & \end{bmatrix}$$

to indicate that the columns of A are the vectors $\mathsf{a}_1, \mathsf{a}_2, \dots, \mathsf{a}_n$.

Again ignoring the problem of brackets, each row of a matrix is the *transpose* of a vector in the following sense.

5.2 Definition. The *transpose* of an $m \times n$ matrix A is the $n \times m$ matrix whose rows are the columns of A in the same order. The transpose of A is denoted A^T.

For example, if $A = \begin{bmatrix} 1 & 2 & 3 \\ 4 & 5 & 6 \\ 7 & 8 & 9 \end{bmatrix}$, then the columns of $A^T = \begin{bmatrix} 1 & 4 & 7 \\ 2 & 5 & 8 \\ 3 & 6 & 9 \end{bmatrix}$ are the three rows of A (in order). If $A = \begin{bmatrix} -1 & 2 \\ 3 & 4 \\ 0 & -5 \end{bmatrix}$, then $A^T = \begin{bmatrix} -1 & 3 & 0 \\ 2 & 4 & -5 \end{bmatrix}$. The transpose of a row matrix is a column matrix (or vector) and the transpose of a vector is a row matrix. Sometimes, we wish to think of a matrix in terms of its rows, so we write

$$A = \begin{bmatrix} \mathsf{a}_1^T & \rightarrow \\ \mathsf{a}_2^T & \rightarrow \\ \vdots & \\ \mathsf{a}_m^T & \rightarrow \end{bmatrix}$$

to indicate that the rows of A are the transposes of the vectors $\mathsf{a}_1, \mathsf{a}_2, \ldots, \mathsf{a}_m$. For example, if

$$A = \begin{bmatrix} 1 & 2 & 3 & 4 \\ 8 & 7 & 6 & 5 \\ 0 & 1 & 1 & 0 \end{bmatrix}, \tag{6}$$

then the columns of A are the vectors

$$\begin{bmatrix} 1 \\ 8 \\ 0 \end{bmatrix}, \begin{bmatrix} 2 \\ 7 \\ 1 \end{bmatrix}, \begin{bmatrix} 3 \\ 6 \\ 1 \end{bmatrix}, \begin{bmatrix} 4 \\ 5 \\ 0 \end{bmatrix}$$

and the rows of A are

$$\begin{bmatrix} 1 \\ 2 \\ 3 \\ 4 \end{bmatrix}^T, \begin{bmatrix} 8 \\ 7 \\ 6 \\ 5 \end{bmatrix}^T, \begin{bmatrix} 0 \\ 1 \\ 1 \\ 0 \end{bmatrix}^T.$$

Occasionally we want to refer to the individual entries of a matrix. With reference to the matrix A in (6), we call the numbers 1, 2, 3, and 4 in the first row the $(1,1)$, $(1,2)$, $(1,3)$, and $(1,4)$ the *entries* of A and label them with a lower case a (to match the name of the matrix A) and two subscripts, denoting row and column. Here

$$a_{11} = 1, \quad a_{12} = 2, \quad a_{13} = 3, \quad a_{14} = 4.$$

In general, the entry in row i and column j of a matrix called A is denoted a_{ij}. So the $(2,3)$ entry of A is 6, the $(3,1)$ entry is 0 and the $(3,3)$ entry is 1.

5.3 Example. Let B be the 2×3 matrix with $b_{ij} = i - j$. The $(1,1)$ entry of B is $b_{11} = 1 - 1 = 0$, the $(1,2)$ entry is $b_{12} = 1 - 2 = -1$, and so on. We obtain $B = \begin{bmatrix} 0 & -1 & -2 \\ 1 & 0 & -1 \end{bmatrix}$. ☺

Matrix Equality

Matrices A and B are *equal* if and only if they have the same number of rows, the same number of columns and corresponding entries are equal.

5.4 Example. Let $A = \begin{bmatrix} -1 & x \\ 2y & -3 \end{bmatrix}$ and $B = \begin{bmatrix} a & -4 \\ 4 & a - b \end{bmatrix}$. If $A = B$, then $-1 = a$, $x = -4$, $2y = 4$, and $-3 = a - b$. Thus $a = -1$, $b = a + 3 = 2$, $x = -4$, and $y = 2$. ☺

Matrix Addition

If matrices $A = \begin{bmatrix} \mathbf{a}_1 & \mathbf{a}_2 & \dots & \mathbf{a}_n \\ \downarrow & \downarrow & & \downarrow \end{bmatrix}$ and $B = \begin{bmatrix} \mathbf{b}_1 & \mathbf{b}_2 & \dots & \mathbf{b}_n \\ \downarrow & \downarrow & & \downarrow \end{bmatrix}$ have the same size, then they can be added in the obvious way:

$$A + B = \begin{bmatrix} \mathbf{a}_1 + \mathbf{b}_1 & \mathbf{a}_2 + \mathbf{b}_2 & \dots & \mathbf{a}_n + \mathbf{b}_n \\ \downarrow & \downarrow & & \downarrow \end{bmatrix}.$$

Of course, this is just adding columnwise, or entry by entry.

5.5 Example. Let $A = \begin{bmatrix} 1 & 2 & 3 \\ 4 & 5 & 6 \end{bmatrix}$, $B = \begin{bmatrix} -2 & 3 & 0 \\ 2 & 1 & -1 \end{bmatrix}$ and $C = \begin{bmatrix} 1 & 0 \\ 0 & -1 \end{bmatrix}$. Then $A+B = \begin{bmatrix} -1 & 5 & 3 \\ 6 & 6 & 5 \end{bmatrix}$ while $A + C$ and $C + B$ are not defined: **You can only add matrices of the same size.** ☺

Scalar Multiplication

Just as we can multiply vectors by scalars, so also we can multiply matrices by scalars. If $A = \begin{bmatrix} \mathbf{a}_1 & \mathbf{a}_2 & \dots & \mathbf{a}_n \\ \downarrow & \downarrow & & \downarrow \end{bmatrix}$ is an $m \times n$ matrix and c is a scalar, then cA is the matrix whose columns are c times the columns of A:

$$cA = \begin{bmatrix} c\mathbf{a}_1 & c\mathbf{a}_2 & \dots & c\mathbf{a}_n \\ \downarrow & \downarrow & & \downarrow \end{bmatrix}.$$

Of course, this is just multiplying entry by entry.

5.6 Example. If $A = \begin{bmatrix} 1 & -1 \\ 2 & 0 \\ -3 & 5 \end{bmatrix}$ and $c = -4$, then $cA = \begin{bmatrix} -4 & 4 \\ -8 & 0 \\ 12 & -20 \end{bmatrix}$. ☺

Most of the basic properties of matrix addition and scalar multiplication have names that you should remember. (Names make conversation a whole lot easier.)

Properties of Matrix Addition and Scalar Multiplication

Let A, B and C be matrices and let c and d be scalars.

1. (Closure under addition) If A and B are $m \times n$ matrices, so is $A + B$.

2. (Addition is commutative) If A and B have the same size, $A + B = B + A$.

3. (Addition is associative) If A, B and C have the same size, $(A + B) + C = A + (B + C)$.

4. (Zero) If A is an $m \times n$ matrix, there is an $m \times n$ zero matrix $\mathbf{0}$ with the property that $A + \mathbf{0} = \mathbf{0} + A = A$.

5. (Additive Inverses) For every $m \times n$ matrix A, there is an $m \times n$ matrix denoted $-A$ (read "minus A") called the additive inverse of A, with the property that $A + (-A) = (-A) + A = \mathbf{0}$ is the $m \times n$ zero matrix.

6. (Closure under scalar multiplication) If c is a scalar and A is an $m \times n$ matrix, then cA is an $m \times n$ matrix.

7. (Scalar associativity) If c and d are scalars, and A is a matrix, then $c(dA) = (cd)A$.

8. (One) $1A = A$.

9. (Distributivity) $c(A + B) = cA + cB$ and $(c + d)A = cA + dA$.

10. If $cA = \mathbf{0}$, then either $c = 0$ or $A = \mathbf{0}$.

By now, you ought to be able immediately to write down what "scalar associativity" and "distributivity" mean because we have encountered them so many times already. Most of the properties listed follow immediately from properties of the real numbers. Commutativity of matrix addition, for instance, follows immediately because ordinary addition of real numbers is a commutative operation.

For any positive integers m and n, the $m \times n$ *zero matrix*, which we denote with bold face type $\mathbf{0}$ ($\underline{0}$ in handwritten work), is the $m \times n$ matrix all of whose entries are 0. It has the property that $A + \mathbf{0} = \mathbf{0} + A = A$ for any $m \times n$ matrix A.

5.7 Example. The 2×2 zero matrix is $\begin{bmatrix} 0 & 0 \\ 0 & 0 \end{bmatrix}$ and it works just like the number zero: $A + \mathbf{0} = A$ for any 2×2 matrix A. For example,

$$\begin{bmatrix} -1 & 2 \\ 3 & 5 \end{bmatrix} + \begin{bmatrix} 0 & 0 \\ 0 & 0 \end{bmatrix} = \begin{bmatrix} -1 & 2 \\ 3 & 5 \end{bmatrix} = \begin{bmatrix} 0 & 0 \\ 0 & 0 \end{bmatrix} + \begin{bmatrix} -1 & 2 \\ 3 & 5 \end{bmatrix}.$$

:)

5.8 Example. If $A = \begin{bmatrix} -1 & 3 \\ -2 & 4 \end{bmatrix}$, then $-A = (-1)\begin{bmatrix} -1 & 3 \\ -2 & 4 \end{bmatrix} = \begin{bmatrix} 1 & -3 \\ 2 & -4 \end{bmatrix}$. Clearly $A + (-A) = \begin{bmatrix} 0 & 0 \\ 0 & 0 \end{bmatrix}$ is $\mathbf{0}$, the 2×2 zero matrix.

:)

Matrix Multiplication

If addition and scalar multiplication of matrices seemed rather dull, matrix multiplication may grab your attention! Would you believe that

$$\begin{bmatrix} 1 & 2 \\ 3 & 4 \end{bmatrix}\begin{bmatrix} 5 & 6 \\ 7 & 8 \end{bmatrix} = \begin{bmatrix} 19 & 22 \\ 43 & 50 \end{bmatrix}?$$

What???

As you might now guess, matrix multiplication is more complicated than matrix addition or scalar multiplication, so we will explain how to multiply matrices in three stages. First we show how to multiply a row matrix by a column matrix, then how to multiply an arbitrary matrix by a column, and finally how to multiply any two matrices (of appropriate sizes).

STAGE 1: Row times column. Suppose $A = [\quad]$ is a row matrix and $B = \begin{bmatrix} \\ \\ \end{bmatrix}$ is a column matrix. Then B can equally be regarded as a vector and the same for A^T. The **matrix product** AB is defined to be the vector **dot product** $A^T B$ (enclosed in square brackets).

$$AB = [A^T \cdot B]. \tag{7}$$

5.9 Examples.

- $\begin{bmatrix} 2 & 4 & -3 \end{bmatrix}\begin{bmatrix} 3 \\ -1 \\ 1 \end{bmatrix} = \begin{bmatrix} 2(3) + 4(-1) - 3(1) \end{bmatrix} = \begin{bmatrix} -1 \end{bmatrix}$

- $\begin{bmatrix} 3 & -1 & 2 & 5 \end{bmatrix}\begin{bmatrix} 4 \\ -1 \\ -1 \\ 2 \end{bmatrix} = \begin{bmatrix} 12 + 1 - 2 + 10 \end{bmatrix} = \begin{bmatrix} 21 \end{bmatrix}$ ☺

5.10

The product of a row and a column is a dot product: $\mathsf{a}^T\mathsf{b} = [\mathsf{a} \cdot \mathsf{b}]$.

STAGE 2: Matrix times column. To find the product of an arbitrary matrix with a column, we work row by row.

Suppose the rows of A are $\mathsf{a}_1^T, \mathsf{a}_2^T, \ldots, \mathsf{a}_m^T$ and b is a column with the same number of components as each of the vectors a_i, say n. This means that A is $m \times n$ and b is $n \times 1$. (Note the repeated n.) The product

$$A\mathsf{b} = \begin{bmatrix} \mathsf{a}_1^T & \rightarrow \\ \mathsf{a}_2^T & \rightarrow \\ \vdots & \\ \mathsf{a}_m^T & \rightarrow \end{bmatrix}\begin{bmatrix} \mathsf{b} \\ \downarrow \\ \\ \\ \end{bmatrix} = \begin{bmatrix} \mathsf{a}_1^T\mathsf{b} \\ \mathsf{a}_2^T\mathsf{b} \\ \vdots \\ \mathsf{a}_m^T\mathsf{b} \end{bmatrix} \tag{8}$$

is a column, the column whose entries are the products of the rows of A with b. Thus the product of an $m \times n$ matrix and an $n \times 1$ matrix is $n \times 1$.

5.11 Example. If $A = \begin{bmatrix} 0 & 1 & 3 \\ -1 & 1 & 2 \\ 2 & 0 & -1 \\ 4 & 5 & -1 \end{bmatrix}$ and $B = \begin{bmatrix} -1 \\ 2 \\ 3 \end{bmatrix}$, then AB is a column whose first entry is the product of the first row of A with B,

$$\begin{bmatrix} 0 & 1 & 3 \end{bmatrix} \begin{bmatrix} -1 \\ 2 \\ 3 \end{bmatrix} = 11,$$

whose second entry is the product of the second row of A with B,

$$\begin{bmatrix} -1 & 1 & 2 \end{bmatrix} \begin{bmatrix} -1 \\ 2 \\ 3 \end{bmatrix} = 9,$$

whose third entry is the product of the third row of A with B,

$$\begin{bmatrix} 2 & 0 & -1 \end{bmatrix} \begin{bmatrix} -1 \\ 2 \\ 3 \end{bmatrix} = -5$$

and whose fourth entry is the product of the fourth row of A with B,

$$\begin{bmatrix} 4 & 5 & -1 \end{bmatrix} \begin{bmatrix} -1 \\ 2 \\ 3 \end{bmatrix} = 3.$$

Thus $AB = \begin{bmatrix} 0 & 1 & 3 \\ -1 & 1 & 2 \\ 2 & 0 & -1 \\ 4 & 5 & -1 \end{bmatrix} \begin{bmatrix} -1 \\ 2 \\ 3 \end{bmatrix} = \begin{bmatrix} 11 \\ 9 \\ -5 \\ 3 \end{bmatrix}$. ☺

5.12 Example. As noted earlier, the system

$$\begin{aligned} 2x_1 + x_2 - x_3 &= 2 \\ 3x_1 - 2x_2 + 6x_3 &= -1 \end{aligned}$$

can be written $A\mathsf{x} = \mathsf{b}$, with $A = \begin{bmatrix} 2 & 1 & -1 \\ 3 & -2 & 6 \end{bmatrix}$, $\mathsf{x} = \begin{bmatrix} x_1 \\ x_2 \\ x_3 \end{bmatrix}$ and $\mathsf{b} = \begin{bmatrix} 2 \\ -1 \end{bmatrix}$. ☺

Remember that the standard basis vectors of R^n are labelled $e_1, \ldots, e_n$:

$$e_1 = \begin{bmatrix} 1 \\ 0 \\ 0 \\ \vdots \\ 0 \end{bmatrix}, \quad e_2 = \begin{bmatrix} 0 \\ 1 \\ 0 \\ \vdots \\ 0 \end{bmatrix}, \quad \ldots, \quad e_n = \begin{bmatrix} 0 \\ 0 \\ \vdots \\ 0 \\ 1 \end{bmatrix}.$$

Suppose $A = \begin{bmatrix} a_1^T & \rightarrow \\ a_2^T & \rightarrow \\ \vdots & \\ a_m^T & \rightarrow \end{bmatrix}$ is an $m \times n$ matrix with rows as indicated. As in (8), we have

$$Ae_1 = \begin{bmatrix} a_1^T & \rightarrow \\ a_2^T & \rightarrow \\ \vdots & \\ a_m^T & \rightarrow \end{bmatrix} \begin{bmatrix} 1 \\ 0 \\ \vdots \\ 0 \end{bmatrix} = \begin{bmatrix} a_1 \cdot e_1 \\ a_2 \cdot e_1 \\ \vdots \\ a_m \cdot e_1 \end{bmatrix},$$

which is just the first column of A, since $a_1 \cdot e_1$ is the first component of a_1, $a_2 \cdot e_1$ is the first component of a_2, and so on. In general, for any i,

5.13 $$\boxed{Ae_i = \text{column } i \text{ of } A.}$$

For example, if $n = 2$ and $A = \begin{bmatrix} 1 & 4 \\ 2 & 5 \\ 3 & 6 \end{bmatrix}$, then

$$Ae_1 = \begin{bmatrix} 1 & 4 \\ 2 & 5 \\ 3 & 6 \end{bmatrix} \begin{bmatrix} 1 \\ 0 \end{bmatrix} = \begin{bmatrix} 1 \\ 2 \\ 3 \end{bmatrix}$$

is the first column of A and

$$Ae_2 = \begin{bmatrix} 1 & 4 \\ 2 & 5 \\ 3 & 6 \end{bmatrix} \begin{bmatrix} 0 \\ 1 \end{bmatrix} = \begin{bmatrix} 4 \\ 5 \\ 6 \end{bmatrix}$$

is the second column of A.

STAGE 3 (End of Story) Matrix by matrix. Suppose that A is an $m \times \mathbf{n}$ matrix and B is an $\mathbf{n} \times p$ matrix. (Notice the repeated n's here.) Let B have columns $b_1, \ldots, b_p$. Then

$$AB = A \begin{bmatrix} b_1 & b_2 & \ldots & b_p \\ \downarrow & \downarrow & & \downarrow \\ & & & \end{bmatrix} = \begin{bmatrix} Ab_1 & Ab_2 & \ldots & Ab_p \\ \downarrow & \downarrow & & \downarrow \\ & & & \end{bmatrix} \tag{9}$$

is the matrix whose columns are the products $Ab_1, Ab_2, \ldots, Ab_p$ of A with the columns of B.

5.14 Example. Let $A = \begin{bmatrix} 2 & 4 & -1 \\ 3 & 5 & 1 \\ 0 & 4 & -2 \end{bmatrix}$ and $B = \begin{bmatrix} 2 & 4 & 0 & 1 & -1 \\ 3 & -2 & 0 & 2 & 4 \\ 1 & 2 & -6 & 2 & 1 \end{bmatrix}$. Then

$$AB = \begin{bmatrix} 2 & 4 & -1 \\ 3 & 5 & 1 \\ 0 & 4 & -2 \end{bmatrix} \begin{bmatrix} 2 & 4 & 0 & 1 & -1 \\ 3 & -2 & 0 & 2 & 4 \\ 1 & 2 & -6 & 2 & 1 \end{bmatrix} = \begin{bmatrix} 15 & -2 & 6 & 8 & 13 \\ 22 & 4 & -6 & 15 & 18 \\ 10 & -12 & 12 & 4 & 14 \end{bmatrix}.$$

The first column of AB is the product of A and the first column of B,

$$A \begin{bmatrix} 2 \\ 3 \\ 1 \end{bmatrix} = \begin{bmatrix} 2 & 4 & -1 \\ 3 & 5 & 1 \\ 0 & 4 & -2 \end{bmatrix} \begin{bmatrix} 2 \\ 3 \\ 1 \end{bmatrix} = \begin{bmatrix} 15 \\ 22 \\ 10 \end{bmatrix},$$

the second column of AB is the product of A and the second column of B,

$$A \begin{bmatrix} 4 \\ -2 \\ 2 \end{bmatrix} = \begin{bmatrix} 2 & 4 & -1 \\ 3 & 5 & 1 \\ 0 & 4 & -2 \end{bmatrix} \begin{bmatrix} 4 \\ -2 \\ 2 \end{bmatrix} = \begin{bmatrix} -2 \\ 4 \\ -12 \end{bmatrix},$$

and so on. ☺

It is useful also to note that the entries of AB are dot products.

5.15 The (i, j) entry of AB is the product of row i of A and column j of B.

5.16 Examples. • With A and B as above, for example, the $(3, 2)$ entry of AB is the product of the third row of A and the second column of B: $-12 = \begin{bmatrix} 0 & 4 & -2 \end{bmatrix} \begin{bmatrix} 4 \\ -2 \\ 2 \end{bmatrix}$.

• $\begin{bmatrix} a_1^T & \rightarrow \\ a_2^T & \rightarrow \end{bmatrix} \begin{bmatrix} b_1 & b_2 & b_3 \\ \downarrow & \downarrow & \downarrow \end{bmatrix} = \begin{bmatrix} a_1 \cdot b_1 & a_1 \cdot b_2 & a_1 \cdot b_3 \\ a_2 \cdot b_1 & a_2 \cdot b_2 & a_2 \cdot b_3 \end{bmatrix}$.

Notice the pattern: the (i, j) entry of AB is the product of row i of A and row j of B, just as we said. ☺

We have used the term **compatible sizes** a couple of times to emphasize that only certain pairs of matrices can be multiplied. If you want to find AB, then the number of entries in each row of A must be the number of entries in each column of B; that is, the number of columns of A must match the number of rows of B. If A is $m \times n$ and we want to form the product AB then B must be $n \times p$ for some p—and then AB will be $m \times p$. Some people remember the little schemata

$$\frac{m}{n}\frac{n}{p} = \frac{m}{p}.$$

5.17 Example. Suppose $A = \begin{bmatrix} 1 & -2 \\ 3 & -4 \end{bmatrix}$ and $B = \begin{bmatrix} -1 & 0 & 1 \\ 3 & 4 & 2 \end{bmatrix}$. Then $AB = \begin{bmatrix} -7 & -8 & -3 \\ -15 & -16 & -5 \end{bmatrix}$ while BA is not defined: B is $2 \times ③$ but A is $② \times 2$. (The circled numbers are different.) ⌣̈

Properties of Matrix Multiplication

As usual after introducing a new algebraic operation, we make a list of its properties. Generally, these are pretty dull (I agree). With matrix multiplication, however, there are some interesting properties too. Be patient!

1. (Associativity) $(AB)C = A(BC)$ whenever all products are defined.
2. (Scalar associativity) $(cA)B = (cA)B = A(cB)$ for any scalar c.
3. (Distributivity) $(A + B)C = AC + BC$ and $A(B + C) = AB + AC$ whenever all products are defined.
4. (Zero) The zero matrix 0 has the property that $A0 = 0$ and $0A = 0$ for a matrix A of appropriate size.
5. (Identity) For any n, there is an $n \times n$ identity matrix I_n that behaves like the number 1, whenever possible: $AI_n = A$ and $I_n B = B$ for matrices A and B of appropriate sizes.

5.18 Definition. The $n \times n$ *identity matrix* is the matrix whose columns are the standard basis vectors $e_1, e_2, \ldots, e_n$ in order. It is denoted I_n (or often, just I).

The 1×1 identity matrix is $[1]$ (which is not much different from the number 1 and just about as interesting). The 2×2 identity matrix is $I_2 = \begin{bmatrix} 1 & 0 \\ 0 & 1 \end{bmatrix}$ and the 3×3 and 4×4 identity matrices are $I_3 = \begin{bmatrix} 1 & 0 & 0 \\ 0 & 1 & 0 \\ 0 & 0 & 1 \end{bmatrix}$ and $I_4 = \begin{bmatrix} 1 & 0 & 0 & 0 \\ 0 & 1 & 0 & 0 \\ 0 & 0 & 1 & 0 \\ 0 & 0 & 0 & 1 \end{bmatrix}$.

Letting $e_1, \ldots, e_n$ be the standard basis vectors, remember that for any $m \times n$ matrix A, Ae_i is column i of A. This explains why $AI_n = A$: the columns of AI_n are the columns of A, in order.

$$AI_n = A \begin{bmatrix} e_1 & e_2 & \ldots & e_n \\ \downarrow & \downarrow & & \downarrow \end{bmatrix} = \begin{bmatrix} Ae_1 & Ae_2 & \ldots & Ae_n \\ \downarrow & \downarrow & & \downarrow \end{bmatrix} = A.$$

One can verify $I_nA = A$ by using a property of the matrix transpose. See Problem 5.27 at the end of this week's notes.

For instance, you should check by doing the actual matrix multiplications that

$$\begin{bmatrix} a & b \\ c & d \end{bmatrix}\begin{bmatrix} 1 & 0 \\ 0 & 1 \end{bmatrix} = \begin{bmatrix} a & b \\ c & d \end{bmatrix} = \begin{bmatrix} 1 & 0 \\ 0 & 1 \end{bmatrix}\begin{bmatrix} a & b \\ c & d \end{bmatrix}.$$

Most of the properties we have discussed so far look pretty routine. In many ways, matrix multiplication (and addition) behaves just the way you'd expect. There are two important ways, however, in which matrix algebra **differs** from ordinary algebra.

5.19 Matrix multiplication is not commutative, in general, $AB \neq BA$.

For example, if $A = \begin{bmatrix} 1 & 0 \\ 0 & 0 \end{bmatrix}$ and $B = \begin{bmatrix} 0 & 1 \\ 0 & 0 \end{bmatrix}$, then $AB = B$ but $BA = 0$.

The fact that matrices do not commute means you have to be very careful in simplifying or expanding matrix products. For instance, in general the product $ABAB$ cannot be simplified, and this has consequences. Here's one.

$$\begin{aligned} (A+B)^2 &= (A+B)(A+B) \\ &= A(A+B) + B(A+B) && \text{using distributivity} \\ &= AA + AB + BA + BB && \text{distributivity again} \\ &= A^2 + AB + BA + B^2 \end{aligned}$$

is usually **not** the same as $A^2 + 2AB + B^2$.

This example serves to illustrate a second place where matrix algebra differs from ordinary algebra. With A and B as above, notice that $BA = 0$ even though $B \neq 0$ and $A \neq 0$.

5.20 $AB = 0$ does not in general imply that $A = 0$ or $B = 0$.

The Significance of Ax

We form *linear combinations* of matrices just as we did with vectors.

5.21 Definition. A *linear combination* of matrices $A_1, A_2, \ldots, A_k$ (all the same size) is a matrix of the form $c_1A_1 + c_2A_2 + \cdots + c_kA_k$, where $c_1, c_2, \ldots, c_k$ are scalars.

For example, $3B - 2C$ is a linear combination of matrices B and C, $c_1B_1 + c_2B_2 + \cdots + c_kB_k$ is a linear combination of matrices B_1, B_2, ..., B_k, and $c_1(AB_1) + c_2(AB_2) + \cdots + c_k(AB_k)$ is a linear combination of matrices AB_1, AB_2, ..., AB_k.

5.22 Matrix multiplication preserves linear combinations. By this, we mean that for any matrix A and any matrices $B_1, B_2, \ldots, B_k$ (of suitable sizes)

$$A(c_1B_1 + c_2B_2 + \cdots + c_kB_k) = c_1(AB_1) + c_2(AB_2) + \cdots + c_k(AB_k).$$

Here's an application of this idea. Suppose A is an $m \times n$ matrix and $\mathsf{x} = \begin{bmatrix} x_1 \\ x_2 \\ \vdots \\ x_n \end{bmatrix}$ is a vector in R^n. Remember—see **1.30**—that x can be written $\mathsf{x} = x_1\mathsf{e}_1 + x_2\mathsf{e}_2 + \cdots + x_n\mathsf{e}_n$. Since matrix multiplication preserves linear combinations, we know that

$$A\mathsf{x} = A(x_1\mathsf{e}_1 + x_2\mathsf{e}_2 + \cdots + x_n\mathsf{e}_n) = x_1A\mathsf{e}_1 + x_2A\mathsf{e}_2 + \cdots + x_nA\mathsf{e}_n.$$

With reference to **5.13**, we note that the expression on the right is x_1 times the first column of A, plus x_2 times the second column of A, and, eventually, plus x_n times the last column of A.

In my opinion, this idea, which I am about to restate, is the single-most important concept in linear algebra. Study it! Understand it! You won't believe how useful it is.

5.23 The product $A\mathsf{x}$ of a matrix A and a vector x is a linear combination of the columns of A, the coefficients being the components of x:

$$\begin{bmatrix} \mathsf{a}_1 & \mathsf{a}_2 & \cdots & \mathsf{a}_n \\ \downarrow & \downarrow & & \downarrow \\ & & & \end{bmatrix}\begin{bmatrix} x_1 \\ x_2 \\ \vdots \\ x_n \end{bmatrix} = x_1\begin{bmatrix} \mathsf{a}_1 \\ \downarrow \\ \\ \end{bmatrix} + x_2\begin{bmatrix} \mathsf{a}_2 \\ \downarrow \\ \\ \end{bmatrix} + \cdots + x_n\begin{bmatrix} \mathsf{a}_n \\ \downarrow \\ \\ \end{bmatrix}.$$

Conversely, any linear combination of vectors is $A\mathsf{x}$ for some matrix A and some vector x.

5.24 Example. Let $A = \begin{bmatrix} 1 & 2 & 3 \\ 4 & 5 & 6 \\ 7 & 8 & 9 \end{bmatrix}$ and $\mathsf{x} = \begin{bmatrix} 2 \\ -1 \\ 4 \end{bmatrix}$. Then

$$\begin{aligned} A\mathsf{x} &= \begin{bmatrix} 1 & 2 & 3 \\ 4 & 5 & 6 \\ 7 & 8 & 9 \end{bmatrix}\begin{bmatrix} 2 \\ -1 \\ 4 \end{bmatrix} = \begin{bmatrix} 12 \\ 27 \\ 42 \end{bmatrix} \\ &= 2\begin{bmatrix} 1 \\ 4 \\ 7 \end{bmatrix} - \begin{bmatrix} 2 \\ 5 \\ 8 \end{bmatrix} + 4\begin{bmatrix} 3 \\ 6 \\ 9 \end{bmatrix} \\ &= \boxed{2} \times \text{ column } 1 + \boxed{-1} \times \text{ column } 2 + \boxed{4} \times \text{ column } 3 \end{aligned}$$

is a linear combination of the columns of A, the coefficients, $2, -1, 4$, being the components of x.

In general, for any $\mathsf{x} = \begin{bmatrix} x_1 \\ x_2 \\ x_3 \end{bmatrix}$, $A\mathsf{x} = x_1\begin{bmatrix} 1 \\ 4 \\ 7 \end{bmatrix} + x_2\begin{bmatrix} 2 \\ 5 \\ 8 \end{bmatrix} + x_3\begin{bmatrix} 3 \\ 6 \\ 9 \end{bmatrix}$. ☺

5.25 Example. Suppose A is an $m \times n$ matrix with columns $\mathrm{a}_1, \mathrm{a}_2, \ldots, \mathrm{a}_n$ and D is the $n \times n$ *diagonal* matrix whose *diagonal entries* are $d_1, \ldots, d_n$; that is,

$$A = \begin{bmatrix} \mathrm{a}_1 & \mathrm{a}_2 & \cdots & \mathrm{a}_n \\ \downarrow & \downarrow & & \downarrow \\ & & & \end{bmatrix} \quad \text{and} \quad D = \begin{bmatrix} d_1 & 0 & 0 & \cdots & 0 \\ 0 & d_2 & 0 & \cdots & 0 \\ 0 & 0 & d_3 & & 0 \\ \vdots & \vdots & & \ddots & \\ 0 & 0 & \cdots & 0 & d_n \end{bmatrix}.$$

What does AD look like? Remember that matrix multiplication takes place column by column—see equation (9). The first column of AD is the product $A\begin{bmatrix} d_1 \\ 0 \\ \vdots \\ 0 \end{bmatrix}$ of A and the first column of D. This product is a linear combination of the columns of A where the coefficients are $d_1, 0, \ldots, 0$. So the product is just $d_1\mathrm{a}_1$, the product of d_1 and the first column of A.

The second column of AD is the product $A\begin{bmatrix} 0 \\ d_2 \\ 0 \\ \vdots \\ 0 \end{bmatrix}$ of A and the second column of D. This product is a linear combination of the columns of A with coefficients $0, d_2, 0, \ldots, 0$; that is, $d_2\mathrm{a}_2$. Similarly, the third column of AD is $d_3\mathrm{a}_3$. In general, column k of AD is $d_k\mathrm{a}_k$; that is, d_k times the kth column of A.

$$AD = \begin{bmatrix} d_1\mathrm{a}_1 & d_2\mathrm{a}_2 & \cdots & d_n\mathrm{a}_n \\ \downarrow & \downarrow & & \downarrow \\ & & & \end{bmatrix}.$$

☺

We have encountered the concept of "matrix transpose." Here are the basic properties of this operation (not all are dull!).

Properties of the Transpose of a Matrix

Let A and B be matrices and let c be a scalar.

1. $(A+B)^T = A^T + B^T$ (assuming A and B have the same size);
2. $(cA)^T = cA^T$;
3. $(A^T)^T = A$;
4. $(AB)^T = B^T A^T$ (assuming AB is defined): the transpose of a product is the product of the transposes, **order reversed**.

The last property is the most interesting. While proofs are straightforward, they are also ugly or a bit advanced, so we'll just content ourselves with some remarks.

Notice first that the "obvious" rule for the transpose of a product, $(AB)^T = A^T B^T$, cannot possibly work since if A is 2×3 and B is 3×4, for example, then A^T is 3×2 and B^T is 4×3, so the product $A^T B^T$ is not even defined. On the other hand, the product $B^T A^T$ is defined and produces a 4×2 matrix, precisely the size of $(AB)^T$ because AB is 2×4. This argument does not **prove** that $(AB)^T = B^T A^T$, of course, but it gives us reason to hope. In a similar spirit, examples suggest that $(AB)^T$ might be always the same as $B^T A^T$.

5.26 Example. Suppose $A = \begin{bmatrix} 1 & 2 & 3 \\ -4 & 0 & -1 \end{bmatrix}$ and $B = \begin{bmatrix} 0 & 1 \\ 1 & -3 \\ 4 & 2 \end{bmatrix}$. Then

$$AB = \begin{bmatrix} 1 & 2 & 3 \\ -4 & 0 & -1 \end{bmatrix} \begin{bmatrix} 0 & 1 \\ 1 & -3 \\ 4 & 2 \end{bmatrix} = \begin{bmatrix} 14 & 1 \\ -4 & -6 \end{bmatrix}.$$

Now $A^T = \begin{bmatrix} 1 & -4 \\ 2 & 0 \\ 3 & -1 \end{bmatrix}$ and $B^T = \begin{bmatrix} 0 & 1 & 4 \\ 1 & -3 & 2 \end{bmatrix}$, so

$$B^T A^T = \begin{bmatrix} 0 & 1 & 4 \\ 1 & -3 & 2 \end{bmatrix} \begin{bmatrix} 1 & -4 \\ 2 & 0 \\ 3 & -1 \end{bmatrix} = \begin{bmatrix} 14 & -4 \\ 1 & -6 \end{bmatrix} = (AB)^T.$$ ☺

5.27 Problem. Suppose A is an $m \times n$ matrix. Use properties of the transpose and our knowledge that $AI_n = A$ to prove that $I_m A = A$ too.

Solution. Let $X = I_m A$. We wish to prove that $X = A$. Now $X^T = (I_m A)^T = A^T I_m^T = A^T I_m$ because $I_m^T = I_m$. Now $A^T I_m = A^T$, so $X^T = A^T$ and $X = (X^T)^T = (A^T)^T = A$. ♘

True/False Questions for Week 5

Decide, with as little calculation as possible, whether each of the following statements is true or false and explain your answer whenever you say "false."

1. The matrix $A = \begin{bmatrix} 1 & 2 \\ 3 & 4 \\ 5 & 6 \end{bmatrix}$ is 3×2.
2. You can only add matrices of the same size.
3. The $(2, 1)$ entry of $\begin{bmatrix} 1 & -1 & 0 \\ 2 & 3 & 4 \end{bmatrix}$ is -1.
4. If A and B and matrices and $2A - B = 0$, the zero matrix, then each entry of A is twice the corresponding entry of B.

5. If $A = \begin{bmatrix} 1 & -2 & 3 \\ 4 & 5 & 0 \end{bmatrix}$, then $A^T = \begin{bmatrix} 1 & 4 \\ -2 & 5 \\ 3 & 0 \end{bmatrix}$.

6. If A and B are matrices such that both AB and BA are defined, then A and B are square.

7. You can only multiply matrices of the same size.

8. If u and v are vectors, then $\mathsf{u} \cdot \mathsf{v}$ is the matrix product $\mathsf{u}^T\mathsf{v}$.

9. If $A = \begin{bmatrix} 1 & 2 & 3 & 4 \\ 5 & 6 & 7 & 8 \\ 9 & 10 & 11 & 12 \end{bmatrix}$ and $\mathsf{b} = \begin{bmatrix} 0 \\ 0 \\ 1 \\ 0 \end{bmatrix}$, then $A\mathsf{b} = \begin{bmatrix} 3 \\ 7 \\ 11 \end{bmatrix}$.

10. If A is a matrix and $A\mathsf{x} = 0$ for every (compatible) vector x, then $A = 0$ is the zero matrix.

11. If A is a 2×3 matrix and B is 3×5, then $(AB)^T$ is 5×2.

12. Suppose A is an $m \times n$ matrix and $A^2 = A$. Then A must be square.

13. The statement $(AB)C = A(BC)$ says that matrix multiplication is commutative.

14. If A and B are matrices, $A \neq 0$, and $AB = 0$, then $B = 0$.

15. If A and B are matrices and $AB = 0$, then $AXB = 0$ for any matrix X (of appropriate size).

16. If $A = \begin{bmatrix} 1 & 2 & 3 & 4 \\ 5 & 6 & 7 & 8 \\ 9 & 10 & 11 & 12 \end{bmatrix}$ and $\mathsf{x} = \begin{bmatrix} -2 \\ 3 \\ 5 \\ -8 \end{bmatrix}$, then $A\mathsf{x} = -2\begin{bmatrix} 1 \\ 5 \\ 9 \end{bmatrix} + 3\begin{bmatrix} 2 \\ 6 \\ 10 \end{bmatrix} + 5\begin{bmatrix} 3 \\ 7 \\ 11 \end{bmatrix} - 8\begin{bmatrix} 4 \\ 8 \\ 12 \end{bmatrix}$.

17. $2\begin{bmatrix} 1 \\ 6 \\ 4 \end{bmatrix} - 7\begin{bmatrix} 3 \\ -1 \\ 2 \end{bmatrix} + 5\begin{bmatrix} -1 \\ 2 \\ 3 \end{bmatrix} = \begin{bmatrix} 1 & 3 & -1 \\ 6 & -1 & 2 \\ 4 & 2 & 3 \end{bmatrix}\begin{bmatrix} 2 \\ -7 \\ 5 \end{bmatrix}$.

Week 5 Test Yourself

Here are a few problems with short answers that you can use to test your understanding of the concepts you have met this week.

1. Write down the 2×4 matrix A with $a_{ij} = i^j$.

2. Write down the 3×2 zero matrix.

3. Given $A = B$ with $A = \begin{bmatrix} 5x - 4y & -x \\ 3x + 3y & -3x - 3y \\ \frac{1}{2}x + y & x + 2y \\ -x + y & x - 2y \end{bmatrix}$ and $B = \begin{bmatrix} 8 & 4 \\ -33 & 33 \\ -9 & -18 \\ -3 & 10 \end{bmatrix}$, find x and y.

4. Write down the 2×3 matrix A for which $a_{ij} = 3i - 2j - 1$.

5. If possible, find $A+B$, $A-B$, $3A^T+B$ and $2A-B$ given $A = \begin{bmatrix} 2 & 1 & 1 \\ -1 & -1 & 4 \end{bmatrix}$; $B = \begin{bmatrix} 2 & -3 & 4 \\ -3 & 1 & 2 \end{bmatrix}$.

6. Find AB and BA (if defined) given $A = \begin{bmatrix} 1 & 2 & -4 \\ -6 & 8 & 1 \end{bmatrix}$; $B = \begin{bmatrix} -3 & 4 \\ 2 & 6 \end{bmatrix}$.

7. If A is any $n \times n$ matrix and x is a vector in R^n, what is the size of $\mathsf{x}^T A \mathsf{x}$ and why?

8. Let $A = \begin{bmatrix} 1 & 2 & 3 \\ 4 & 5 & 6 \\ 7 & 8 & 9 \end{bmatrix}$, $\mathsf{x} = \begin{bmatrix} -4 \\ 1 \\ 7 \end{bmatrix}$, and $\mathsf{b} = \begin{bmatrix} 19 \\ 31 \\ 43 \end{bmatrix}$. Given that $A\mathsf{x} = \mathsf{b}$, write b as a linear combination of the columns of A.

9. Verify that $(AB)^T = B^T A^T \neq A^T B^T$ with $A = \begin{bmatrix} 1 & 3 \\ 2 & 1 \end{bmatrix}$ and $B = \begin{bmatrix} 1 & 3 \\ 0 & 1 \end{bmatrix}$.

10. If A and B are matrices of the appropriate size and c is a scalar, then $c(AB) = (cA)B$ (scalar associativity). This is more useful than it perhaps looks. Illustrate by computing the product XY with $X = \frac{1}{3}\begin{bmatrix} -3 & 5 & -2 \\ 4 & -21 & 16 \\ -1 & 14 & -8 \end{bmatrix}$ and $Y = \begin{bmatrix} 0 & 2 & 3 \\ 4 & 5 & 6 \\ 7 & 8 & 9 \end{bmatrix}$.

11. Let $A = \begin{bmatrix} 1 & 0 \\ 2 & -1 \end{bmatrix}$ and $B = \begin{bmatrix} 2 & 4 \\ -3 & 1 \end{bmatrix}$. Check that $AB \neq BA$ and that $(A+B)^2 \neq A^2 + 2AB + B^2$.

12. What is the name of the matrix $\begin{bmatrix} 1 & 0 & 0 \\ 0 & 1 & 0 \\ 0 & 0 & 1 \end{bmatrix}$?

13. Explain the significance of the matrix $A = \begin{bmatrix} 1 & 0 \\ 0 & 1 \end{bmatrix}$.

14. Let $A = \begin{bmatrix} 1 & 2 & 3 \\ 4 & 5 & 6 \end{bmatrix}$ and $\mathsf{x} = \begin{bmatrix} -2 \\ 3 \\ 5 \end{bmatrix}$. Write $A\mathsf{x}$ as a linear combination of the columns of A.

15. Write $4\begin{bmatrix} 1 \\ 2 \\ 3 \end{bmatrix} - 3\begin{bmatrix} 5 \\ 6 \\ 7 \end{bmatrix}$ in the form $A\mathsf{x}$.

16. Write down the matrix product $\begin{bmatrix} 1 & 2 \\ 3 & 4 \\ 5 & 6 \end{bmatrix} \begin{bmatrix} 10 & 0 & 0 & 0 \\ 0 & 20 & 0 & 0 \end{bmatrix}$ without calculation.

17. Suppose E is a matrix and $E^2 = E$. Why must E be square?

18. Suppose $A = LU$ is the product of a matrix L and and a matrix U the same size as A. Explain why L must be square.

Week 6

Systems of Linear Equations

A *linear equation* in variables $x_1, x_2, \ldots$ is an equation like $2x_1 - 3x_2 + x_3 - \sqrt{2}x_4 = 17$, where each variable appears by itself (and to the first power) and the coefficients are scalars. A set of one or more linear equations is called a *linear system*. *To solve* a linear system means to find values of the variables that make each equation true.

Many people could solve the system

$$\begin{aligned} x + y &= 7 \\ x - 2y &= -2 \end{aligned} \tag{10}$$

via some means or other and, in fact, I have assumed you can solve simple systems at several parts of this course already. But suppose you were faced with a system of eight equations in 17 unknowns. How would you go about finding a solution in this case? In this section, we describe a **procedure** for solving systems of linear equations called "Gaussian Elimination." First we need the concept of equivalence.

Two linear systems are said to be *equivalent* if they have the same solutions. For example, system (10) is equivalent to

$$\begin{aligned} x + y &= 7 \\ 3y &= 9 \end{aligned} \tag{11}$$

because each system has the same solution, $x = 4$, $y = 3$.

Gaussian elimination involves transforming one system to another system that is equivalent but easier to solve. For example, system (11) makes it immediately clear that $y = 3$, so it is easier to solve than (10).

There are three ways to change a system of linear equations to an equivalent system. Maybe there are more, but these three have a long history!

First, we can change the order of the equations. For example, the numbers x, y, and z that satisfy

$$\begin{aligned} -2x + y + z &= -9 \\ x + y + z &= 0 \\ x + 3y &= -3 \end{aligned} \tag{12}$$

are the same as the numbers that satisfy

$$\begin{aligned} x + y + z &= 0 \\ -2x + y + z &= -9 \\ x + 3y &= -3, \end{aligned}$$

since the equations in each system are the same: we simply interchanged the first two equations of (12).

Second, in any system of equations, we can multiply any equation by a number (different from 0) without changing the solution. For example,

$$x + 3y = -3 \quad \text{if and only if} \quad \tfrac{1}{3}x + y = -1.$$

Third, and this is less obvious, in any system of equations, we can always add or subtract a multiple of one equation (say E') from another (say E) without changing the solution. (I remember this as $E \to E - cE'$.) Suppose, for instance, that we replace the first equation in system (12) by the first equation plus twice the third, giving

$$\begin{array}{rcl} -2x + y + z + 2(x + 3y) &=& -9 + 2(-3) \\ x + y + z &=& 0 \\ x + 3y &=& -3; \end{array}$$

that is,

$$\begin{array}{rcl} 7y + z &=& -3 \\ x + y + z &=& 0 \\ x + 3y &=& -3. \end{array} \tag{13}$$

Any values of x, y, and z that satisfy this system must also satisfy system (12) and, conversely, any values of x, y, and z that satisfy the original system (12) also satisfy (13). The systems (12) and (13) are equivalent.

So we have described three ways to change one system of equations into an equivalent system:

1. interchange two equations;
2. multiply an equation by any number other than 0;
3. $E \to E - cE'$; that is, replace an equation by that equation minus any multiple of another.

By the way, it makes no difference whether we say "minus" or "plus" in Statement 3 since, for example, adding twice an equation is the same as subtracting minus twice that equation, but we prefer the language of subtraction. (I'll tell you why later.)

Here is how these operations might be employed to solve system (12). (This is our first example of Gaussian elimination.)

First, we interchange the first two equations to get

$$\begin{array}{rcl} x + y + z &=& 0 \\ -2x + y + z &=& -9 \\ x + 3y &=& -3. \end{array}$$

Then replace the second equation by that equation plus twice the first—we denote this as $E2 \to E2 + 2(E1)$—to obtain

$$\begin{array}{rcl} x + y + z &=& 0 \\ 3y + 3z &=& -9 \\ x + 3y &=& -3. \end{array} \tag{14}$$

We describe the transformation from system (12) to system (14) like this:

$$\begin{array}{rcr} -2x + y + z & = & -9 \\ x + y + z & = & 0 \\ x + 3y & = & -3 \end{array} \quad \xrightarrow{E1 \leftrightarrow E2} \quad \begin{array}{rcr} x + y + z & = & 0 \\ -2x + y + z & = & -9 \\ x + 3y & = & -3 \end{array}$$

$$\xrightarrow{E2 \to E2 + 2(E1)} \quad \begin{array}{rcr} x + y + z & = & 0 \\ 3y + 3z & = & -9 \\ x + 3y & = & -3. \end{array}$$

We continue with our solution, replacing equation 3 with equation 3 minus equation 1:

$$\begin{array}{rcr} x + y + z & = & 0 \\ 3y + 3z & = & -9 \\ x + 3y & = & -3 \end{array} \quad \xrightarrow{E3 \to E3 - E1} \quad \begin{array}{rcr} x + y + z & = & 0 \\ 3y + 3z & = & -9 \\ 2y - z & = & -3. \end{array}$$

Now we multiply the equation 2 by $\frac{1}{3}$ and replace equation 3 with equation 3 minus twice equation 2:

$$\begin{array}{rcr} x + y + z & = & 0 \\ 3y + 3z & = & -9 \\ 2y - z & = & -3 \end{array} \quad \xrightarrow{E2 \to \frac{1}{3}E2} \quad \begin{array}{rcr} x + y + z & = & 0 \\ y + z & = & -3 \\ 2y - z & = & -3 \end{array}$$

$$\xrightarrow{E3 \leftarrow E3 - 2(E2)} \quad \begin{array}{rcr} x + y + z & = & 0 \\ y + z & = & -3 \\ -3z & = & 3. \end{array}$$

We have reached our goal, which was to transform the original system to one that is *upper triangular*, like this: ◹ .

At this point, we can obtain a solution in a straightforward manner by using a technique called *back substitution*, which means solving the equations from the bottom up. The third equation gives $z = -1$. The second, $y + z = -3$, tells us $y = -3 - z = -2$ and the first, $x + y + z = 0$, gives $x = -y - z = 3$.

It is easy to check that $x = 3$, $y = -2$, $z = -1$ satisfies the original system (12), so we have our solution. **Always check your answer!**

We can make the process just described look a lot simpler if we notice that we have only to keep track of the coefficients and the numbers to the right of the equals signs. We can remember where the x, y, z and $=$ go (I hope!). So we solve system (12) again, writing down only what is important.

In matrix form, our system was

$$\begin{bmatrix} -2 & 1 & 1 \\ 1 & 1 & 1 \\ 1 & 3 & 0 \end{bmatrix} \begin{bmatrix} x \\ y \\ z \end{bmatrix} = \begin{bmatrix} -9 \\ 0 \\ 3 \end{bmatrix},$$

which has the form $A\mathsf{x} = \mathsf{b}$. The matrix $A = \begin{bmatrix} -2 & 1 & 1 \\ 1 & 1 & 1 \\ 1 & 3 & 0 \end{bmatrix}$, is called (naturally) the *matrix of coefficients.* The vector $\mathsf{x} = \begin{bmatrix} x \\ y \\ z \end{bmatrix}$ holds the unknowns and $\mathsf{b} = \begin{bmatrix} -9 \\ 0 \\ -3 \end{bmatrix}$ is the vector whose components are the numbers on the right side of the equals signs. The *augmented matrix,* which we denote symbolically $[A|\mathsf{b}]$ is A followed by a vertical line and then b:

$$[A|\mathsf{b}] = \left[\begin{array}{ccc|c} -2 & 1 & 1 & -9 \\ 1 & 1 & 1 & 0 \\ 1 & 3 & 0 & -3 \end{array}\right].$$

Just keeping track of the numbers, with augmented matrices, our solution to system (12) now looks like this:

$$\left[\begin{array}{ccc|c} -2 & 1 & 1 & -9 \\ 1 & 1 & 1 & 0 \\ 1 & 3 & 0 & -3 \end{array}\right] \overset{R1 \leftrightarrow R2}{\longrightarrow} \left[\begin{array}{ccc|c} 1 & 1 & 1 & 0 \\ -2 & 1 & 1 & -9 \\ 1 & 3 & 0 & -3 \end{array}\right]$$

$$\overset{\substack{R2 \to R2 + 2(R1) \\ R3 \to R3 - R1}}{\longrightarrow} \left[\begin{array}{ccc|c} 1 & 1 & 1 & 0 \\ 0 & 3 & 3 & -9 \\ 0 & 2 & -1 & -3 \end{array}\right] \overset{R2 \to \frac{1}{3}R2}{\longrightarrow} \left[\begin{array}{ccc|c} 1 & 1 & 1 & 0 \\ 0 & 1 & 1 & -3 \\ 0 & 2 & -1 & -3 \end{array}\right] \qquad (15)$$

$$\overset{R3 \to R3 - 2(R2)}{\longrightarrow} \left[\begin{array}{ccc|c} 1 & 1 & 1 & 0 \\ 0 & 1 & 1 & -3 \\ 0 & 0 & -3 & 3 \end{array}\right] \qquad (16)$$

Here we have used R for *row* rather than E for *equation* to describe the changes to each system, writing, for example, $R3 \to R3 - 2(R2)$ at the last step, rather than $E3 \to E3 - 2(E2)$. The three basic operations that we have been performing on equations now become operations on rows, called the *elementary row operations.*

6.1 The Elementary Row Operations.

1. Interchange two rows.
2. Multiply a row by any scalar except 0.
3. $R \to R - cR'$; that is, replace a row by that row minus a multiple of another.

We could equally well have stated the third elementary row operation as

3. replace a row by that row *plus* a multiple of another,

since $R1 + c(R2) = R1 - (-c)(R2)$, but, as we said before, there are advantages to using the language of subtraction. In particular, when doing so, the multiple has a special name. It's called a *multiplier,* $-c$ in our example, a concept that will be important to us later.

We described an upper triangular matrix with a picture and we do so again in Figure 12. Here's the precise definition.

6.2 Definitions. The *main diagonal* of an $m \times n$ matrix A is the list of elements $a_{11}, a_{22}, \ldots$. A matrix is *diagonal* if its only nonzero entries lie on the main diagonal, *upper triangular* if all entries below the main diagonal are 0, and *lower triangular* if all entries above the main diagonal are 0.

Diagonal Upper triangular Lower triangular

Figure 12: Schematic representations of a diagonal, lower triangular, and upper triangular matrix.

6.3 Examples. The main diagonal of $\begin{bmatrix} 1 & 2 & 3 \\ 4 & 5 & 6 \\ 7 & 8 & 9 \end{bmatrix}$ is $1, 5, 9$, and the main diagonal of $\begin{bmatrix} 7 & 9 & 3 \\ 6 & 5 & 4 \end{bmatrix}$ is $7, 5$. The matrix $\begin{bmatrix} -3 & 4 & 1 & 1 \\ 0 & 8 & 2 & -3 \end{bmatrix}$ is upper triangular while the matrix $\begin{bmatrix} 1 & 0 \\ -2 & 1 \end{bmatrix}$ is lower triangular. The matrix $\begin{bmatrix} 1 & 0 & 0 \\ 0 & -7 & 0 \\ 0 & 0 & 2 \end{bmatrix}$ is diagonal.

The goal of a procedure called "Gaussian elimination" that we have illustrated once and shall soon describe is to change a system of equations to an equivalent system, one where the matrix of coefficients is upper triangular.

6.4 Example. The system

$$\begin{aligned} x + 2y &= 1 \\ 4x + 8y &= 3 \end{aligned}$$

is $A\mathsf{x} = \mathsf{b}$ with $A = \begin{bmatrix} 1 & 2 \\ 4 & 8 \end{bmatrix}$, $\mathsf{x} = \begin{bmatrix} x \\ y \end{bmatrix}$, and $\mathsf{b} = \begin{bmatrix} 1 \\ 3 \end{bmatrix}$. The augmented matrix is $[A|\mathsf{b}] = \left[\begin{array}{cc|c} 1 & 2 & 1 \\ 4 & 8 & 3 \end{array}\right]$. To solve, we first observe that the $(1,1)$ entry of the augmented matrix is not zero, so no row interchanges are necessary. The first pivot is that 1 in the $(1,1)$ position. Gaussian elimination on the augmented matrix requires only one step.

$$\left[\begin{array}{cc|c} \textcircled{1} & 2 & 1 \\ 4 & 8 & 3 \end{array}\right] \xrightarrow{R2 \to R2 - 4(R1)} \left[\begin{array}{cc|c} 1 & 2 & 1 \\ 0 & 0 & -1 \end{array}\right].$$

The equations that correspond to this triangular matrix are

$$\begin{aligned} x + 2y &= 1 \\ 0 &= -1. \end{aligned}$$

The last equation is not true, so there is no solution, a fact we might have noted earlier. If $x + 2y = 1$, then certainly $4x + 8y = 4(x + 2y) = 4$, not 3. So if the first equation is satisfied, the second one is not. A system like this, with no solution, is called *inconsistent*.

:-)

6.5 Gaussian Elimination. To move a matrix to one that is upper triangular,

1. If the matrix is all 0, do nothing. Otherwise, change the order of the rows, if necessary, to ensure that the first non-zero entry in the first row is a leftmost nonzero entry in the matrix. This number is called a *pivot* and the column containing this pivot is the first *pivot column.*

2. Use the third elementary row operation to get 0s in the pivot column below the pivot.

3. Repeat steps 1 and 2 on the submatrix immediately to the right of and below the pivot.

4. Repeat step 3 until you reach the last nonzero row. The first nonzero entry in this row (assuming this row has never been multiplied by a scalar) is the last pivot.

If this process looks complicated, this is only because it has to be written to cover all possible situations, and there are many possible situations. In the easiest situations, the pivots are all on the diagonal. Unless the first column is all 0, we can always arrange the rows so that the $(1, 1)$-entry is nonzero. Then we use the third elementary row operation to make the rest of column one all 0. Then rearrange the rows if necessary to get the $(2, 2)$-entry nonzero and use this element to make the rest of column two (below the $(2, 2)$-entry) all 0. Now continue column by column sweeping across the matrix from left to right each time trying to get something nonzero in the diagonal position and using this element to get 0s below.

6.6 Example. We use Gaussian elimination to bring $A = \begin{bmatrix} -1 & 2 & 3 \\ 4 & 1 & 0 \\ -2 & 5 & 8 \end{bmatrix}$ to an upper triangular matrix. The -1 in the $(1, 1)$-position is just what we need. It's our first pivot and we use this and the third elementary operation to put 0s in column 1 below the pivot, like this.

$$A \to \begin{bmatrix} -1 & 2 & 3 \\ 0 & 9 & 12 \\ 0 & 1 & 2 \end{bmatrix} \tag{17}$$

Now we have two choices. Working by hand, we might interchange rows two and three in order to put the 1 in the $(2,2)$-position:

$$\begin{bmatrix} -1 & 2 & 3 \\ 0 & 9 & 12 \\ 0 & 1 & 2 \end{bmatrix} \to \begin{bmatrix} -1 & 2 & 3 \\ 0 & 1 & 2 \\ 0 & 9 & 12 \end{bmatrix}$$

and use this 1 (the second pivot) to put a 0 where the 9 is:

$$\begin{bmatrix} -1 & 2 & 3 \\ 0 & 1 & 2 \\ 0 & 9 & 12 \end{bmatrix} \to \begin{bmatrix} -1 & 2 & 3 \\ 0 & 1 & 2 \\ 0 & 0 & -6 \end{bmatrix}.$$

Or, returning to (17), if we do not interchange rows two and three, we could leave the second row as it is (or divide by 9) and use the $(2,2)$-entry to put a 0 below. So, eitherc

$$\begin{bmatrix} -1 & 2 & 3 \\ 0 & 9 & 12 \\ 0 & 1 & 2 \end{bmatrix} \to \begin{bmatrix} -1 & 2 & 3 \\ 0 & 9 & 12 \\ 0 & 0 & 2-\frac{12}{9} \end{bmatrix} = \begin{bmatrix} -1 & 2 & 3 \\ 0 & 9 & 12 \\ 0 & 0 & \frac{2}{3} \end{bmatrix}$$

or

$$\begin{bmatrix} -1 & 2 & 3 \\ 0 & 9 & 12 \\ 0 & 1 & 2 \end{bmatrix} \to \begin{bmatrix} -1 & 2 & 3 \\ 0 & 1 & \frac{4}{3} \\ 0 & 1 & 2 \end{bmatrix} \to \begin{bmatrix} -1 & 2 & 3 \\ 0 & 1 & \frac{4}{3} \\ 0 & 0 & \frac{2}{3} \end{bmatrix}.$$

There many ways to skin a cat! ⌣̈

6.7 Example. Sometimes we have to fiddle to make a desired diagonal entry nonzero. Suppose $A = \begin{bmatrix} -2 & 1 & 0 & 1 \\ 3 & -3 & 1 & 1 \\ 1 & -1 & 1 & 5 \end{bmatrix}$. Here's a start at Gaussian elimination:

$$A \to \begin{bmatrix} 1 & -1 & 1 & 5 \\ 3 & -3 & 1 & 1 \\ -2 & 1 & 0 & 1 \end{bmatrix}$$

(interchanging rows one and three because it is easier to work by hand with a 1 in the top left corner)

$$\to \begin{bmatrix} 1 & -1 & 1 & 5 \\ 0 & 0 & -2 & -14 \\ 0 & -1 & 2 & 10 \end{bmatrix}$$

putting 0 in the rest of column one. To work on column two, we would like the $(2,2)$-entry to be nonzero. Here then, we "fiddle," interchanging rows two and three and continuing, like this

$$\to \begin{bmatrix} 1 & -1 & 1 & 5 \\ 0 & -1 & 2 & 10 \\ 0 & 0 & -2 & -14 \end{bmatrix}.$$

Column two is fine and—hey! this is upper triangular—so we are finished. ⌣̈

6.8 Example. It's not always possible to "fiddle" as in the previous example. Here's the start of Gaussian elimination on $A = \begin{bmatrix} 1 & -2 & -1 & 1 \\ -3 & 6 & 5 & 1 \\ -1 & 2 & 4 & 3 \\ 1 & -2 & 1 & 1 \end{bmatrix}$.

$$A \to \begin{bmatrix} 1 & -2 & -1 & 1 \\ 0 & 0 & 2 & 4 \\ 0 & 0 & 3 & 4 \\ 0 & 0 & 2 & 0 \end{bmatrix}.$$

There is no way to make the $(2,2)$-entry a 0 (without making an interchange with row one, which is not allowed). Gaussian elimination keeps moving to the right and down. So we move to column three. We could leave the 2 4 alone or divide by 2.

$$\to \begin{bmatrix} 1 & -2 & -1 & 1 \\ 0 & 0 & 1 & 2 \\ 0 & 0 & 3 & 4 \\ 0 & 0 & 2 & 0 \end{bmatrix} \to \begin{bmatrix} 1 & -2 & -1 & 1 \\ 0 & 0 & 1 & 2 \\ 0 & 0 & 0 & -2 \\ 0 & 0 & 0 & -4 \end{bmatrix},$$

which is upper triangular. ☺

6.9 To Solve a System of Linear Equations $A\mathsf{x} = \mathsf{b}$

1. Write down the *augmented matrix* $[A|\mathsf{b}]$.
2. Use Gaussian elimination to reduce the augmented matrix to an upper triangular matrix.
3. Write down the equations that correspond to the triangular matrix and solve by back substitution.

We have already seen that a system of equations may have no solution, and in situations where there is a solution, the solution may not be unique: There may be more than one solution.

6.10 Example. The system

$$\begin{aligned} 2x - y - 7z &= 6 \\ x + y - 2z &= -3 \end{aligned}$$

is $A\mathsf{x} = \mathsf{b}$, with

$$A = \begin{bmatrix} 2 & -1 & -7 \\ 1 & 1 & -2 \end{bmatrix}, \quad \mathsf{x} = \begin{bmatrix} x \\ y \\ z \end{bmatrix} \quad \text{and} \quad \mathsf{b} = \begin{bmatrix} 6 \\ -3 \end{bmatrix}. \tag{18}$$

The augmented matrix is

$$[A|\mathsf{b}] = \left[\begin{array}{ccc|c} 2 & -1 & -7 & 6 \\ 1 & 1 & -2 & -3 \end{array}\right].$$

While there is no need for initial row interchanges, it is easier to work with a 1 rather than a 2 in the $(1, 1)$ position, so we interchange rows one and two (the 1 becomes the first pivot), and then apply the third elementary row operation to get a 0 below the pivot.

$$\left[\begin{array}{ccc|c} 2 & -1 & -7 & 6 \\ 1 & 1 & -2 & -3 \end{array}\right] \xrightarrow{R1 \leftrightarrow R2} \left[\begin{array}{ccc|c} \textcircled{1} & 1 & -2 & -3 \\ 2 & -1 & -7 & 6 \end{array}\right] \xrightarrow{R2 \to R2-2(R1)} \left[\begin{array}{ccc|c} 1 & 1 & -2 & -3 \\ 0 & \textcircled{-3} & -3 & 12 \end{array}\right]$$

Now we focus on the submatrix immediately to the right of and below the first pivot. This is $[-3 \ -3 \ 12]$. The second pivot is the -3. The matrix is upper triangular. We can stop here, or divide the last row by the -3 (at the price of the disappearance of the second pivot), obtaining

$$\left[\begin{array}{ccc|c} 1 & 1 & -2 & -3 \\ 0 & 1 & 1 & -4 \end{array}\right].$$

The equations corresponding to this matrix are

$$\begin{aligned} x + y - 2z &= -3 \\ y + z &= -4. \end{aligned}$$

Applying back substitution, the last equation says $y = -4 - z$ and the first says $x + y - 2z = -3$. So $x = -3 - y + 2z = -3 - (-4 - z) + 2z = 1 + 3z$. The solution is

$$\begin{aligned} x &= 1 + 3z \\ y &= -4 - z \\ z &= z. \end{aligned}$$

This is an example of a system that has infinitely many solutions. Our solution shows that z can be any number at all, but once we have chosen z, both x and y are determined: $x = 1 + 3z$, $y = -4 - z$. To emphasize the idea that z can be any number, we introduce the *parameter* t and let $z = t$. The solution is then written

$$\begin{aligned} x &= 1 + 3t \\ y &= -4 - t \\ z &= t. \end{aligned}$$

The variable z is called *free* because, as the equation $z = t$ tries to show, z can be any number. The system of equations in this problem has infinitely many solutions, one for each value of t. We write the solution as a vector, like this:

$$\begin{bmatrix} x \\ y \\ z \end{bmatrix} = \begin{bmatrix} 1 + 3t \\ -4 - t \\ t \end{bmatrix} = \begin{bmatrix} 1 \\ -4 \\ 0 \end{bmatrix} + t \begin{bmatrix} 3 \\ -1 \\ 1 \end{bmatrix}.$$

Such an equation should look familiar. It is the equation of a line, the one through $(1, -4, 0)$ with direction $\mathsf{d} = \begin{bmatrix} 3 \\ -1 \\ 1 \end{bmatrix}$. Geometrically, this is what we expect. Each of the two given equations describes a plane in 3-space. When we solve the system, we are finding the points that lie on both planes. Since the planes are not parallel, the points that lie on both form a line. :)

Any solution to the linear system in Example 6.10 is the sum of the vector $\begin{bmatrix} 1 \\ -4 \\ 0 \end{bmatrix}$ and a scalar multiple of $\begin{bmatrix} 3 \\ -1 \\ 1 \end{bmatrix}$. The solution $\begin{bmatrix} 1 \\ -4 \\ 0 \end{bmatrix}$ is called a *particular solution* and we ask the reader to check that the vector $\begin{bmatrix} 3 \\ -1 \\ 1 \end{bmatrix}$ is a solution to the system $A\mathsf{x} = \mathsf{0}$. It is characteristic of linear systems that whenever there is more than one solution, each solution is the sum of a particular solution and the solution to $A\mathsf{x} = \mathsf{0}$.

6.11 Example. The system

$$\begin{aligned} 3x_2 - 12x_3 - 7x_4 &= -15 \\ x_2 - 4x_3 &= 2 \\ -2x_1 + 4x_2 + 5x_4 &= -1. \end{aligned}$$

is $A\mathsf{x} = \mathsf{b}$ with

$$A = \begin{bmatrix} 0 & 3 & -12 & -7 \\ 0 & 1 & -4 & 0 \\ -2 & 4 & 0 & 5 \end{bmatrix}, \quad \mathsf{x} = \begin{bmatrix} x_1 \\ x_2 \\ x_3 \\ x_4 \end{bmatrix}, \quad \mathsf{b} = \begin{bmatrix} -15 \\ 2 \\ -1 \end{bmatrix}.$$

We form the augmented matrix and immediately interchange rows one and three in order to get a nonzero entry in the $(1,1)$ position.

$$\left[\begin{array}{cccc|c} 0 & 3 & -12 & -7 & -15 \\ 0 & 1 & -4 & 0 & 2 \\ -2 & 4 & 0 & 5 & -1 \end{array}\right] \xrightarrow{R1 \leftrightarrow R3} \left[\begin{array}{cccc|c} \boxed{-2} & 4 & 0 & 5 & -1 \\ 0 & \boxed{1} & -4 & 0 & 2 \\ 0 & 3 & -12 & -7 & -15 \end{array}\right].$$

The first pivot is -2 and the rest of the column below this pivot is already 0. So now we look at the submatrix immediately to the right of and below this pivot. The second pivot is the 1 in the $(2,2)$ position. We use the third elementary row operation to put a 0 below it

$$\xrightarrow{R3 \to R3 - 3(R2)} \left[\begin{array}{cccc|c} -2 & 4 & 0 & 5 & -1 \\ 0 & 1 & -4 & 0 & 2 \\ 0 & 0 & 0 & -7 & -21 \end{array}\right],$$

obtaining an upper triangular matrix. Here is back substitution:

$$\begin{aligned} -7x_4 &= -21 && \text{so } x_4 = 3, \\ x_2 - 4x_3 &= 2, && \text{which we write as } x_2 = 2 + 4x_3. \end{aligned}$$

We see that x_3 is free, so we set $x_3 = t$ and obtain $x_2 = 2 + 4t$. Finally,

$$-2x_1 + 4x_2 + 5x_4 = -1,$$

so $-2x_1 = -1 - 4x_2 - 5x_4 = -1 - 4(2+4t) - 15 = -24 - 16t$ and $x = 12 + 8t$. Our solution vector is

$$\mathsf{x} = \begin{bmatrix} 12+8t \\ 2+4t \\ t \\ 3 \end{bmatrix} = \begin{bmatrix} 12 \\ 2 \\ 0 \\ 3 \end{bmatrix} + t \begin{bmatrix} 8 \\ 4 \\ 1 \\ 0 \end{bmatrix}.$$

As before, the vector $\begin{bmatrix} 12 \\ 2 \\ 0 \\ 3 \end{bmatrix}$ is a particular solution to $A\mathsf{x} = \mathsf{b}$ and $\begin{bmatrix} 8 \\ 4 \\ 1 \\ 0 \end{bmatrix}$ is a solution to $A\mathsf{x} = \mathsf{0}$.

☺

In general, there are three possible outcomes when we attempt to solve a system of linear equations.

6.12 A system of linear equations may have

1. a unique solution,
2. no solution, or
3. infinitely many solutions.

The solution to system (12) with which we began our discussion of systems of linear equations was unique. In Example 6.4, we met an inconsistent system (no solution) and, in Examples 6.10 and 6.11, we met systems with infinitely many solutions.

Row Echelon Form

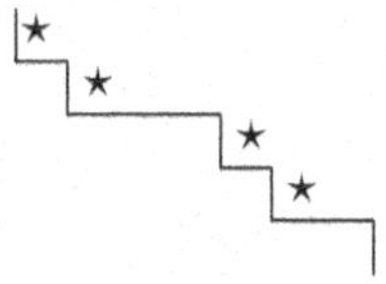

At this point, we confess that the goal of Gaussian elimination is to transform a matrix not simply into any upper triangular matrix, but into a special kind of upper triangular matrix called a *row echelon matrix.* A row echelon matrix takes its name from the French word "échelon" meaning "step." When a matrix is in row echelon form, the path formed by the pivots resembles a staircase.

6.13 Definition. A *row echelon matrix* is an upper triangular matrix with the following properties:

1. If there are any rows consisting entirely of 0s, these are at the bottom.

2. The pivots step from left to right as you read down the matrix (a *pivot* is the first nonzero entry in a nonzero row).

3. All the entries in a column below a pivot are 0.

A *row echelon form* of a matrix A is a row echelon matrix U to which A can be transformed via elementary row operations.

6.14 Definition. If U is a row echelon matrix, a *pivot* of U is the first nonzero element in a nonzero row and a *pivot column* is a column containing a pivot. If A is any matrix, a *pivot column* of A is a pivot column of a row echelon form U of A. If U was obtained without any multiplication of rows by nonzero scalars, then the *pivots of* A are the pivots of U.

6.15 Examples. · The matrix $U = \begin{bmatrix} 0 & 0 & \textcircled{2} & 1 & 5 \\ 0 & 0 & 0 & 0 & \textcircled{1} \\ 0 & 0 & 0 & 0 & 0 \end{bmatrix}$ is a row echelon matrix. The pivots are the circled entries. The pivot columns are columns three and five.

· The matrix $U = \begin{bmatrix} \textcircled{7} & 1 & 3 & 4 & 5 \\ 0 & 0 & \textcircled{-2} & 3 & 1 \\ 0 & 0 & 0 & 0 & \textcircled{4} \end{bmatrix}$ is a row echelon matrix. The pivots are circled. Since these are in columns one, three, and five, the pivot columns are columns one, three, and five.

· $U = \begin{bmatrix} 0 & 0 & 2 & 1 & 5 \\ 0 & 0 & 0 & 0 & 1 \\ 0 & 0 & 0 & 0 & 3 \end{bmatrix}$ is an upper triangular matrix that is not row echelon. If row echelon, the 1 in the $(2,5)$ position would be a pivot, but then the number below it should be 0, not 3.

· $U = \begin{bmatrix} 0 & 7 & 1 & 3 & 4 \\ 0 & 0 & 2 & 1 & 0 \\ 1 & 0 & -1 & 4 & 2 \end{bmatrix}$ is not row echelon because the leading nonzero entries (the wanna-be pivots) do not step from left to right.

6.16 Examples. · Let $A = \begin{bmatrix} 3 & -2 & 1 & -4 \\ 6 & -4 & 8 & -7 \\ 9 & -6 & 9 & -12 \end{bmatrix}$. Then

$$A \xrightarrow[R3 \to R3 - 3(R1)]{R2 \to R2 - 2(R1)} \begin{bmatrix} 3 & -2 & 1 & -4 \\ 0 & 0 & 6 & 1 \\ 0 & 0 & 6 & 0 \end{bmatrix} \xrightarrow{R3 \to R3 - 6(R2)} \begin{bmatrix} \textcircled{3} & -2 & 1 & -4 \\ 0 & 0 & \textcircled{6} & 1 \\ 0 & 0 & 0 & \textcircled{-1} \end{bmatrix},$$

which is row echelon, so the pivots of A are 3, 6 and -1, and the pivot columns of A are columns one, three, and four.

- Let $A = \begin{bmatrix} -2 & 3 & 4 \\ 1 & 0 & 1 \\ 4 & -5 & 6 \end{bmatrix}$. Then $A \xrightarrow{R1 \leftrightarrow R2} \begin{bmatrix} 1 & 0 & 1 \\ -2 & 3 & 4 \\ 4 & -5 & 6 \end{bmatrix}$

$$\xrightarrow[R3 \to R3 - 4(R1)]{R2 \to R2 + 2(R1)} \begin{bmatrix} 1 & 0 & 1 \\ 0 & 3 & 6 \\ 0 & -5 & 2 \end{bmatrix} \xrightarrow{R3 \to R3 + \frac{5}{3}(R2)} \begin{bmatrix} \textcircled{1} & 0 & 1 \\ 0 & \textcircled{3} & 6 \\ 0 & 0 & \textcircled{12} \end{bmatrix},$$

which is row echelon, so the pivots of A are 1, 3, and 12, and the pivot columns of A are columns one, two, and three.

Perhaps a more natural way to have achieved row echelon form here is like this:

$$A \to \begin{bmatrix} 1 & 0 & 1 \\ -2 & 3 & 4 \\ 4 & -5 & 6 \end{bmatrix} \to \begin{bmatrix} 1 & 0 & 1 \\ 0 & 3 & 6 \\ 0 & -5 & 2 \end{bmatrix} \xrightarrow{*} \begin{bmatrix} 1 & 0 & 1 \\ 0 & 1 & 2 \\ 0 & -5 & 2 \end{bmatrix} \to \begin{bmatrix} 1 & 0 & 1 \\ 0 & 1 & 2 \\ 0 & 0 & 12 \end{bmatrix}.$$

This is another row echelon form of A, but we lost the pivots when we multiplied the second row by $\frac{1}{3}$ at the step labelled $*$. This example also illustrates that row echelon form is not unique. Moreover, pivots are not unique. Had we proceeded without interchanging rows one and two at the beginning, like this,

$$A \to \begin{bmatrix} -2 & 3 & 4 \\ 0 & \frac{3}{2} & 3 \\ 0 & 1 & 14 \end{bmatrix} \to \begin{bmatrix} -2 & 3 & 4 \\ 0 & \frac{3}{2} & 3 \\ 0 & 0 & 12 \end{bmatrix},$$

the pivots would have been -2, $\frac{3}{2}$, and 12, not $1, 3, 12$ as they were before. Interesting, while pivots are not unique, the products of two different sets of pivots must be the same, apart from sign. ♧

6.17 Remark. As we just said, row echelon form is not unique, since different sequences of elementary row operations can transform a matrix A into different row echelon matrices. Thus we say **a** row echelon form, not **the** row echelon form. On the other hand, the columns in which the pivots eventually appear—the pivot columns—are unique. This is a fact we simply record without a proof. In the last example, for instance,no matter how our Gaussian elimination proceeded, every column was a pivot column.

6.18 Remark. The definition of row echelon form varies from author to author. It was long the custom, and some authors still require, that the leading nonzero entry in a nonzero row of a row echelon matrix be a 1. Such a matrix (with all pivots 1) is said to be in *reduced row echelon* form, a concept to which we return briefly in Week 7—see 7.17. Making leading nonzero entries 1 makes a lot of sense when doing calculations by hand, but it is much less important today with our ready access to calculators and computers. Having said this, in the rest of this section, we will regularly divide each row containing a pivot by that pivot, thus turning the first nonzero entry in that row into a 1. Putting 0s below a 1 by hand is much easier than putting 0s below $\frac{2}{3}$, for instance (and I make fewer mistakes!).

Free Variables

The pivot columns of a matrix are important for several reasons. The variable that corresponds to a pivot column is never free. For example, look at the pivot in column three of

$$\left[\begin{array}{cccc|c} 1 & 1 & 3 & 4 & 5 \\ 0 & 0 & \textcircled{1} & 3 & 1 \\ 0 & 0 & 0 & 1 & -3 \end{array}\right]. \tag{19}$$

The second equation reads $x_3 + 3x_4 = 1$, so $x_3 = 1 - x_4$ isn't free; it depends on x_4. On the other hand, a variable that corresponds to a column that is not a pivot column is always free. The equations that correspond to the rows of the matrix in (19) are

$$\begin{aligned} x_1 + x_2 + 3x_3 + 4x_4 &= 5 \\ x_3 + 3x_4 &= 1 \\ x_4 &= -3. \end{aligned}$$

The variables x_1, x_3, and x_4 are not free since

$$\begin{aligned} x_4 &= -3 \\ x_3 &= 1 - 3x_4 = 10 \\ x_1 &= 5 - x_2 - 3x_3 - 4x_4 = 5 - x_2 - 3(10) - 4(-3) = -13 - x_2, \end{aligned}$$

but there is no equation of the form $x_2 = *$, so x_2 is free and, to so signify, we set $x_2 = t$, a parameter. The solution is $\begin{bmatrix} x_1 \\ x_2 \\ x_3 \\ x_4 \end{bmatrix} = \begin{bmatrix} -13-t \\ t \\ 10 \\ -3 \end{bmatrix} = \begin{bmatrix} -13 \\ 0 \\ 10 \\ -3 \end{bmatrix} + t\begin{bmatrix} -1 \\ 1 \\ 0 \\ 0 \end{bmatrix}$. To summarize,

6.19

> Free variables are variables that correspond to columns that are not pivot columns.

6.20 Example. The system

$$\begin{aligned} 2x_1 - 2x_2 - x_3 + 4x_4 &= 9 \\ -x_1 + x_2 + 2x_3 + x_4 &= -3 \end{aligned}$$

is $A\mathsf{x} = \mathsf{b}$ where

$$A = \begin{bmatrix} 2 & -2 & -1 & 4 \\ -1 & 1 & 2 & 1 \end{bmatrix}, \quad \mathsf{x} = \begin{bmatrix} x_1 \\ x_2 \\ x_3 \\ x_4 \end{bmatrix}, \quad \text{and} \quad \mathsf{b} = \begin{bmatrix} 9 \\ -3 \end{bmatrix}.$$

The augmented matrix is

$$[A|\mathsf{b}] = \left[\begin{array}{cccc|c} 2 & -2 & -1 & 4 & 9 \\ -1 & 1 & 2 & 1 & -3 \end{array}\right].$$

Here is one instance of Gaussian elimination:

$$\left[\begin{array}{cccc|c} 2 & -2 & -1 & 4 & 9 \\ -1 & 1 & 2 & 1 & -3 \end{array}\right] \xrightarrow{R1\leftrightarrow R2} \left[\begin{array}{cccc|c} -1 & 1 & 2 & 1 & -3 \\ 2 & -2 & -1 & 4 & 9 \end{array}\right]$$

$$\xrightarrow{R2\to R2+2(R1)} \left[\begin{array}{cccc|c} -1 & 1 & 2 & 1 & -3 \\ 0 & 0 & 3 & 6 & 3 \end{array}\right] \xrightarrow{R2\to\frac{1}{3}(R2)} \left[\begin{array}{cccc|c} -1 & 1 & 2 & 1 & -3 \\ 0 & 0 & 1 & 2 & 1 \end{array}\right].$$

This is a row echelon matrix. The pivot columns are columns one and three. The columns which are not pivot columns are columns two and four. The corresponding variables are free and, to make this point, we set $x_2 = t$ and $x_4 = s$. The equations corresponding to the rows of the row echelon matrix are

$$\begin{aligned} -x_1 + x_2 + 2x_3 \;+\; x_4 &= -3 \\ 3x_3 \;+\; 6x_4 &= 3. \end{aligned}$$

The second equation says $3x_3 = 3 - 6x_4 = 3 - 6s$, so $x_3 = 1 - 2s$.
The first equation says
$$\begin{aligned} -x_1 &= -x_2 - 2x_3 - x_4 - 3 \\ &= -t - 2(1-2s) - s - 3 \\ &= -t + 3s - 5, \quad \text{so } x_1 = t - 3s + 5. \end{aligned}$$

The solution is

$$\begin{aligned} x_1 &= t - 3s + 5 \\ x_2 &= t \\ x_3 &= 1 - 2s \\ x_4 &= s \end{aligned}$$

which, in vector form, is

$$\mathsf{x} = \begin{bmatrix} x_1 \\ x_2 \\ x_3 \\ x_4 \end{bmatrix} = \begin{bmatrix} t - 3s + 5 \\ t \\ 1 - 2s \\ s \end{bmatrix} = \begin{bmatrix} 5 \\ 0 \\ 1 \\ 0 \end{bmatrix} + t\begin{bmatrix} 1 \\ 1 \\ 0 \\ 0 \end{bmatrix} + s\begin{bmatrix} -3 \\ 0 \\ -2 \\ 1 \end{bmatrix}.$$

Note that $\begin{bmatrix} 5 \\ 0 \\ 1 \\ 0 \end{bmatrix}$ is a particular solution to $A\mathsf{x} = \mathsf{b}$ while $t\begin{bmatrix} 1 \\ 1 \\ 0 \\ 0 \end{bmatrix} + s\begin{bmatrix} -3 \\ 0 \\ -2 \\ 1 \end{bmatrix}$ is a solution to $A\mathsf{x} = \mathsf{0}$.

☺

6.21 Example. Is the vector $\begin{bmatrix} 9 \\ -3 \end{bmatrix}$ a linear combination of the columns of $A = \begin{bmatrix} 2 & -2 & -1 & 4 \\ -1 & 1 & 2 & 1 \end{bmatrix}$?
The question asks if there are scalars a, b, c, and d so that

$$\begin{bmatrix} 9 \\ -3 \end{bmatrix} = a\begin{bmatrix} 2 \\ -1 \end{bmatrix} + b\begin{bmatrix} -2 \\ 1 \end{bmatrix} + c\begin{bmatrix} -1 \\ 2 \end{bmatrix} + d\begin{bmatrix} 4 \\ 1 \end{bmatrix}.$$

By **5.23**, this is just $\begin{bmatrix} 9 \\ -3 \end{bmatrix} = A\begin{bmatrix} a \\ b \\ c \\ d \end{bmatrix}$, the system considered in Example 6.20. There are many solutions, one of which is $\begin{bmatrix} a \\ b \\ c \\ d \end{bmatrix} = \begin{bmatrix} 5 \\ 0 \\ 1 \\ 0 \end{bmatrix}$. So, for instance, $\begin{bmatrix} 9 \\ -3 \end{bmatrix} = 5\begin{bmatrix} 2 \\ -1 \end{bmatrix} + 0\begin{bmatrix} -2 \\ 1 \end{bmatrix} + 1\begin{bmatrix} -1 \\ 2 \end{bmatrix} + 0\begin{bmatrix} 4 \\ 1 \end{bmatrix}$, the coefficients being the components of the vector $\begin{bmatrix} 5 \\ 0 \\ 1 \\ 0 \end{bmatrix}$. ☺

We conclude this section with a couple of examples of the sort that some students don't like. They involve letters as well as numbers! I hope that what follows will convince even the skeptics that letters aren't so bad.

6.22 Example. Let me (try to) convince you that the system

$$\begin{aligned} x + y + z &= a \\ x + 2y - z &= b \\ 2x - 2y + z &= c \end{aligned}$$

has a unique solution no matter what a, b, and c might be. We apply Gaussian elimination to the augmented matrix. It's a bit of a pain working with letters, but not **that** much of a pain!

$$\left[\begin{array}{ccc|c} 1 & 1 & 1 & a \\ 1 & 2 & -1 & b \\ 2 & -2 & 1 & c \end{array}\right] \to \left[\begin{array}{ccc|c} 1 & 1 & 1 & a \\ 0 & 1 & -2 & b-a \\ 0 & -4 & -1 & c-2a \end{array}\right] \to \left[\begin{array}{ccc|c} 1 & 1 & 1 & a \\ 0 & 1 & -2 & b-a \\ 0 & 0 & -9 & (c-2a)+4(b-a) \end{array}\right].$$

The equation which corresponds to the last row is $-9z = -6a + 4b + c$ which has exactly one solution z no matter the values of a, b, c. The equation corresponding to the second row is $y = 2z + b - a$. Since we know z, we know y and there is just one y because there is just one z. Finally the first row determines just one x to match y and z. Not bad, eh? ☺

6.23 Problem. Consider the system

$$\begin{aligned} x + y + z &= -2 \\ x + 2y - z &= 1 \\ 2x + ay + bz &= 2. \end{aligned}$$

Find conditions on a and b that (i) guarantee a unique solution; (ii) assure there are no solutions and (iii) assure there are infinitely many solutions.

Solution. We apply Gaussian elimination to the augmented matrix (exactly as if the letters were numbers).

$$\left[\begin{array}{ccc|c} 1 & 1 & 1 & -2 \\ 1 & 2 & -1 & 1 \\ 2 & a & b & 2 \end{array}\right] \to \left[\begin{array}{ccc|c} 1 & 1 & 1 & -2 \\ 0 & 1 & -2 & 3 \\ 0 & a-2 & b-2 & 6 \end{array}\right]$$

$$\to \left[\begin{array}{ccc|c} 1 & 1 & 1 & -2 \\ 0 & 1 & -2 & 3 \\ 0 & 0 & (b-2)+2(a-2) & 6-3(a-2) \end{array}\right] = \left[\begin{array}{ccc|c} 1 & 1 & 1 & -2 \\ 0 & 1 & -2 & 3 \\ 0 & 0 & 2a+b-6 & -3a+12 \end{array}\right].$$

Suppose $2a + b - 6 \neq 0$. The equation corresponding to the last row gives us a unique z just as it did in Example 6.22. And then, just as in that example, we get just one y and just one x. There is a unique solution if $2a + b - 6 \neq 0$.

Suppose $2a + b - 6 = 0$.

Case 1. If $-3a + 12 \neq 0$ ($a \neq 4$), the last equation says that 0 is not 0. There is no solution.

Case 2. If $-3a + 12 = 0$ ($a = 4$, $b = -2$), the last equation reads $0 = 0$, which is fine. The variable z is free and there are infinitely many solutions. ♘

True/False Questions for Week 6

Decide, with as little calculation as possible, whether each of the following statements is true or false and explain your answer whenever you say "false."

1. Changing the order of the rows of a 3×3 matrix to $R2$, $R3$, $R1$ is an elementary row operation.

2. If a linear system $A\mathsf{x} = \mathsf{b}$ has a solution, then b is a linear combination of the columns of A.

3. The systems

$$\begin{aligned} x - y - z &= 0 \\ -4x + y + 2z &= -2 \\ 2x + y - z &= 3 \end{aligned} \quad \text{and} \quad \begin{aligned} x - y - z &= 0 \\ -4x + y + 2z &= -2 \\ 3y + z &= 3 \end{aligned}$$

   are equivalent.

4. Suppose we have a system of linear equations and we multiply one of the equations by $\frac{1}{2}$. Then the new system is equivalent to the first.

5. It is possible for a system of linear equations to have exactly two (different) solutions.

6. The matrix $\begin{bmatrix} 0 & 4 & 3 & -4 & 1 \\ 0 & 0 & 0 & 5 & 1 \\ 0 & 0 & 0 & 0 & -1 \end{bmatrix}$ is in row echelon form.

7. The first nonzero entry of a nonzero row of a row echelon matrix is called a pivot.

8. The first nonzero entry of a nonzero row of an upper triangular matrix is called a pivot.

9. The pivots of the matrix $\begin{bmatrix} 4 & 0 & 2 \\ 0 & 7 & 3 \\ 0 & 0 & 1 \end{bmatrix}$ are 4, 7 and 1.

10. The pivots of the matrix $\begin{bmatrix} 4 & 0 & 2 \\ 0 & 7 & 1 \\ 0 & 1 & 1 \end{bmatrix}$ are 4, 7 and $\frac{6}{7}$.

11. The pivots of the matrix $\begin{bmatrix} 0 & 0 & 2 & 1 & 5 \\ 0 & 0 & 0 & 0 & 1 \\ 0 & 0 & 0 & 0 & 3 \end{bmatrix}$ are 2, 1 and 3.

12. In general, if you want to know the pivots of a matrix, you must reduce to row echelon form without multiplying (or dividing) any row by a nonzero scalar.

13. If Gaussian elimination on the augmented matrix of a system of linear equations leads to the matrix $\left[\begin{array}{cccc|c} 0 & 1 & 3 & -4 & 1 \\ 0 & 0 & 0 & 1 & 1 \\ 0 & 0 & 0 & 0 & 1 \end{array}\right]$, the system has a unique solution.

14. In the solution to a system of linear equations, free variables are those that correspond to columns of row echelon form that contain the pivots.

Week 6 Test Yourself

Here are a few problems with short answers that you can use to test your understanding of the concepts you have met this week.

1. Write the system
$$\begin{aligned} -2x_1 + x_2 + 5x_3 &= -10 \\ -8x_1 + 7x_2 + 19x_3 &= -42 \end{aligned}$$
in the form $A\mathsf{x} = \mathsf{b}$. What is A? What is x? What is b?

2. What does it mean to say that two systems of linear equations are "equivalent?"

3. What are the elementary row operations?

4. What is the main diagonal of the matrix $\begin{bmatrix} 1 & 2 \\ 3 & 4 \\ 5 & 6 \\ 7 & 8 \end{bmatrix}$?

5. Which of the following matrices are **not** in row echelon form? Give a reason in each case.

(a) $\begin{bmatrix} 0 & 1 \\ 1 & 0 \end{bmatrix}$ (b) $\begin{bmatrix} 1 & 2 & 3 \\ 0 & 1 & 0 \\ 0 & 0 & 1 \end{bmatrix}$ (c) $\begin{bmatrix} 1 & 5 \\ 0 & 0 \\ 0 & 1 \\ 0 & 0 \\ 0 & 0 \end{bmatrix}$ (d) $\begin{bmatrix} 1 & 2 & 3 & 4 \\ 0 & 0 & -1 & 10 \\ 0 & 0 & 0 & 0 \end{bmatrix}$

6. Reduce $\begin{bmatrix} 1 & -1 & -2 \\ 2 & -3 & -5 \\ -1 & 4 & 5 \end{bmatrix}$ to row echelon form. What are the pivots? What are the pivot columns?

7. In solving the system $A\mathsf{x} = \mathsf{b}$, Gaussian elimination on the augmented matrix $[A|\mathsf{b}]$ led to the row echelon matrix $\left[\begin{array}{cccc|c} 1 & -2 & 3 & -1 & 5 \end{array}\right]$. The variables were x_1, x_2, x_3, x_4.

   (a) Circle the pivots. Identify the free variables.

   (b) Write the solution to the given system as a vector or as a linear combination of vectors, as appropriate.

8. Here's a matrix in row echelon form:

$$\left[\begin{array}{cccc|c} 1 & 0 & 0 & 3 & 2 \\ 0 & 1 & 1 & 2 & 3 \\ 0 & 0 & 0 & 1 & \frac{1}{3} \\ 0 & 0 & 0 & 0 & 0 \end{array}\right]$$

It's the row echelon matrix obtained after Gaussian elimination was applied to the augmented matrix $[A|\mathsf{b}]$. Find the solution, if any, to $A\mathsf{x} = \mathsf{b}$ (in vector form). If there is a solution, state whether this is unique or whether there are infinitely many solutions. The variables are x_1, x_2, x_3, x_4.

9. Let $A = \begin{bmatrix} 2 & 5 \\ 1 & 3 \end{bmatrix}$ and $\mathsf{b} = \begin{bmatrix} b_1 \\ b_2 \end{bmatrix}$.

   (a) Solve the system $A\mathsf{x} = \mathsf{b}$ for $\mathsf{x} = \begin{bmatrix} x_1 \\ x_2 \end{bmatrix}$.

   (b) Write b as a linear combination of the columns of A.

10. Determine whether $\begin{bmatrix} 1 \\ 6 \\ -4 \end{bmatrix}$ is a linear combination of the columns of $A = \begin{bmatrix} 2 & 3 & 4 \\ 4 & 7 & 5 \\ 6 & -1 & 9 \end{bmatrix}$.

11. Find all solutions, if any, to each of the given systems of equations expressing solutions as a linear combination of vectors.

   (a) $$\begin{aligned} 2x + 2y + 2z &= 2 \\ 4x + 6y + 6z &= 4 \\ 6x + 6y + 10z &= 2 \end{aligned}$$

   (b) $$\begin{aligned} -2x_1 + x_2 + 5x_3 &= -10 \\ -8x_1 + 7x_2 + 19x_3 &= -42 \end{aligned}$$

   (c) $$\begin{aligned} 2x - y + z &= 2 \\ 3x + y - 6z &= -9 \\ -x + 2y - 5z &= -4 \end{aligned}$$

12. Consider the system $$\begin{aligned} 5x + 2y &= a \\ -15x - 6y &= b. \end{aligned}$$

   (a) Under what conditions (on a and b), if any, does this system fail to have a solution?

   (b) Under what conditions, if any, does the system have a unique solution? Find any unique solution that may exist.

   (c) Under what conditions are there infinitely many solutions? Find these solutions if and when they exist.

Week 7

Homogeneous Systems

A *homogeneous* system has the form $A\mathsf{x} = \mathsf{0}$: The vector b to the right of the equals sign is the zero vector. For example, the linear system

$$\begin{aligned} x_1 + 2x_2 + x_3 &= 0 \\ x_1 + 3x_2 &= 0 \\ 2x_1 + x_2 + 5x_3 &= 0 \end{aligned} \tag{20}$$

is homogeneous. Unlike linear systems in general, a homogeneous system always has a solution, set all the variables equal to 0. We call this the *trivial* solution. So the real interest in a homogeneous system is whether or not there is a *nontrivial* solution, that is, a solution $\mathsf{x} \neq \mathsf{0}$.

When we speak of the homogeneous system *corresponding to* the system $A\mathsf{x} = \mathsf{b}$, we mean the system $A\mathsf{x} = \mathsf{0}$, with the same matrix of coefficients A. For example, system (20) is the homogeneous system corresponding to

$$\begin{aligned} x_1 + 2x_2 + x_3 &= 2 \\ x_1 + 3x_2 &= -1 \\ 2x_1 + x_2 + 5x_3 &= 13. \end{aligned} \tag{21}$$

When applying Gaussian elimination to a homogeneous system, we never augment the matrix of coefficients with a column of 0s because the elementary row operations don't change these 0s. So to solve the homogeneous system (20), we apply Gaussian elimination directly to the matrix of coefficients,

$$\begin{bmatrix} 1 & 2 & 1 \\ 1 & 3 & 0 \\ 2 & 1 & 5 \end{bmatrix}, \tag{22}$$

like this:

$$\begin{bmatrix} 1 & 2 & 1 \\ 1 & 3 & 0 \\ 2 & 1 & 5 \end{bmatrix} \to \begin{bmatrix} 1 & 2 & 1 \\ 0 & 1 & -1 \\ 0 & -3 & 3 \end{bmatrix} \to \begin{bmatrix} 1 & 2 & 1 \\ 0 & 1 & -1 \\ 0 & 0 & 0 \end{bmatrix}.$$

The last matrix is in row echelon form. There are two pivot columns. Column three is the only nonpivot column and we set the corresponding variable $x_3 = t$ to indicate that x_3 is free. The equations corresponding to the first two rows are

$$\begin{aligned} x_1 + 2x_2 + x_3 &= 0 \\ x_2 - x_3 &= 0. \end{aligned}$$

By back substitution, we get $x_2 = x_3 = t$ and $x_1 = -2x_2 - x_3 = -3t$. In vector form, the solution to the homogeneous system (20) is

$$\mathsf{x} = \begin{bmatrix} x_1 \\ x_2 \\ x_3 \end{bmatrix} = \begin{bmatrix} -3t \\ t \\ t \end{bmatrix} = t \begin{bmatrix} -3 \\ 1 \\ 1 \end{bmatrix}. \tag{23}$$

Interestingly, this solution is closely related to the solution of the *nonhomogeneous system* (21). To solve (21), we employ the same sequence of steps in Gaussian elimination as we did before, except that this time we start with the augmented matrix

$$[A|\mathsf{b}] = \left[\begin{array}{ccc|c} 1 & 2 & 1 & 2 \\ 1 & 3 & 0 & -1 \\ 2 & 1 & 5 & 13 \end{array}\right].$$

We proceed

$$\left[\begin{array}{ccc|c} 1 & 2 & 1 & 2 \\ 1 & 3 & 0 & -1 \\ 2 & 1 & 5 & 13 \end{array}\right] \to \left[\begin{array}{ccc|c} 1 & 2 & 1 & 2 \\ 0 & 1 & -1 & -3 \\ 0 & -3 & 3 & 9 \end{array}\right] \to \left[\begin{array}{ccc|c} 1 & 2 & 1 & 2 \\ 0 & 1 & -1 & -3 \\ 0 & 0 & 0 & 0 \end{array}\right].$$

This is row echelon form. As before, $x_3 = t$ is the lone free variable, but this time back substitution gives $x_2 = -3 + x_2 = -3 + t$, $x_1 = 2 - 2x_2 - x_3 = 2 - 3t$, so the solution

$$\mathsf{x} = \begin{bmatrix} x_1 \\ x_2 \\ x_3 \end{bmatrix} = \begin{bmatrix} 2-3t \\ -3+t \\ t \end{bmatrix} = \begin{bmatrix} 2 \\ -3 \\ 0 \end{bmatrix} + t \begin{bmatrix} -3 \\ 1 \\ 1 \end{bmatrix}. \tag{24}$$

Part of this solution should seem familiar. The vector $\mathsf{x}_h = t \begin{bmatrix} -3 \\ 1 \\ 1 \end{bmatrix}$ is the solution to the homogeneous system (20). The first vector in (24), $\mathsf{x}_p = \begin{bmatrix} 2 \\ -3 \\ 0 \end{bmatrix}$, is a *particular solution* to system (21); we get $\mathsf{x} = \begin{bmatrix} 2 \\ -3 \\ 0 \end{bmatrix}$ when $t = 0$. The general solution is $\mathsf{x}_p + \mathsf{x}_h$, the sum of x_p and a solution to the corresponding homogeneous system. This is one instance of a general pattern that we have already seen in several examples.

7.1 **Proposition.** *Suppose* x_p *is a particular solution to the system* $A\mathsf{x} = \mathsf{b}$ *and* x *is any other solution. Then* $\mathsf{x} = \mathsf{x}_p + \mathsf{x}_h$ *is the sum of* x_p *and a solution* x_h *to the homogeneous system* $A\mathsf{x} = 0$.

Proof. Since x_p and x are both solutions to $A\mathsf{x} = \mathsf{b}$, using distributivity of matrix multiplication, $A(\mathsf{x} - \mathsf{x}_p) = A\mathsf{x} - A\mathsf{x}_p = \mathsf{b} - \mathsf{b} = 0$. So $\mathsf{x} - \mathsf{x}_p = \mathsf{x}_h$ is a solution to the homogeneous system $A\mathsf{x} = 0$. This gives $\mathsf{x} = \mathsf{x}_p + \mathsf{x}_h$. ■

In Example 6.10, we considered the linear system

$$\begin{aligned} 2x - y - 7z &= 6 \\ x + y - 2z &= -3, \end{aligned}$$

and showed that the solutions were all of the form

$$\begin{bmatrix} x \\ y \\ z \end{bmatrix} = \begin{bmatrix} 1+3t \\ -4-t \\ t \end{bmatrix} = \begin{bmatrix} 1 \\ -4 \\ 0 \end{bmatrix} + t \begin{bmatrix} 3 \\ -1 \\ 1 \end{bmatrix}.$$

We noted that the vector $\begin{bmatrix} 1 \\ -4 \\ 0 \end{bmatrix}$ is one particular solution to the given system and that vectors of the form $t\begin{bmatrix} 3 \\ -1 \\ 1 \end{bmatrix}$ are solutions to the corresponding homogeneous system. Here $\mathsf{x}_p = \begin{bmatrix} 1 \\ -4 \\ 0 \end{bmatrix}$ and $\mathsf{x}_h = t\begin{bmatrix} 3 \\ -1 \\ 1 \end{bmatrix}$.

More on Linear Independence

Recall that vectors $\mathsf{x}_1, \mathsf{x}_2, \ldots, \mathsf{x}_n$ are linearly independent if and only if

$$c_1\mathsf{x}_1 + c_2\mathsf{x}_2 + \cdots + c_n\mathsf{x}_n = \mathsf{0} \implies c_1 = 0, c_2 = 0, \ldots, c_n = 0. \tag{25}$$

Remembering the oh-so-important **5.23**, we recognize $c_1\mathsf{x}_1 + c_2\mathsf{x}_2 + \cdots + c_n\mathsf{x}_n$ as $A\mathsf{c}$, where

$$A = \begin{bmatrix} \mathsf{x}_1 & \mathsf{x}_2 & \ldots & \mathsf{x}_n \\ \downarrow & \downarrow & & \downarrow \end{bmatrix}$$

is the matrix whose columns are the vectors $\mathsf{x}_1, \mathsf{x}_2, \ldots, \mathsf{x}_n$ and $\mathsf{c} = \begin{bmatrix} c_1 \\ c_2 \\ \vdots \\ c_n \end{bmatrix}$. So there is a connection between linear independence and homogeneous systems.

7.2 Vectors $\mathsf{x}_1, \ldots, \mathsf{x}_n$ are linearly independent if and only if $A\mathsf{c} = \mathsf{0}$ implies $\mathsf{c} = \mathsf{0}$, where A the matrix whose columns are the vectors; that is, the homogeneous system $A\mathsf{x} = \mathsf{0}$ has only the trivial solution.

7.3 Example. Let $\mathsf{x}_1 = \begin{bmatrix} 2 \\ 3 \\ 1 \\ 4 \end{bmatrix}$, $\mathsf{x}_2 = \begin{bmatrix} -1 \\ 1 \\ 2 \\ 3 \end{bmatrix}$, and $\mathsf{x}_3 = \begin{bmatrix} 4 \\ 0 \\ -1 \\ 1 \end{bmatrix}$. To decide whether these vectors are linearly independent or linearly dependent, we form the matrix $A = \begin{bmatrix} 2 & -1 & 4 \\ 3 & 1 & 0 \\ 1 & 2 & -1 \\ 4 & 3 & 1 \end{bmatrix}$ whose columns are the given vectors and solve the homogeneous system $A\mathsf{x} = \mathsf{0}$. One possible

Gaussian elimination is

$$\begin{bmatrix} 2 & -1 & 4 \\ 3 & 1 & 0 \\ 1 & 2 & -1 \\ 4 & 3 & 1 \end{bmatrix} \to \begin{bmatrix} 1 & 2 & -1 \\ 3 & 1 & 0 \\ 2 & -1 & 4 \\ 4 & 3 & 1 \end{bmatrix} \to \begin{bmatrix} 1 & 2 & -1 \\ 0 & -5 & 3 \\ 0 & -5 & 6 \\ 0 & -5 & 5 \end{bmatrix} \to \begin{bmatrix} 1 & 2 & -1 \\ 0 & -5 & 3 \\ 0 & 0 & 3 \\ 0 & 0 & 2 \end{bmatrix} \to \begin{bmatrix} 1 & 2 & -1 \\ 0 & -5 & 3 \\ 0 & 0 & 1 \\ 0 & 0 & 0 \end{bmatrix}.$$

Assuming the variables are x_1, x_2, x_3, back substitution gives $x_3 = 0$, then $-5x_2 + 3x_3 = 0$, so $x_2 = 0$ and, finally, $x_1 + 2x_2 - x_3 = 0$, so $x_1 = 0$. Thus $\mathsf{x} = 0$ is the only solution to $A\mathsf{x} = 0$. The given vectors are linearly independent. ☺

7.4 Example. Suppose we want to know whether or not $\mathsf{x}_1 = \begin{bmatrix} -2 \\ 2 \\ 3 \\ 5 \end{bmatrix}$, $\mathsf{x}_2 = \begin{bmatrix} -3 \\ 5 \\ 11 \\ 3 \end{bmatrix}$, $\mathsf{x}_3 = \begin{bmatrix} 0 \\ 2 \\ 1 \\ -1 \end{bmatrix}$ and $\mathsf{x}_4 = \begin{bmatrix} 1 \\ -1 \\ 4 \\ -6 \end{bmatrix}$ and linearly independent. We form the matrix $A = \begin{bmatrix} -2 & -3 & 0 & 1 \\ 2 & 5 & 2 & -1 \\ 3 & 11 & 1 & 4 \\ 5 & 3 & -1 & -6 \end{bmatrix}$ whose columns are the given vectors and solve the homogeneous system $A\mathsf{x} = 0$. One possible Gaussian elimination is

$$\begin{bmatrix} -2 & -3 & 0 & 1 \\ 2 & 5 & 2 & -1 \\ 3 & 11 & 1 & 4 \\ 5 & 3 & -1 & -6 \end{bmatrix} \to \begin{bmatrix} -2 & -3 & 0 & 1 \\ 0 & 2 & 2 & 0 \\ 0 & \frac{13}{2} & 1 & \frac{11}{2} \\ 0 & -\frac{9}{2} & -1 & -\frac{7}{2} \end{bmatrix} \to \begin{bmatrix} -2 & -3 & 0 & 1 \\ 0 & 1 & 1 & 0 \\ 0 & \frac{13}{2} & 1 & \frac{11}{2} \\ 0 & -\frac{9}{2} & -1 & -\frac{7}{2} \end{bmatrix}$$

$$\to \begin{bmatrix} -2 & -3 & 0 & 1 \\ 0 & 1 & 1 & 0 \\ 0 & 0 & -\frac{11}{2} & \frac{11}{2} \\ 0 & 0 & \frac{7}{2} & -\frac{7}{2} \end{bmatrix} \to \begin{bmatrix} -2 & -3 & 0 & 1 \\ 0 & 1 & 1 & 0 \\ 0 & 0 & 1 & -1 \\ 0 & 0 & 0 & 0 \end{bmatrix}.$$

Assuming the variables are x_1, x_2, x_3, x_4, we have $x_4 = t$ free. Then back substitution gives $x_3 = x_4 = t$, then $x_2 = -x_3 = -t$ and $-2x_1 = 3x_2 - x_4 = -4t$, so $x_1 = 2t$. The solution is $\begin{bmatrix} 2t \\ -t \\ t \\ t \end{bmatrix}$. There are many nontrivial solutions, so the vectors are linearly dependent. For instance (with $t = 1$), $2\mathsf{x}_1 - \mathsf{x}_2 + \mathsf{x}_3 + \mathsf{x}_4 = 0$. ☺

The Inverse of a Matrix

Remember that an identity matrix is a square matrix with 1s on the main diagonal and 0s everywhere else. The $n \times n$ identity matrix is denoted I_n or sometimes just I (if we know what n is, or if we don't care). For example, the 2×2 identity matrix is $\begin{bmatrix} 1 & 0 \\ 0 & 1 \end{bmatrix}$. When we discussed the properties of matrix multiplication—p. 64—we noted that the identity matrix acts like the number 1 whenever it has a chance: $IA = A$ for any matrix A, provided IA is defined, just as $1 \cdot a = a$ for any real number a. How far does this analogy extend? If a is a nonzero real number, there is another number b with $ab = 1$. The number $b = \frac{1}{a}$ is called the (multiplicative) inverse of a and often written a^{-1}. This idea motivates the idea of the **inverse** of a matrix.

7.5 Definition. A matrix A is *invertible* or is said to *have an inverse* if there is another matrix B such that $AB = I$ and $BA = I$. The matrix B is called the *inverse* of A.

A matrix can have at most one inverse because if $AB = BA = I$ and $AC = CA = I$ too, then $B = BI = B(AC) = (BA)C = IC = C$. Thus we refer to **the** inverse of A and, by analogy with numbers, use the notation A^{-1}.

7.6 Example. The matrix $A = \begin{bmatrix} 1 & -2 \\ 2 & -3 \end{bmatrix}$ is invertible and $A^{-1} = B = \begin{bmatrix} -3 & 2 \\ -2 & 1 \end{bmatrix}$ since

$$AB = \begin{bmatrix} 1 & -2 \\ 2 & -3 \end{bmatrix}\begin{bmatrix} -3 & 2 \\ -2 & 1 \end{bmatrix} = \begin{bmatrix} 1 & 0 \\ 0 & 1 \end{bmatrix} = I$$

and

$$BA = \begin{bmatrix} -3 & 2 \\ -2 & 1 \end{bmatrix}\begin{bmatrix} 1 & -2 \\ 2 & -3 \end{bmatrix} = \begin{bmatrix} 1 & 0 \\ 0 & 1 \end{bmatrix} = I$$

too. ☺

7.7 Example. The matrix $A = \begin{bmatrix} 1 & 2 \\ 2 & 4 \end{bmatrix}$ is not invertible, and here's why. Let $B = \begin{bmatrix} x & z \\ y & w \end{bmatrix}$ and compute AB:

$$AB = \begin{bmatrix} 1 & 2 \\ 2 & 4 \end{bmatrix}\begin{bmatrix} x & y \\ z & w \end{bmatrix} = \begin{bmatrix} x + 2z & y + 2w \\ 2x + 4z & 2y + 4w \end{bmatrix}.$$

The second row is twice the first. This is not the case with $I = \begin{bmatrix} 1 & 0 \\ 0 & 1 \end{bmatrix}$, so AB can never be I, so there can be no inverse for A. We note in passing that the matrix A is not 0, so while any nonzero number has an inverse, not so with matrices. ☺

7.8 Remarks. 1. It can be shown (Math 2051) that only a square matrix can have an inverse. Moreover, if A is square and $AB = I$, then $BA = I$ too, so we didn't really have to check $BA = I$ in Example 7.6.

2. The equations $AB = BA = I$ say not only that A is invertible with $A^{-1} = B$, but that B is invertible with $B^{-1} = A$. Thus $(A^{-1})^{-1} = B^{-1} = A$.

3. If A and B are invertible matrices of the same size, then AB is invertible and $(AB)^{-1} = B^{-1}A^{-1}$. This follows because $X = B^{-1}A^{-1}$ is a matrix satisfying $(AB)X = I$ and $X(AB) = I$. These equations say $X = (AB)^{-1}$.

   Note the analogy with transpose. Just as $(AB)^T = B^T A^T$, so $(AB)^{-1} = B^{-1}A^{-1}$ (assuming A and B are both invertible).

Matrix Equations

Just as the solution to $ax = b$ is $x = a^{-1}b$ (if a is invertible), so the solution to the matrix equation $AX = B$ is $X = A^{-1}B$ (if A is invertible). To see why, let $X = A^{-1}B$ and compute $AX = A(A^{-1}B) = IB = B$.

To see how we found this X, start with the equation $AX = B$. We want just X on the left, so we must multiply the left side of $AX = B$ by A^{-1} **on the left**. Thus we must multiply the right side of $AX = B$ **on the left** by A^{-1} as well. This gives $X = A^{-1}B$. Since matrix multiplication is not commutative, this is not (usually) the same as $X = BA^{-1}$. Again we note a difference from ordinary algebra where, assuming $a \neq 0$, the solution to $ax = b$ can be written $x = a^{-1}b$ or $x = ba^{-1}$.

7.9 Example. We use this idea to solve the system $\begin{aligned} x - 2y &= 5 \\ 2x - 3y &= 7. \end{aligned}$

This is $\mathsf{A}\mathsf{x} = \mathsf{b}$, where $A = \begin{bmatrix} 1 & -2 \\ 2 & -3 \end{bmatrix}$ is the matrix of coefficients, $\mathsf{x} = \begin{bmatrix} x \\ y \end{bmatrix}$ is the vector of unknowns, and $\mathsf{b} = \begin{bmatrix} 5 \\ 7 \end{bmatrix}$. Now A has inverse $A^{-1} = \begin{bmatrix} -3 & 2 \\ -2 & 1 \end{bmatrix}$ as we saw in Example 7.6, so

$$\mathsf{x} = A^{-1}\mathsf{b} = \begin{bmatrix} -3 & 2 \\ -2 & 1 \end{bmatrix} \begin{bmatrix} 5 \\ 7 \end{bmatrix} = \begin{bmatrix} -1 \\ -3 \end{bmatrix}.$$

The solution is $x = -1$, $y = -3$. ☺

This is perhaps a good place to tabulate some of the similarities between ordinary algebra and matrix algebra when it comes to solving equations.

	Ordinary Algebra	Matrix Algebra
multiplicative identity	1	I
inverse	a^{-1}	A^{-1}
equations	$ax = b \implies x = a^{-1}b$ if $a \neq 0$	$AX = B \implies X = A^{-1}B$ if A is invertible
condition for invertibility	$a \neq 0$	$\det A \neq 0$ (to come later)

There are differences as well as similarities between algebra with matrices and algebra with numbers. One major difference is that matrix multiplication is, in general, not commutative. So, for example, a matrix of the form $A^{-1}XA$ cannot (usually) be simplified.

7.10 Example. Suppose A, B and X are invertible matrices and $AX^{-1}A^{-1} = B^{-1}$. To find X, we first multiply the given equation on the right by A, obtaining $AX^{-1}A^{-1}A = B^{-1}A$. Since $A^{-1}A = I$, we obtain $AX^{-1} = B^{-1}A$. Now multiply on the left by A^{-1}. This gives $X^{-1} = A^{-1}B^{-1}A$. Finally $X = (X^{-1})^{-1} = (A^{-1}B^{-1}A)^{-1} = A^{-1}(B^{-1})^{-1}(A^{-1})^{-1}$ (remembering to reverse order) $= A^{-1}BA$. ☺

Finding the Inverse of a Matrix

It's easy to determine whether or not a 2×2 matrix is invertible and to find the inverse of an invertible 2×2 matrix. Do you remember what is meant by the "determinant" of a 2×2 matrix? Those are the numbers you calculate when finding the cross product of a vector. The determinant of $\begin{bmatrix} a & b \\ c & d \end{bmatrix}$ is the number $ad - bc$, and this is often denoted $\det A$.

7.11 Proposition. *The matrix* $A = \begin{bmatrix} a & b \\ c & d \end{bmatrix}$ *is invertible if and only if* $\det A \neq 0$. *If* $ad - bc \neq 0$, *then* $A^{-1} = \frac{1}{ad-bc}\begin{bmatrix} d & -b \\ -c & a \end{bmatrix}$.

Proof. We begin by checking that

$$A\begin{bmatrix} d & -b \\ -c & a \end{bmatrix} = (ad - bc)I. \tag{26}$$

($\Leftarrow$) Now suppose $ad - bc \neq 0$. Dividing (26) by $ad - bc$ gives $A\left(\frac{1}{ad-bc}\begin{bmatrix} d & -b \\ -c & a \end{bmatrix}\right) = I$, so A is invertible and $A^{-1} = \frac{1}{ad-bc}\begin{bmatrix} d & -b \\ -c & a \end{bmatrix}$.

($\Rightarrow$) Assume A is invertible but $ad - bc = 0$. Then (26) says $A\begin{bmatrix} d & -b \\ -c & a \end{bmatrix} = 0$. Multiplying this equation on the left by A^{-1} then gives $\begin{bmatrix} d & -b \\ -c & a \end{bmatrix} = \begin{bmatrix} 0 & 0 \\ 0 & 0 \end{bmatrix}$, but this gives $d = b = c = a = 0$, so $A = 0$, contradicting that A is invertible. ∎

It's not hard to remember the formula for the inverse of $A = \begin{bmatrix} a & b \\ c & d \end{bmatrix}$. The inverse is $\frac{1}{ad-bc}$ times $\begin{bmatrix} d & -b \\ -c & a \end{bmatrix}$. This matrix is obtained from A by switching the diagonal entries and changing the sign of the off-diagonal elements. For example, if $A = \begin{bmatrix} 1 & -2 \\ 3 & 5 \end{bmatrix}$, then $ad-bc = 1(5)-(-2)(3) = 11$ and $A^{-1} = \frac{1}{11}\begin{bmatrix} 5 & 2 \\ -3 & 1 \end{bmatrix}$.

7.12 **Remark.** Remember that we earlier called $ad-bc$ the *determinant* of $A = \begin{bmatrix} a & b \\ c & d \end{bmatrix}$, so the proposition says a 2×2 matrix is invertible if and only if its determinant is not 0, and then there is a formula for A^{-1} which begins $\frac{1}{\det A}$. All this holds for larger matrices, as we shall soon see.

So we know how to find the inverse of (invertible) 2×2 matrices, but what about larger matrices?

Let A be an $n \times n$ matrix. If A is invertible, there is a matrix B with

$$AB = I = \begin{bmatrix} 1 & 0 & \dots & 0 \\ 0 & 1 & \dots & 0 \\ \vdots & \vdots & \ddots & \vdots \\ 0 & 0 & \dots & 1 \end{bmatrix} = \begin{bmatrix} \mathsf{e}_1 & \mathsf{e}_2 & \dots & \mathsf{e}_n \\ \downarrow & \downarrow & & \downarrow \\ & & & \end{bmatrix}.$$

Letting $\mathsf{x}_1, \mathsf{x}_2, \dots, \mathsf{x}_n$ be the columns of B, so that $B = \begin{bmatrix} \mathsf{x}_1 & \mathsf{x}_2 & \dots & \mathsf{x}_n \\ \downarrow & \downarrow & & \downarrow \end{bmatrix}$, we have $AB = \begin{bmatrix} A\mathsf{x}_1 & A\mathsf{x}_2 & \dots & A\mathsf{x}_n \\ \downarrow & \downarrow & & \downarrow \end{bmatrix}$. We want $AB = I$, so we need $A\mathsf{x}_1 = \mathsf{e}_1$, $A\mathsf{x}_2 = \mathsf{e}_2$, ..., $A\mathsf{x}_n = \mathsf{e}_n$. Thus we must solve n systems of equations $A\mathsf{x} = \mathsf{e}_i$ for the columns $\mathsf{x}_1, \mathsf{x}_2, \dots, \mathsf{x}_n$ of B.

7.13 **Example.** Suppose $A = \begin{bmatrix} 1 & 4 \\ 2 & 7 \end{bmatrix}$. We seek the matrix $B = \begin{bmatrix} \mathsf{b}_1 & \mathsf{b}_2 \\ \downarrow & \downarrow \end{bmatrix}$ with $AB = I$. As seen, this requires solving two systems of equations. First we solve $A\mathsf{x} = \mathsf{e}_1$ for b_1. Gaussian elimination proceeds

$$\left[\begin{array}{cc|c} 1 & 4 & 1 \\ 2 & 7 & 0 \end{array}\right] \to \left[\begin{array}{cc|c} 1 & 4 & 1 \\ 0 & -1 & -2 \end{array}\right] \to \left[\begin{array}{cc|c} 1 & 4 & 1 \\ 0 & \textcircled{1} & 2 \end{array}\right]$$

so, if $\mathsf{b}_1 = \begin{bmatrix} x \\ y \end{bmatrix}$, we have $y = 2$, $x + 4y = 1$, so $x = 1 - 4y = -7$. The first column of the inverse matrix is $\mathsf{b}_1 = \begin{bmatrix} -7 \\ 2 \end{bmatrix}$. This solution can also be obtained by continuing the Gaussian elimination, using the circled pivot as indicated to get a 0 **above** it:

$$\left[\begin{array}{cc|c} 1 & 4 & 1 \\ 0 & \textcircled{1} & 2 \end{array}\right] \to \left[\begin{array}{cc|c} 1 & 0 & -7 \\ 0 & 1 & 2 \end{array}\right],$$

thus $x = -7$, $y = 2$ comes directly and the desired column b_1 appears as the final column of the augmented matrix. Solving the second system this way,

$$\left[\begin{array}{cc|c} 1 & 4 & 0 \\ 2 & 7 & 1 \end{array}\right] \to \left[\begin{array}{cc|c} 1 & 4 & 0 \\ 0 & -1 & 1 \end{array}\right] \to \left[\begin{array}{cc|c} 1 & 4 & 0 \\ 0 & 1 & -1 \end{array}\right] \to \left[\begin{array}{cc|c} 1 & 0 & 4 \\ 0 & 1 & -1 \end{array}\right],$$

and so b_2, the second column of B, appears as the final column of the matrix. We conclude that $B = A^{-1} = \begin{bmatrix} -7 & 4 \\ 2 & -1 \end{bmatrix}$. ⌣̈

I suppose you noticed that in the two systems of linear equations we just solved, the sequence of row operations required to bring each matrix to row echelon form were precisely the same. With a little care, we can solve these two systems at the same time, like this:

$$\left[\begin{array}{cc|cc} 1 & 4 & 1 & 0 \\ 2 & 7 & 0 & 1 \end{array}\right] \to \left[\begin{array}{cc|cc} 1 & 4 & 1 & 0 \\ 0 & -1 & -2 & 1 \end{array}\right] \to \left[\begin{array}{cc|cc} 1 & 4 & 1 & 0 \\ 0 & 1 & 2 & -1 \end{array}\right] \to \left[\begin{array}{cc|cc} 1 & 0 & -7 & 4 \\ 0 & 1 & 2 & -1 \end{array}\right].$$

Schematically, we started with $[A \mid I]$ and finished with $[I \mid A^{-1}]$.

This example illustrates a clever way to find the inverse of a matrix.

7.14 If A can be carried to the identity matrix I by a sequence of elementary row operations, then A is invertible and, after carrying $[A|I]$ to $[I|B]$, the matrix $B = A^{-1}$; otherwise, A is not invertible.

Cool?

7.15 Example. Suppose $A = \begin{bmatrix} 2 & 7 & 0 \\ 1 & 4 & -1 \\ 1 & 3 & 0 \end{bmatrix}$.

Gaussian elimination proceeds $[A|I] = \left[\begin{array}{ccc|ccc} 2 & 7 & 0 & 1 & 0 & 0 \\ 1 & 4 & -1 & 0 & 1 & 0 \\ 1 & 3 & 0 & 0 & 0 & 1 \end{array}\right]$

$$\to \left[\begin{array}{ccc|ccc} 1 & 4 & -1 & 0 & 1 & 0 \\ 2 & 7 & 0 & 1 & 0 & 0 \\ 1 & 3 & 0 & 0 & 0 & 1 \end{array}\right] \to \left[\begin{array}{ccc|ccc} 1 & 4 & -1 & 0 & 1 & 0 \\ 0 & -1 & 2 & 1 & -2 & 0 \\ 0 & -1 & 1 & 0 & -1 & 1 \end{array}\right]$$

$$\to \left[\begin{array}{ccc|ccc} 1 & 4 & -1 & 0 & 1 & 0 \\ 0 & 1 & -2 & -1 & 2 & 0 \\ 0 & -1 & 1 & 0 & -1 & 1 \end{array}\right] \to \left[\begin{array}{ccc|ccc} 1 & 0 & 7 & 4 & -7 & 0 \\ 0 & 1 & -2 & -1 & 2 & 0 \\ 0 & 0 & -1 & -1 & 1 & 1 \end{array}\right]$$

$$\to \left[\begin{array}{ccc|ccc} 1 & 0 & 7 & 4 & -7 & 0 \\ 0 & 1 & -2 & -1 & 2 & 0 \\ 0 & 0 & 1 & 1 & -1 & -1 \end{array}\right] \to \left[\begin{array}{ccc|ccc} 1 & 0 & 0 & -3 & 0 & 7 \\ 0 & 1 & 0 & 1 & 0 & -2 \\ 0 & 0 & 1 & 1 & -1 & -1 \end{array}\right].$$

If we haven't made a mistake, A^{-1} should be the matrix to the right of the vertical line in this last matrix: $B = \begin{bmatrix} -3 & 0 & 7 \\ 1 & 0 & -2 \\ 1 & -1 & -1 \end{bmatrix}$. We check that

$$AB = \begin{bmatrix} 2 & 7 & 0 \\ 1 & 4 & -1 \\ 1 & 3 & 0 \end{bmatrix} \begin{bmatrix} -3 & 0 & 7 \\ 1 & 0 & -2 \\ 1 & -1 & -1 \end{bmatrix} = \begin{bmatrix} 1 & 0 & 0 \\ 0 & 1 & 0 \\ 0 & 0 & 1 \end{bmatrix} = I,$$

so $B = A^{-1}$. ☺

7.16 Example. Suppose $A = \begin{bmatrix} 1 & 2 & 0 \\ 3 & -1 & 2 \\ -2 & 3 & -2 \end{bmatrix}$. Gaussian elimination applied to $[A|I]$ proceeds

$$\left[\begin{array}{rrr|rrr} 1 & 2 & 0 & 1 & 0 & 0 \\ 3 & -1 & 2 & 0 & 1 & 0 \\ -2 & 3 & -2 & 0 & 0 & 1 \end{array}\right] \to \left[\begin{array}{rrr|rrr} 1 & 2 & 0 & 1 & 0 & 0 \\ 0 & -7 & 2 & -3 & 1 & 0 \\ 0 & 7 & -2 & 2 & 0 & 1 \end{array}\right] \to \left[\begin{array}{rrr|rrr} 1 & 2 & 0 & 1 & 0 & 0 \\ 0 & -7 & 2 & -3 & 1 & 0 \\ 0 & 0 & 0 & -1 & 1 & 1 \end{array}\right].$$

There is no point in continuing. Those three 0s in the third row can't be changed. We cannot make the left 3×3 block the identity matrix. The given matrix A has no inverse. ☺

Reduced Row Echelon Form

To see if a matrix A has an inverse, we extended the Gaussian elimination process, using it to carry A not just to row echelon form, but to **reduced** row echelon form.

7.17 Definition. A matrix is in *reduced row echelon form* if

1. it is in row echelon form;
2. each pivot is a 1;
3. every entry above and below a pivot is 0.

7.18 Examples. The following matrices are all in reduced row echelon form:

$$\begin{bmatrix} 1 & 0 \\ 0 & 1 \end{bmatrix}, \quad \begin{bmatrix} 1 & 0 & -2 \\ 0 & 1 & 1 \end{bmatrix}, \quad \begin{bmatrix} 1 & 0 & 0 & 0 \\ 0 & 0 & 1 & 0 \\ 0 & 0 & 0 & 1 \end{bmatrix}, \quad \begin{bmatrix} 1 & 0 & 0 & 0 & 0 & 4 \\ 0 & 1 & 0 & 5 & 0 & -2 \\ 0 & 0 & 0 & 0 & 1 & 3 \end{bmatrix}.$$

7.19 Example. Continuing the Gaussian elimination process on matrix $A = \begin{bmatrix} 1 & 2 & 0 \\ 3 & -1 & 2 \\ -2 & 3 & -2 \end{bmatrix}$ in Example 7.16, we have

$$A \to \begin{bmatrix} 1 & 2 & 0 \\ 0 & 1 & -\frac{2}{7} \\ 0 & 0 & 0 \end{bmatrix} \to \begin{bmatrix} 1 & 0 & \frac{4}{7} \\ 0 & 1 & -\frac{2}{7} \\ 0 & 0 & 0 \end{bmatrix},$$

so the reduced row echelon form of A is $U = \begin{bmatrix} 1 & 0 & \frac{4}{7} \\ 0 & 1 & -\frac{2}{7} \\ 0 & 0 & 0 \end{bmatrix}$. ☺

To bring a matrix to **reduced** row echelon form, we use Gaussian elimination in the usual way, sweeping across the columns of the matrix from left to right making sure all nonzero leading entries are 1, and then, each time we get a leading 1, using this to get 0s in the rest of the corresponding column, below **and above** the 1.

7.20 Example. $\begin{bmatrix} \textcircled{1} & 2 & -5 \\ 3 & 1 & 5 \\ -2 & 3 & 4 \end{bmatrix} \to \begin{bmatrix} 1 & 2 & -5 \\ 0 & -5 & 20 \\ 0 & 7 & -6 \end{bmatrix} \to \begin{bmatrix} 1 & 2 & -5 \\ 0 & \textcircled{1} & -4 \\ 0 & 7 & -6 \end{bmatrix} \to \begin{bmatrix} 1 & 0 & 3 \\ 0 & 1 & -4 \\ 0 & 0 & 22 \end{bmatrix} \to \begin{bmatrix} 1 & 0 & 3 \\ 0 & 1 & -4 \\ 0 & 0 & \textcircled{1} \end{bmatrix}$

$\to \begin{bmatrix} 1 & 0 & 0 \\ 0 & 1 & 0 \\ 0 & 0 & 1 \end{bmatrix}$, so the reduced row echelon form of $\begin{bmatrix} 1 & 2 & -5 \\ 3 & 1 & 5 \\ -2 & 3 & 4 \end{bmatrix}$ is $\begin{bmatrix} 1 & 0 & 0 \\ 0 & 1 & 0 \\ 0 & 0 & 1 \end{bmatrix}$. ☺

Our interest in reduced row echelon form derives from the method we have described for finding the inverse of a matrix, when there is an inverse. If the reduced row echelon form of A is the identity matrix, then A has an inverse.

7.21 Example. Suppose we are asked for the conditions on a, b, c in order that $A = \begin{bmatrix} 1 & 2 & 0 \\ 3 & -1 & 2 \\ a & b & c \end{bmatrix}$ have an inverse. We want the reduced row echelon form of A to be I. Gaussian elimination proceeds

$$A \to \begin{bmatrix} 1 & 2 & 0 \\ 0 & -7 & 2 \\ 0 & b-2a & c \end{bmatrix} \to \begin{bmatrix} 1 & 2 & 0 \\ 0 & 1 & -\frac{2}{7} \\ 0 & b-2a & c \end{bmatrix} \to \begin{bmatrix} 1 & 0 & \frac{4}{7} \\ 0 & 1 & -\frac{2}{7} \\ 0 & 0 & c-(b-2a)(-\frac{2}{7}) \end{bmatrix}.$$

Reduced row echelon form will be I if and only if we can make the $(3,3)$-entry of this last matrix a 1 (which can then be used to put 0s in the rest of column three). So A has an inverse if and only if $c-(b-2a)(-\frac{2}{7}) \neq 0$, that is, if and only if $-4a+2b+7c \neq 0$. ☺

True/False Questions for Week 7

Decide, with as little calculation as possible, whether each of the following statements is true or false and explain your answer whenever you say "false."

1. The system $\begin{aligned} -2x_1 \qquad\quad - x_3 &= 0 \\ 5x_2 + 6x_3 &= 0 \\ 7x_1 + 5x_2 - 4x_3 &= 0 \end{aligned}$ is homogeneous.

2. If vectors u, v, and w are linearly dependent and A is a matrix whose columns are these vectors, then the homogeneous system $A\mathsf{x} = 0$ has a nontrivial solution.

3. If A is a square nonzero matrix, then A is invertible.

4. If $A^3 = I$, then $A^{-1} = A^2$.

5. If A and B are matrices and A is invertible, then $A^{-1}BA = B$.

6. If A and B are invertible matrices and $XA = B$, then $X = A^{-1}B$.

7. If B is invertible, the solution to $XB = A$ is $X = AB^{-1}$.

8. If A is invertible, then the columns of A are linearly independent.

9. The matrix $\begin{bmatrix} 1 & 0 & 3 \\ 0 & 1 & 0 \end{bmatrix}$ is in reduced row echelon form.

10. The matrix $\begin{bmatrix} 1 & 3 & 0 \\ 0 & 1 & 0 \end{bmatrix}$ is in reduced row echelon form.

11. The matrix $\begin{bmatrix} 0 & 1 & 3 & 0 & -4 \\ 0 & 0 & 0 & 1 & 0 \\ 0 & 0 & 0 & 0 & 1 \end{bmatrix}$ is in reduced row echelon form.

Week 7 Test Yourself

Here are a few problems with short answers that you can use to test your understanding of the concepts you have met this week.

1. Each system below has at least one solution. Why? Now solve each system expressing your answer as a vector or as a linear combination of vectors.

   (a) $$\begin{aligned} x - 2y + z &= 0 \\ 3x - 7y + 2z &= 0 \end{aligned}$$

   (b) $$\begin{aligned} -x_1 + x_2 + 2x_3 + x_4 + 3x_5 &= 0 \\ x_1 - x_2 + x_3 + 3x_4 - 4x_5 &= 0 \\ -8x_1 + 8x_2 + 7x_3 - 4x_4 + 36x_5 &= 0 \end{aligned}$$

2. Consider the following system of linear equations.

$$\begin{aligned} x_1 - 2x_2 + 3x_3 + x_4 + 3x_5 + 4x_6 &= -1 \\ -3x_1 + 6x_2 - 8x_3 + 2x_4 - 11x_5 - 15x_6 &= 2 \\ x_1 - 2x_2 + 2x_3 - 4x_4 + 6x_5 + 9x_6 &= 3 \\ -2x_1 + 4x_2 - 6x_3 - 2x_4 - 6x_5 - 7x_6 &= 1. \end{aligned}$$

   (a) Write this system in the form $A\mathsf{x} = \mathsf{b}$.

   (b) Solve the system expressing your answer as a vector $\mathsf{x} = \mathsf{x}_p + \mathsf{x}_h$, where x_p is a particular solution and x_h is a solution to the corresponding homogeneous system.

   (c) Express $\begin{bmatrix} -1 \\ 2 \\ 3 \\ 1 \end{bmatrix}$ as a linear combination of the columns of A.

3. State some ways in which matrix algebra is different from ordinary algebra with numbers.

4. Determine whether each of the following pairs of matrices are inverses.

(a) $A = \begin{bmatrix} 0 & 1 \\ 1 & 0 \end{bmatrix}$, $B = A$

(b) $A = \begin{bmatrix} 1 & 0 & 0 \\ 2 & 1 & 0 \\ -1 & 2 & 1 \end{bmatrix}$, $B = \begin{bmatrix} 1 & 0 & 0 \\ -2 & 1 & 1 \\ 5 & -1 & -1 \end{bmatrix}$

5. Let $A = \begin{bmatrix} 1 & 3 \\ -2 & 5 \end{bmatrix}$ and $B = \begin{bmatrix} 2 & -1 \\ 1 & -1 \end{bmatrix}$. Find AB. Find $(AB)^{-1}$, A^{-1} and B^{-1} in each case using the formula given in Proposition 7.11. Then check that $(AB)^{-1} = B^{-1}A^{-1}$ but **not** $A^{-1}B^{-1}$.

6. If A is an invertible $n \times n$ matrix and B is a $t \times n$ matrix, show that the equation $XA = B$ has a solution by finding it. What is the size of X?

7. Let A be a 3×3 matrix such that $A\begin{bmatrix} 1 \\ 2 \\ 3 \end{bmatrix} = \begin{bmatrix} 0 \\ 0 \\ 0 \end{bmatrix}$. Can A be invertible? Explain.

8. Let $A = \begin{bmatrix} 1 & -7 & 1 \\ 2 & -9 & 1 \end{bmatrix}$ and $B = \begin{bmatrix} -1 & 3 \\ 0 & 1 \\ 2 & 4 \end{bmatrix}$.

(a) Compute AB and BA.

(b) Is A invertible? Explain.

9. Let $A = \begin{bmatrix} 1 & 1 \\ 2 & 4 \end{bmatrix}$ and $C = \begin{bmatrix} 5 & 3 \\ 2 & 2 \end{bmatrix}$.

Given that B is a 2×2 matrix and that $ABC^{-1} = I$, find B.

10. Suppose that B and $I - B^{-1}$ are invertible matrices and the matrix equation $(X - A)^{-1} = BX^{-1}$ has a solution X. Find X.

11. (a) Suppose A and B are invertible matrices and X is a matrix such that $AXB = A + B$. What is X?

(b) Suppose A, B, C, and X are matrices, X and C invertible, such that $X^{-1}A = C - X^{-1}B$. What is X?

(c) Suppose A, B, and X are invertible matrices such that $BAX = XABX$. What is X?

12. Let $A = \begin{bmatrix} -1 & 2 & 2 \\ 3 & 0 & 5 \\ 2 & -1 & 0 \end{bmatrix}$.

Use the fact that $A^{-1} = \frac{1}{9}\begin{bmatrix} 5 & -2 & 10 \\ 10 & -4 & 11 \\ -3 & 3 & -6 \end{bmatrix}$ to solve the system
$$\begin{aligned} -x + 2y + 2z &= 12 \\ 3x \qquad\;\; + 5z &= -1 \\ 2x - \;\; y \qquad\;\; &= -8. \end{aligned}$$

13. Determine whether or not each of the following matrices has an inverse. Find the inverse whenever this exists.

(a) $\begin{bmatrix} 3 & 1 \\ -6 & -3 \end{bmatrix}$ (b) $\begin{bmatrix} 2 & 4 & 2 \\ 1 & 2 & 3 \\ 3 & 2 & 1 \end{bmatrix}$ (c) $\begin{bmatrix} 0 & -1 & 2 \\ 2 & 1 & 4 \\ 1 & -1 & 5 \end{bmatrix}$

14. Determine whether or not the columns of each of the matrices in Exercise 13 are linearly independent.

15. What is the reduced row echelon form of the matrix $A = \begin{bmatrix} 1 & 2 & 0 \\ 3 & -1 & 2 \\ -2 & 3 & -2 \end{bmatrix}$?

Week 8

Elementary Matrices

I hope you remember the three elementary row operations on a matrix.

1. the interchange of two rows;
2. multiplication of a row by a nonzero scalar;
3. $(R \to R - cR')$ replacing a row by that row less a multiple of another row.

We start this week's work by showing that each of these operations can be achieved by matrix multiplication.

8.1 Definition. An *elementary matrix* is a square matrix obtained from the identity matrix by a single elementary row operation.

8.2 Examples. • $E_1 = \begin{bmatrix} 0 & 1 & 0 \\ 1 & 0 & 0 \\ 0 & 0 & 1 \end{bmatrix}$ is an elementary matrix, obtained from the 3×3 identity matrix I by interchanging rows one and two.

- The matrix $E_2 = \begin{bmatrix} 1 & 0 & 0 \\ 0 & 3 & 0 \\ 0 & 0 & 1 \end{bmatrix}$ is elementary; it is obtained from I by multiplying row two by 3.

- The matrix $E_3 = \begin{bmatrix} 1 & 0 & 0 \\ 0 & 1 & 0 \\ -4 & 0 & 1 \end{bmatrix}$ is elementary; it is obtained from I by replacing row three by row three minus 4 times row one. ☺

Here is one big reason why elementary matrices are important. Multiplication by an elementary matrix performs an elementary row operation.

8.3 If A is a matrix and E is an elementary matrix and EA is defined, then EA is that matrix obtained by applying to A the elementary operation that produced E.

For some reason, this reminds me of an old proverb. Remember the one that begins, "Do unto others ..."? Well, after an elementary matrix E is born, E goes out into the world and does unto others what was done unto the identity to form E.

To illustrate, let $A = \begin{bmatrix} a & d \\ b & e \\ c & f \end{bmatrix}$ and let E_1, E_2 and E_3 be as before. The matrix E_1A is

$$E_1A = \begin{bmatrix} 0 & 1 & 0 \\ 1 & 0 & 0 \\ 0 & 0 & 1 \end{bmatrix} \begin{bmatrix} a & d \\ b & e \\ c & f \end{bmatrix} = \begin{bmatrix} b & e \\ a & d \\ c & f \end{bmatrix},$$

which is just A with rows one and two interchanged. The matrix E_2A is

$$E_2A = \begin{bmatrix} 1 & 0 & 0 \\ 0 & 3 & 0 \\ 0 & 0 & 1 \end{bmatrix} \begin{bmatrix} a & d \\ b & e \\ c & f \end{bmatrix} = \begin{bmatrix} a & d \\ 3b & 3e \\ c & f \end{bmatrix},$$

which is A with its second row multiplied by 3, and

$$E_3A = \begin{bmatrix} 1 & 0 & 0 \\ 0 & 1 & 0 \\ -4 & 0 & 1 \end{bmatrix} \begin{bmatrix} a & d \\ b & e \\ c & f \end{bmatrix} = \begin{bmatrix} a & d \\ b & e \\ -4a + c & -4d + f \end{bmatrix},$$

which is A, but with row three replaced by row three minus 4 times row one.

Another important property of an elementary matrix is that its inverse is also elementary; moreover, we can say precisely what elementary matrix.

8.4 An elementary matrix is invertible and the inverse of an elementary matrix is also an elementary matrix: If E is elementary, E^{-1} is the elementary matrix that reverses (or undoes) what E does.

We illustrate with a few examples.

8.5 Example. $E_1 = \begin{bmatrix} 0 & 1 & 0 \\ 1 & 0 & 0 \\ 0 & 0 & 1 \end{bmatrix}$ is the elementary matrix that interchanges rows one and two of a matrix. To undo such an interchange, we interchange rows one and two again. According to **8.4** we should have $E_1^{-1} = E_1$, and this is indeed the case because

$$E_1E_1 = \begin{bmatrix} 0 & 1 & 0 \\ 1 & 0 & 0 \\ 0 & 0 & 1 \end{bmatrix} \begin{bmatrix} 0 & 1 & 0 \\ 1 & 0 & 0 \\ 0 & 0 & 1 \end{bmatrix} = \begin{bmatrix} 1 & 0 & 0 \\ 0 & 1 & 0 \\ 0 & 0 & 1 \end{bmatrix} = I.$$

☺

8.6 Example. To multiply row two of a matrix by 3, multiply (on the left) by the elementary matrix $E_2 = \begin{bmatrix} 1 & 0 & 0 \\ 0 & 3 & 0 \\ 0 & 0 & 1 \end{bmatrix}$. To undo the effect of this operation, we should multiply row two by

$\frac{1}{3}$. According to **8.4**, E_2^{-1} should be $B = \begin{bmatrix} 1 & 0 & 0 \\ 0 & \frac{1}{3} & 0 \\ 0 & 0 & 1 \end{bmatrix}$, the elementary matrix that multiplies row two of a matrix by $\frac{1}{3}$ and again this is true because

$$E_2B = \begin{bmatrix} 1 & 0 & 0 \\ 0 & 3 & 0 \\ 0 & 0 & 1 \end{bmatrix} \begin{bmatrix} 1 & 0 & 0 \\ 0 & \frac{1}{3} & 0 \\ 0 & 0 & 1 \end{bmatrix} = \begin{bmatrix} 1 & 0 & 0 \\ 0 & 1 & 0 \\ 0 & 0 & 1 \end{bmatrix} = I.$$ ☺

8.7 Example. Multiplying a matrix by the elementary matrix $E_3 = \begin{bmatrix} 1 & 0 & 0 \\ 0 & 1 & 0 \\ -4 & 0 & 1 \end{bmatrix}$ replaces row three of that matrix by row three minus 4 times row one. To undo this operation, we must add 4 times row one back to row three; that is, replace row three by row three plus 4 times row one. Apparently then, the inverse of E_3 is $B = \begin{bmatrix} 1 & 0 & 0 \\ 0 & 1 & 0 \\ 4 & 0 & 1 \end{bmatrix}$ and, once again, we check this is the case:

$$E_3B = \begin{bmatrix} 1 & 0 & 0 \\ 0 & 1 & 0 \\ -4 & 0 & 1 \end{bmatrix} \begin{bmatrix} 1 & 0 & 0 \\ 0 & 1 & 0 \\ 4 & 0 & 1 \end{bmatrix} = \begin{bmatrix} 1 & 0 & 0 \\ 0 & 1 & 0 \\ 0 & 0 & 1 \end{bmatrix} = I.$$ ☺

LU Factorization

What can we do with these ideas about elementary matrices?

Let A be the 2×4 matrix $A = \begin{bmatrix} 2 & 4 & 1 & 1 \\ -6 & 0 & 0 & 5 \end{bmatrix}$. We can move A to row echelon form in one step by adding three times row one to row two:

$$A = \begin{bmatrix} 2 & 4 & 1 & 1 \\ -6 & 0 & 0 & 5 \end{bmatrix} \to \begin{bmatrix} 2 & 4 & 1 & 1 \\ 0 & 12 & 3 & 8 \end{bmatrix} = U.$$

The elementary row operation—add three times row one to row two—is accomplished by multiplying A on the left by $E = \begin{bmatrix} 1 & 0 \\ 3 & 1 \end{bmatrix}$. Thus $U = EA$. Since E is invertible, $A = E^{-1}U$. The inverse of E is the elementary matrix which "undoes" E: $E^{-1} = \begin{bmatrix} 1 & 0 \\ -3 & 1 \end{bmatrix}$. Setting $L = E^{-1}$, we have a factorization of the matrix A:

$$\begin{bmatrix} 2 & 4 & 1 & 1 \\ -6 & 0 & 0 & 5 \end{bmatrix} = \begin{bmatrix} 1 & 0 \\ -3 & 1 \end{bmatrix} \begin{bmatrix} 2 & 4 & 1 & 1 \\ 0 & 12 & 3 & 8 \end{bmatrix} \qquad (27)$$

$$A = L \quad U.$$

The symbols are chosen because L is a **lower triangular** matrix and U is **upper triangular** (any row echelon matrix is upper triangular).

8.8 Definition. An *LU factorization* of a matrix A is a factorization $A = LU$ of A where L is a square lower triangular matrix with 1s on the diagonal and U is a row echelon form of A.

The "U" stands for "upper triangular:" remember that a row echelon matrix is a particular kind of upper triangular matrix.

8.9 Example. $\begin{bmatrix} 1 & 2 \\ 4 & 6 \end{bmatrix} = \begin{bmatrix} 1 & 0 \\ 4 & 1 \end{bmatrix} \begin{bmatrix} 1 & 2 \\ 0 & -2 \end{bmatrix}$ is an LU factorization of $A = \begin{bmatrix} 1 & 2 \\ 4 & 6 \end{bmatrix}$. :)

8.10 Example. Suppose we want an LU factorization of $A = \begin{bmatrix} -1 & 1 & -2 & 1 \\ 2 & 1 & 7 & 0 \\ -3 & 0 & 1 & 5 \end{bmatrix}$. We reduce A to row echelon form making note of the elementary matrices that effect the row operations we use. To begin, we add 2 times row one to row two. This can be achieved by multiplying A (on the left) by a certain elementary matrix, which must be 3×3 since A is 3×4. The elementary matrix we need is $E_1 = \begin{bmatrix} 1 & 0 & 0 \\ 2 & 1 & 0 \\ 0 & 0 & 1 \end{bmatrix}$:

$$E_1 A = \begin{bmatrix} 1 & 0 & 0 \\ 2 & 1 & 0 \\ 0 & 0 & 1 \end{bmatrix} \begin{bmatrix} -1 & 1 & -2 & 1 \\ 2 & 1 & 7 & 0 \\ -3 & 0 & 1 & 5 \end{bmatrix} = \begin{bmatrix} -1 & 1 & -2 & 1 \\ 0 & 3 & 3 & 2 \\ -3 & 0 & 1 & 5 \end{bmatrix}.$$

Now we replace the third row of $E_1 A$ by row three less 3 times row one. This is achieved by multiplying by $E_2 = \begin{bmatrix} 1 & 0 & 0 \\ 0 & 1 & 0 \\ -3 & 0 & 1 \end{bmatrix}$:

$$E_2(E_1 A) = \begin{bmatrix} 1 & 0 & 0 \\ 0 & 1 & 0 \\ -3 & 0 & 1 \end{bmatrix} \begin{bmatrix} -1 & 1 & -2 & 1 \\ 0 & 3 & 3 & 2 \\ -3 & 0 & 1 & 5 \end{bmatrix} = \begin{bmatrix} -1 & 1 & -2 & 1 \\ 0 & 3 & 3 & 2 \\ 0 & -3 & 7 & 2 \end{bmatrix}.$$

Finally, we add row two to row three. This corresponds to multiplication by $E_3 = \begin{bmatrix} 1 & 0 & 0 \\ 0 & 1 & 0 \\ 0 & 1 & 1 \end{bmatrix}$:

$$E_3(E_2 E_1 A) = \begin{bmatrix} 1 & 0 & 0 \\ 0 & 1 & 0 \\ 0 & 1 & 1 \end{bmatrix} \begin{bmatrix} -1 & 1 & -2 & 1 \\ 0 & 3 & 3 & 2 \\ 0 & -3 & 7 & 2 \end{bmatrix} = \begin{bmatrix} -1 & 1 & -2 & 1 \\ 0 & 3 & 3 & 2 \\ 0 & 0 & 10 & 4 \end{bmatrix} = U.$$

We have shown that $(E_3 E_2 E_1)A = U$. Since each of E_3, E_2, and E_1 is invertible, so is their product, and $(E_3 E_2 E_1)^{-1} = E_1^{-1} E_2^{-1} E_3^{-1}$ because the inverse of the product of invertible matrices is the product of the inverses, **order reversed**. Therefore,

$$(E_3E_2E_1)^{-1} = E_1^{-1}E_2^{-1}E_3^{-1}$$

$$= \begin{bmatrix} 1 & 0 & 0 \\ -2 & 1 & 0 \\ 0 & 0 & 1 \end{bmatrix}\begin{bmatrix} 1 & 0 & 0 \\ 0 & 1 & 0 \\ 3 & 0 & 1 \end{bmatrix}\begin{bmatrix} 1 & 0 & 0 \\ 0 & 1 & 0 \\ 0 & -1 & 1 \end{bmatrix} = \begin{bmatrix} 1 & 0 & 0 \\ -2 & 1 & 0 \\ 3 & -1 & 1 \end{bmatrix},$$

and $(E_3E_2E_1)A = U$ implies

$$A = (E_3E_2E_1)^{-1}U = \begin{bmatrix} 1 & 0 & 0 \\ -2 & 1 & 0 \\ 3 & -1 & 1 \end{bmatrix}\begin{bmatrix} -1 & 1 & -2 & 1 \\ 0 & 3 & 3 & 2 \\ 0 & 0 & 10 & 4 \end{bmatrix},$$

an equation I hope you'll check. It's not the sort of thing you would find by accident! ☺

Our primary purpose in introducing elementary matrices was to show you how the L in an LU factorization is obtained, and why this is lower triangular. The matrix L is the product of elementary matrices and if these are all lower triangular, then L will be lower triangular too. (This is easy to see.) Notice that the elementary matrix associated with a row interchange is **not** lower triangular— $\begin{bmatrix} 1 & 0 & 0 \\ 0 & 0 & 1 \\ 1 & 0 & 0 \end{bmatrix}$, for instance—so if such a matrix is involved in the product giving L, then it will be almost impossible for L to be lower triangular. Similarly, an application of the type $R_j \mapsto R_j - cR_i$ with $i < j$ is not allowed, since it also corresponds to an elementary matrix which is not lower triangular.

8.11 Problem. Does $A = \begin{bmatrix} -2 & 6 & 0 & 5 \\ 1 & -3 & 3 & 1 \\ -1 & 0 & 1 & 9 \end{bmatrix}$ have an LU factorization?

Solution. We try to bring A to row echelon form without row interchanges. We begin

$$A = \begin{bmatrix} -2 & 6 & 0 & 5 \\ 1 & -3 & 3 & 1 \\ -1 & 0 & 1 & 9 \end{bmatrix} \to \begin{bmatrix} -2 & 6 & 0 & 5 \\ 0 & 0 & 3 & \frac{7}{2} \\ 0 & -3 & 1 & \frac{13}{2} \end{bmatrix}$$

Now we need either a row interchange, or an operation like $R_2 \mapsto R_2 + R_3$ which is not allowed, so we conclude that this matrix does **not** have an LU factorization. ♤

These examples illustrate the following fact.

8.12 Theorem. *If a matrix A can be reduced to a row echelon matrix without row interchanges, then A has an LU factorization.*

As we have seen, the matrix U mentioned in this theorem is just any row echelon form of A and L is the product of certain elementary matrices. This will be lower triangular as long as each of the elementary matrices is lower triangular, and this will be so as long as we avoided row interchanges.

L without effort

Our primary purpose in introducing elementary matrices is theoretical. They show why Theorem 8.12 is true: the lower triangular L in an LU factorization is the product of lower triangular elementary matrices. In practice, this is **not** how L is determined, however. **If you use only the third elementary row operation in the Gaussian elimination process** (only an operation of the type $R_1 \to R_1 - cR_2$), then L can be found "without effort."

Suppose $A = \begin{bmatrix} -1 & 1 & -2 & 1 \\ 2 & 1 & 7 & 0 \\ -3 & 0 & 1 & 5 \end{bmatrix}$. In Example 8.10, we reduced A to a row echelon matrix U using only the third elementary row operation, like this:

$$A = \begin{bmatrix} -1 & 1 & -2 & 1 \\ 2 & 1 & 7 & 0 \\ -3 & 0 & 1 & 5 \end{bmatrix} \xrightarrow{R2 \to R2-(-2)(R1)} \begin{bmatrix} -1 & 1 & -2 & 1 \\ 0 & 3 & 3 & 2 \\ -3 & 0 & 1 & 5 \end{bmatrix}$$

$$\xrightarrow{R3 \to R3-3(R1)} \begin{bmatrix} -1 & 1 & -2 & 1 \\ 0 & 3 & 3 & 2 \\ 0 & -3 & 7 & 2 \end{bmatrix} \xrightarrow{R3 \to R3-(-1)(R2)} \begin{bmatrix} -1 & 1 & -2 & 1 \\ 0 & 3 & 3 & 3 \\ 0 & 0 & 10 & 4 \end{bmatrix}$$

and factored $A = LU$ with

$$L = \begin{bmatrix} 1 & 0 & 0 \\ -2 & 1 & 0 \\ 3 & -1 & 1 \end{bmatrix} \quad \text{and} \quad U = \begin{bmatrix} -1 & 1 & -2 & 1 \\ 0 & 3 & 3 & 2 \\ 0 & 0 & 10 & 4 \end{bmatrix}.$$

Previously, we got this L by multiplying together certain elementary matrices, but this wasn't necessary. In this situation, we could have simply written L down without any calculation whatsoever!

Remember my preference for the language of subtraction to describe the third elementary row operation? I said I'd show you why I like to say "replace a row by that row **less** a multiple of another row" and not "**plus**" a multiple of another. And now that's what I'm going to do.

The matrix L is square and lower triangular with diagonal entries all 1. That's easy to remember. The entries below the main diagonal, the -2, the 3, the -1, are the *multipliers* that were needed in the Gaussian elimination **when we expressed the third elementary operation in terms of subtraction.** Our operations were

$$\begin{aligned} R2 &\to R2-(\mathbf{-2})(R1) \\ R3 &\to R3-(\mathbf{3})(R1) \\ R3 &\to R3-(\mathbf{-1})(R2). \end{aligned}$$

The scalars -2, 3, -1 are called *multipliers* and these are the entries below the main diagonal in L:

-2 in row two, column one,
3 in row three, column one,
-1 in row three, column two.

Notice that the positions of the multipliers in L correspond to the positions in U where that multiplier was used to get a 0. The -2, for instance, was used to get a 0 in the $(2,1)$ position of U; so -2 goes into the $(2,1)$ position of L, and similarly for the multipliers 3 and -1.

In general, if A can be reduced to a row echelon matrix using only the third elementary row operation, then the lower triangular matrix L is square with 1s on the main diagonal and the multipliers used by the third elementary row operation below the main diagonal.

8.13

> To find an LU factorization of a matrix A, try to reduce A to a row echelon matrix U from the top down using only the third elementary row operation in the form $R_j \mapsto R_j - c(R_i)$ with $j > i$.

"Top down" means that once a pivot has been identified, we put 0s below that pivot in order starting with the entry immediately below that pivot.

8.14 Example. We move $A = \begin{bmatrix} 1 & 2 & 3 \\ 5 & 14 & 7 \\ 9 & 10 & 0 \\ 0 & 4 & 3 \end{bmatrix}$ to row echelon form using only the third row operation.

$$A \xrightarrow[R3 \to R3 - 9(R1)]{R2 \to R2 - 5(R1)} \begin{bmatrix} 1 & 2 & 3 \\ 0 & 4 & -8 \\ 0 & -8 & -27 \\ 0 & 4 & 3 \end{bmatrix}$$

$$\xrightarrow[R4 \to R4 - 1(R2)]{R3 \to R3 - (-2)(R1)} \begin{bmatrix} 1 & 2 & 3 \\ 0 & 4 & -8 \\ 0 & 0 & -43 \\ 0 & 0 & 11 \end{bmatrix} \xrightarrow{R4 \to R4 - (-\frac{11}{43})(R3)} \begin{bmatrix} 1 & 2 & 3 \\ 0 & 4 & -8 \\ 0 & 0 & -43 \\ 0 & 0 & 0 \end{bmatrix} = U.$$

So $A = LU$ with $L = \begin{bmatrix} 1 & 0 & 0 & 0 \\ 5 & 1 & 0 & 0 \\ 9 & -2 & 1 & 0 \\ 0 & 1 & -\frac{11}{43} & 1 \end{bmatrix}$. ☺

8.15 Remark. Why did we say "from the top down" in Definition 8.13? Suppose $A = \begin{bmatrix} 1 & 2 & 3 \\ 1 & 1 & 1 \\ 2 & 2 & 3 \end{bmatrix}$ and we apply the following sequence of third elementary row operations:

$$A = \begin{bmatrix} 1 & 2 & 3 \\ 1 & 1 & 1 \\ 2 & 2 & 3 \end{bmatrix} \xrightarrow{R3 \to R3 - 2(R2)} \begin{bmatrix} 1 & 2 & 3 \\ 1 & 1 & 1 \\ 0 & 0 & 1 \end{bmatrix} \xrightarrow{R2 \to R2 - R1} \begin{bmatrix} 1 & 2 & 3 \\ 0 & -1 & -2 \\ 0 & 0 & 1 \end{bmatrix} = U.$$

Thus the 0s below the first pivot were not produced from the top down. The matrix recording the multipliers is $L = \begin{bmatrix} 1 & 0 & 0 \\ 1 & 1 & 0 \\ 0 & 2 & 1 \end{bmatrix}$, but $A \neq LU$. For this observation and example, the author is indebted to one of the reviewers of this book.

Forcing ourselves to use only the third row operation can lead to messy Gaussian elimination when working by hand, but remember that such a calculation is easy by machine and it leads to enormous savings in the calculation of L. Henceforth,

Why LU?

One use for LU factorization is that it simplifies the solution of systems of linear equations. Suppose we want to solve the system $A\mathsf{x} = \mathsf{b}$ and we know that $A = LU$. So the system is $LU\mathsf{x} = \mathsf{b}$, which we think of as $L(U\mathsf{x}) = \mathsf{b}$. Let $\mathsf{y} = U\mathsf{x}$. The system is now $L\mathsf{y} = \mathsf{b}$, which is easily solved for y by "forward substitution" because L is lower triangular. Then we solve $U\mathsf{x} = \mathsf{y}$ for x, which is easy by back substitution because U is upper triangular.

Here's an example.

8.16 Example. Suppose we want to solve the system

$$\begin{aligned} -3x_1 \qquad\quad + x_3 + \quad x_4 &= 2 \\ -12x_1 + 2x_2 + 2x_3 + 11x_4 &= -2 \\ 9x_1 + 4x_2 + x_3 + 7x_4 &= -10 \\ -3x_1 - 2x_2 + 5x_3 - 6x_4 &= 14. \end{aligned}$$

This is $A\mathsf{x} = \mathsf{b}$ with $A = \begin{bmatrix} -3 & 0 & 1 & 1 \\ -12 & 2 & 2 & 11 \\ 9 & 4 & 1 & 7 \\ -3 & -2 & 5 & -6 \end{bmatrix}$, $\mathsf{x} = \begin{bmatrix} x_1 \\ x_2 \\ x_3 \\ x_4 \end{bmatrix}$ and $\mathsf{b} = \begin{bmatrix} 2 \\ -2 \\ -10 \\ 14 \end{bmatrix}$. Suppose we know $A = LU$, with

$$L = \begin{bmatrix} 1 & 0 & 0 & 0 \\ 4 & 1 & 0 & 0 \\ -3 & 2 & 1 & 0 \\ 1 & -1 & \frac{1}{4} & 1 \end{bmatrix} \quad \text{and } U = \begin{bmatrix} -3 & 0 & 1 & 1 \\ 0 & 2 & -2 & 7 \\ 0 & 0 & 8 & -4 \\ 0 & 0 & 0 & 1 \end{bmatrix}.$$

To solve $A\mathsf{x} = \mathsf{b}$, which is $L(U\mathsf{x}) = \mathsf{b}$, we first solve $L\mathsf{y} = \mathsf{b}$ for $\mathsf{y} = \begin{bmatrix} y_1 \\ y_2 \\ y_3 \\ y_4 \end{bmatrix}$. This is the system

$$\begin{aligned} y_1 \qquad\qquad\qquad\qquad &= 2 \\ 4y_1 + y_2 \qquad\qquad\quad &= -2 \\ -3y_1 + 2y_2 + y_3 \qquad &= -10 \\ y_1 - y_2 + \tfrac{1}{4}y_3 + y_4 &= 14. \end{aligned}$$

By *forward substitution* we have

$y_1 = 2,$

$4y_1 + y_2 = -2$, so $y_2 = -2 - 8 = -10$,

$-3y_1 + 2y_2 + y_3 = -10$, so $y_3 = -10 + 6 + 20 = 16$,

$y_1 - y_2 + \frac{1}{4}y_3 + y_4 = 14$, so $y_4 = 14 - 2 - 10 - 4 = -2$.

So $\mathsf{y} = \begin{bmatrix} 2 \\ -10 \\ 16 \\ -2 \end{bmatrix}$. Now we solve $U\mathsf{x} = \mathsf{y}$. The corresponding equations are

$$\begin{aligned} -3x_1 \qquad\quad + x_3 + x_4 &= 2 \\ 2x_2 - 2x_3 + 7x_4 &= -10 \\ 8x_3 - 4x_4 &= 16 \\ x_4 &= -2 \end{aligned}$$

so, by back substitution,

$x_4 = -2$,

$8x_3 - 4x_4 = 16$, so $8x_3 = 16 - 8 = 8$ and $x_3 = -1$,

$2x_2 - 2x_3 + 7x_4 = -10$, so $2x_2 = -10 + 2 + 14 = 6$ and $x_2 = 3$,

$-3x_1 + x_3 + x_4 = 2$, so $-3x_1 = 2 - 1 + 2 = 3$ and $x_1 = -1$.

We get $\mathsf{x} = \begin{bmatrix} -1 \\ 3 \\ 1 \\ -2 \end{bmatrix}$ and check that this is correct by verifying that $A\mathsf{x} = \mathsf{b}$:

$$\begin{bmatrix} -3 & 0 & 1 & 1 \\ -12 & 2 & 2 & 11 \\ 9 & 4 & 1 & 7 \\ -3 & -2 & 5 & -6 \end{bmatrix} \begin{bmatrix} -1 \\ 3 \\ 1 \\ -2 \end{bmatrix} = \begin{bmatrix} 2 \\ -2 \\ -10 \\ 14 \end{bmatrix}.$$

:-)

Concluding Remark. While we don't go through the calculations here, it is more efficient to solve two triangular systems than one general system of equations. By this we mean, that the number of additions and multiplications required to solve two triangular $n\times$ systems is of the order of n^2, whereas it requires about n^3 operations to solve a general system. The reduction from n^3 to n^2 is significant when n is large.

PLU Factorization; Row Interchanges

To find an LU factorization of A, you might have noticed in **8.13** that I said we **try** to move A to a row echelon matrix without interchanging rows. This suggests this might not always be possible and in fact it isn't. Look again at Problem 8.11, or consider the matrix $\begin{bmatrix} 2 & 4 & 2 \\ 1 & 2 & 3 \\ 3 & 2 & 1 \end{bmatrix}$.

If we try to move this to row echelon form, we begin

$$\begin{bmatrix} 2 & 4 & 2 \\ 1 & 2 & 3 \\ 3 & 2 & 1 \end{bmatrix} \to \begin{bmatrix} 2 & 4 & 2 \\ 0 & 0 & 2 \\ 0 & -4 & -2 \end{bmatrix},$$

but now are forced to interchange rows two and three. In an attempt to avoid this problem, we change A at the outset by interchanging its second and third rows. This gives a matrix A' that we can reduce to row echelon form without row interchanges, like this:

$$A' = \begin{bmatrix} 2 & 4 & 2 \\ 3 & 2 & 1 \\ 1 & 2 & 3 \end{bmatrix} \rightarrow \begin{bmatrix} 2 & 4 & 2 \\ 0 & -4 & -2 \\ 0 & 0 & 2 \end{bmatrix} = U.$$

So

$$A' = LU = \begin{bmatrix} 1 & 0 & 0 \\ \frac{3}{2} & 1 & 0 \\ \frac{1}{2} & 0 & 1 \end{bmatrix} \begin{bmatrix} 2 & 4 & 2 \\ 0 & -4 & -2 \\ 0 & 0 & 2 \end{bmatrix}. \tag{28}$$

Of course, this is a factorization of A', not of A, but there is a nice connection between A' and A, a connection we have seen before. The matrix A' is obtained from A by interchanging rows two and three, so $A' = PA$ where $P = \begin{bmatrix} 1 & 0 & 0 \\ 0 & 0 & 1 \\ 0 & 1 & 0 \end{bmatrix}$ is the elementary matrix obtained from the 3×3 identity matrix by interchanging rows two and three. Thus (28) is an equation of the form $PA = LU$, and this implies that $A = P^{-1}LU$.

Now the inverse of an elementary matrix like P is another elementary matrix, the one that "undoes" P. In this case, $P^{-1} = P$, because if we interchange rows two and three and then immediately want to undo this, we must interchange rows two and three again. In this example, $P^{-1} = P$ and our factorization of A is $A = PLU$.

In general, Gaussian elimination may involve several interchanges of rows so that the initial matrix A' is A with its rows scrambled. In this case, $A' = PA$ where P is the product of a number of row interchange elementary matrices. It is not hard to see that P is obtained from the identity matrix by reordering its rows just as the rows of A were reordered to give A'. Such a matrix is called a *permutation matrix.*

8.17 Definition. A *permutation matrix* is a matrix whose rows are the rows of the identity matrix in some order.

8.18 Example. $P_1 = \begin{bmatrix} 0 & 1 \\ 1 & 0 \end{bmatrix}$, $P_2 = \begin{bmatrix} 0 & 1 & 0 \\ 0 & 0 & 1 \\ 1 & 0 & 0 \end{bmatrix}$, $P_3 = \begin{bmatrix} 0 & 0 & 1 & 0 \\ 0 & 1 & 0 & 0 \\ 0 & 0 & 0 & 1 \\ 1 & 0 & 0 & 0 \end{bmatrix}$ are all permutation matrices.

Multiplying a matrix A on the left by a permutation matrix changes the rows of A in a predictable way. With reference to the three permutation matrices just cited, the rows of P_1 are the rows of the identity in the order 2, 1. Thus, whenever P_1A is defined, P_1A is A with its rows arranged in the same order, 2, 1. For example, $P_1 \begin{bmatrix} 1 & 2 \\ 3 & 4 \end{bmatrix} = \begin{bmatrix} 3 & 4 \\ 1 & 2 \end{bmatrix}$. The rows of P_2 are those of the identity matrix in the order 2, 3, 1 so, whenever P_2A is defined, this matrix

is the matrix A with its rows appearing in the order 2, 3, 1. For example, $P_2 \begin{bmatrix} 1 & 2 \\ 3 & 4 \\ 5 & 6 \end{bmatrix} = \begin{bmatrix} 3 & 4 \\ 5 & 6 \\ 1 & 2 \end{bmatrix}$.
Similarly, whenever P_3A is defined, this matrix is just A with rows in order 3, 2, 4, 1.

Of great help to us is that fact that it's easy to find the inverse of a permutation matrix.

8.19 If P is a permutation matrix, then $P^{-1} = P^T$.

This is easily seen, because the i, j entry of PP^T is the dot product of row i of P and column j of P^T, but this is the dot product of row i of P and row j of P (definition of P^T) and this is 1 if and only if $i = j$ (because the dot product of two standard basis vectors is 0 unless the vectors are the same, in which case it is 1). We conclude that the i, j entry of PP^T is 0 unless $i = j$ in which case it is 1. But this defines the identity matrix.

For example, the inverses of the permutation matrices in Example 8.18 are

$$P_1^{-1} = P_1^T = \begin{bmatrix} 0 & 1 \\ 1 & 0 \end{bmatrix}, \quad P_2^{-1} = P_2^T = \begin{bmatrix} 0 & 0 & 1 \\ 1 & 0 & 0 \\ 0 & 1 & 0 \end{bmatrix} \text{ and } P_3^{-1} = P_3^T = \begin{bmatrix} 0 & 0 & 0 & 1 \\ 0 & 1 & 0 & 0 \\ 1 & 0 & 0 & 0 \\ 0 & 0 & 1 & 0 \end{bmatrix}.$$

8.20 In general, we cannot expect to factor a given matrix A in the form $A = LU$, but we can always achieve $P'A = LU$ for some permutation matrix P', so that $A = (P')^{-1}LU = PLU$, with $P = (P')^{-1} = (P')^T$ also a permutation matrix.

8.21 **Example.** Suppose we want an LU factorization of $A = \begin{bmatrix} 0 & 1 & 1 & 0 \\ 0 & -1 & -1 & 4 \\ -2 & 3 & 2 & 1 \\ -6 & 4 & 2 & -8 \end{bmatrix}$. The first step in moving A to a row echelon matrix requires a row interchange. There is no LU factorization, but there is a PLU factorization. After experimenting with the rows of A, we discover that if we write the rows in order 3, 1, 4, 2, then we can carry the new matrix to row echelon form without row interchanges. So we start with

$$PA = \begin{bmatrix} -2 & 3 & 2 & 1 \\ 0 & 1 & 1 & 0 \\ -6 & 4 & 2 & -8 \\ 0 & -1 & -1 & 4 \end{bmatrix},$$

where $P = \begin{bmatrix} 0 & 0 & 1 & 0 \\ 1 & 0 & 0 & 0 \\ 0 & 0 & 0 & 1 \\ 0 & 1 & 0 & 0 \end{bmatrix}$ is the permutation matrix formed from the identity by reordering the rows of the identity as we have reordered those of A, in the order 3, 1, 4, 2. Gaussian elimination, using only the third elementary row operation, yields

$$PA = \begin{bmatrix} -2 & 3 & 2 & 1 \\ 0 & 1 & 1 & 0 \\ -6 & 4 & 2 & -8 \\ 0 & -1 & -1 & 4 \end{bmatrix} \longrightarrow \begin{bmatrix} -2 & 3 & 2 & 1 \\ 0 & 1 & 1 & 0 \\ 0 & -5 & -4 & -11 \\ 0 & -1 & -1 & 4 \end{bmatrix}$$

$$\longrightarrow \begin{bmatrix} -2 & 3 & 2 & 1 \\ 0 & 1 & 1 & 0 \\ 0 & 0 & 1 & -11 \\ 0 & 0 & 0 & 4 \end{bmatrix} = U$$

with $L = \begin{bmatrix} 1 & 0 & 0 & 0 \\ 0 & 1 & 0 & 0 \\ 3 & -5 & 1 & 0 \\ 0 & -1 & 0 & 1 \end{bmatrix}$.

We have $PA = LU$, so $A = P^{-1}LU$, with $P^{-1} = P^T = \begin{bmatrix} 0 & 1 & 0 & 0 \\ 0 & 0 & 0 & 1 \\ 1 & 0 & 0 & 0 \\ 0 & 0 & 1 & 0 \end{bmatrix}$. :)

We conclude this week by illustrating some other ways in which elementary matrices are of use.

8.22 Theorem. *A matrix is invertible if and only if it can be written as the product of elementary matrices.*

Proof. Suppose A is the product of elementary matrices. An elementary matrix is invertible. Also, the product of invertible matrices is invertible. Thus A is invertible.

On the other hand, suppose A is invertible. Then we can transform A to the identity matrix by a sequence of elementary row operations. The Gaussian elimination process looks like this:

$$A \to E_1A \to E_2(E_1A) \to \cdots \to (E_kE_{k-1} \cdots E_2E_1)A = I,$$

where $E_1, E_2, \ldots, E_k$ are elementary matrices.

So $A = (E_kE_{k-1} \cdots E_2E_1)^{-1} = E_1^{-1}E_2^{-1} \cdots E_{k-1}^{-1}E_k^{-1}$. Since the inverse of an elementary matrix is elementary, $A = E_1^{-1}E_2^{-1} \cdots E_k^{-1}$ is an expression of A as the product of elementary matrices. ■

8.23 Problem. Express $A = \begin{bmatrix} 2 & 3 \\ 1 & 1 \end{bmatrix}$ as the product of elementary matrices.

Solution. We move A to the identity matrix by a sequence of elementary row operations.

$$A = \begin{bmatrix} 2 & 3 \\ 1 & 1 \end{bmatrix} \to \begin{bmatrix} 1 & 1 \\ 2 & 3 \end{bmatrix} = E_1A$$

$$\to \begin{bmatrix} 1 & 1 \\ 0 & 1 \end{bmatrix} = E_2E_1A \to \begin{bmatrix} 1 & 0 \\ 0 & 1 \end{bmatrix} = E_3E_2E_1A = I$$

with

$$E_1 = \begin{bmatrix} 0 & 1 \\ 1 & 0 \end{bmatrix}, \quad E_2 = \begin{bmatrix} 1 & 0 \\ -2 & 1 \end{bmatrix}, \quad \text{and} \quad E_3 = \begin{bmatrix} 1 & -1 \\ 0 & 1 \end{bmatrix}.$$

So $(E_3E_2E_1)A = I$ and $A = (E_3E_2E_1)^{-1} = E_1^{-1}E_2^{-1}E_3^{-1}$;

$$\begin{bmatrix} 2 & 3 \\ 1 & 1 \end{bmatrix} = \begin{bmatrix} 0 & 1 \\ 1 & 0 \end{bmatrix}\begin{bmatrix} 1 & 0 \\ 2 & 1 \end{bmatrix}\begin{bmatrix} 1 & 1 \\ 0 & 1 \end{bmatrix}.$$

Finally, we justify a property of square matrices we have used on a number of occasions.

8.24 Theorem. *If A and B are square $n \times n$ matrices and $AB = I$ (the $n \times n$ identity matrix), then A is invertible. In particular, $BA = I$ too.*

Proof. If $AB = I$, then the system $B\mathsf{x} = 0$ has only the solution $\mathsf{x} = 0$: from $B\mathsf{x} = 0$, we have $AB\mathsf{x} = A0$, so $I\mathsf{x} = 0$ and $\mathsf{x} = 0$. Thus, when we try to solve $B\mathsf{x} = 0$ by reducing B to row echelon form, there are no free variables (free variables imply infinitely many solutions), every column contains a pivot, and it follows that we can transform B to I with elementary row operations. As in Theorem 8.22, this means that $EB = I$ for some product E of elementary matrices. Since E is invertible, $B = E^{-1}$ is also invertible. Now $AB = I$ gives $AE^{-1} = I$, so $A = E$ is the product of elementary matrices and so invertible by Theorem 8.22. ∎

True/False Questions for Week 8

Decide, with as little calculation as possible, whether each of the following statements is true or false and explain your answer whenever you say "false."

1. The matrix $\begin{bmatrix} 1 & 0 & 0 \\ 0 & 2 & 1 \\ 0 & 0 & 1 \end{bmatrix}$ is elementary.

2. If E and F are elementary matrices, then EF is elementary.

3. If E and F are elementary matrices, then EF is invertible.

4. If $E = \begin{bmatrix} 1 & 0 & 0 \\ 0 & 1 & -2 \\ 0 & 0 & 1 \end{bmatrix}$, then $E^{-1} = \begin{bmatrix} 1 & 0 & 0 \\ 0 & 1 & 2 \\ 0 & 0 & 1 \end{bmatrix}$.

5. If $E = \begin{bmatrix} 1 & 0 & -3 \\ 0 & 1 & 0 \\ 0 & 0 & 1 \end{bmatrix}$ and A is a $3 \times n$ matrix, rows two and three of EA are the same as rows two and three of A.

6. The product of lower triangular matrices is lower triangular.

7. The product of elementary matrices is lower triangular.

8. The product of elementary matrices corresponding to elementary row operations of types 2 and 3 is lower triangular.

9. Any matrix that can be reduced to row echelon form without row interchanges has an LU factorization.

10. If $A = LU$ is an LU factorization of A, we can solve $A\mathsf{x} = \mathsf{b}$ by first solving $U\mathsf{y} = \mathsf{b}$ for y and then $L\mathsf{x} = \mathsf{y}$ for x.

11. If $E = \begin{bmatrix} 0 & 1 & 0 \\ 1 & 0 & 0 \\ 0 & 0 & 1 \end{bmatrix}$, then $E^{-1} = E$.

12. If P is a permutation matrix, then $P^T P = I$.

13. The matrix $A = \begin{bmatrix} 0 & 5 & 1 \\ 4 & 10 & 1 \\ 0 & 1 & 11 \end{bmatrix}$ has an LU factorization.

Week 8 Test Yourself

Here are a few problems with short answers that you can use to test your understanding of the concepts you have met this week.

1. Let $E = \begin{bmatrix} 1 & 0 & 0 \\ 0 & 1 & 4 \\ 0 & 0 & 1 \end{bmatrix}$ and let A be an arbitrary $3 \times n$ matrix, $n \geq 1$.

   (a) Explain the connection between A and EA.

   (b) What is the name for a matrix such as E?

   (c) Is E invertible? If so, what is its inverse?

2. Let $A = \begin{bmatrix} 1 & 2 & 3 \\ 4 & 5 & 6 \\ 6 & 8 & 9 \end{bmatrix}$. Write down an elementary matrix E so that

   (a) $EA = \begin{bmatrix} 1 & 2 & 3 \\ 4 & 5 & 6 \\ 0 & -4 & -9 \end{bmatrix}$ (b) $EA = \begin{bmatrix} 1 & 2 & 3 \\ 0 & -3 & -6 \\ 6 & 8 & 9 \end{bmatrix}$

   (c) $EA = \begin{bmatrix} 1 & 2 & 3 \\ 4 & 5 & 6 \\ 0 & \frac{1}{2} & 0 \end{bmatrix}$ (d) $EA = \begin{bmatrix} 1 & 2 & 3 \\ 6 & 8 & 9 \\ 4 & 5 & 6 \end{bmatrix}$.

3. Write down the inverse of each of the following matrices without calculation. Explain your reasoning.

   (a) $A = \begin{bmatrix} 1 & 3 \\ 0 & 1 \end{bmatrix}$ (b) $A = \begin{bmatrix} 1 & 0 \\ -5 & 1 \end{bmatrix}$ (c) $A = \begin{bmatrix} 0 & 0 & 1 \\ 0 & 1 & 0 \\ 1 & 0 & 0 \end{bmatrix}$

4. What number x will force an interchange of rows when Gaussian elimination is applied to $\begin{bmatrix} 2 & 5 & 1 \\ 4 & x & 1 \\ 0 & 1 & -1 \end{bmatrix}$?

5. Is the product of elementary matrices invertible? Explain.

6. Does $A = \begin{bmatrix} 2 & 5 & 1 \\ 4 & 10 & 1 \\ 0 & 1 & -1 \end{bmatrix}$ have an LU factorization? Explain.

7. What is an "LU factorization" of a matrix A?

8. If $A = LU$ is an LU factorization, then L has to be square. Why?

9. Suppose A is an $m \times n$ matrix that can be reduced by Gaussian elimination to a row echelon matrix U without the use of row interchanges. How does this lead to an LU factorization of A? Explain clearly how the lower triangular matrix L arises.

10. If possible, find an LU factorization of $A = \begin{bmatrix} -1 & 1 & -2 \\ 2 & 1 & 7 \end{bmatrix}$ and express L as the product of elementary matrices.

11. Find an LU factorization of $A = \begin{bmatrix} -2 & 1 & 3 & 4 \\ 4 & 0 & -9 & -7 \\ -6 & 10 & 4 & 15 \end{bmatrix}$ "without effort."

12. Let $A = \begin{bmatrix} 2 & 1 \\ 6 & 8 \end{bmatrix} = \begin{bmatrix} 2 & 0 \\ 6 & 5 \end{bmatrix} \begin{bmatrix} 1 & \frac{1}{2} \\ 0 & 1 \end{bmatrix} = LU$.

    (a) Use this factorization of A to solve the system $A\begin{bmatrix} x \\ y \end{bmatrix} = \begin{bmatrix} -2 \\ 9 \end{bmatrix}$.

    (b) Express $\begin{bmatrix} -2 \\ 9 \end{bmatrix}$ as a linear combination of the columns of A.

13. Write down the inverse of $A = \begin{bmatrix} 0 & 0 & 1 \\ 1 & 0 & 0 \\ 0 & 1 & 0 \end{bmatrix}$ without calculation.

14. Let $A = \begin{bmatrix} 0 & 1 & -2 \\ 0 & 0 & 7 \\ 2 & 4 & 9 \end{bmatrix}$. Write down a permutation matrix P such that PA is row echelon.

Week 9

UNIT THREE: DETERMINANTS

Minors and Cofactors

In order to define the cross product of two vectors in R^3, we introduced a number called the "determinant" of a 2×2 matrix. I hope you remember that if $A = \begin{bmatrix} a & b \\ c & d \end{bmatrix}$, then

$$\det A = |A| = \begin{vmatrix} a & b \\ c & d \end{vmatrix} = ad - bc.$$

(We denote the determinant either with det or vertical bars.) We encountered this number again when trying to find a formula for the inverse of a 2×2 matrix. Letting $B = \begin{bmatrix} d & -b \\ -c & a \end{bmatrix}$, the product $AB = \begin{bmatrix} a & b \\ c & d \end{bmatrix} \begin{bmatrix} d & -b \\ -c & a \end{bmatrix} = \begin{bmatrix} ad-bc & 0 \\ 0 & ad-bc \end{bmatrix} = (\det A)I$. So, if $\det A \neq 0$, then A has an inverse, namely, $\frac{1}{ad-bc}B$—see Proposition 7.11 in Unit 2.

These ideas generalize to square matrices of any size. Any square matrix A has a determinant, which is a number denoted $\det A$ or $|A|$. To any square matrix, we can associate another matrix B called the "adjoint" of A, the product of A and its adjoint is $(\det A)I$ and if $\det A \neq 0$, then $A^{-1} = \frac{1}{\det A} \operatorname{adj} A$. It's time to slow down!

9.1 Definitions. If $A = [a]$ is a 1×1 matrix, we define $\det A = a$, just the number. In general, if $n > 1$ and A is an $n \times n$ matrix, the (i,j) *minor* of A is the determinant of the $(n-1)\times(n-1)$ matrix obtained from A by deleting row i and column j. It's denoted m_{ij}. The (i,j) *cofactor* of A is $(-1)^{i+j} m_{ij}$. This is denoted c_{ij}. The *adjoint* of A, denoted $\operatorname{adj} A$, is the transpose of its matrix of cofactors.

These definitions probably look horrible, but you'll be computing minors and cofactors and adjoints before you know it. I promise! Here's an example.

9.2 Example. Let $A = \begin{bmatrix} a & b \\ c & d \end{bmatrix}$. The $(1,1)$ minor is the determinant of the matrix obtained by deleting row 1 and column 1 of A. The remaining matrix is just $[d]$, whose determinant is $m_{11} = d$. The $(1,2)$ minor is the determinant of the matrix obtained by deleting row 1 and column 2. The remaining matrix is $[c]$, whose determinant is $m_{12} = c$. The matrix of minors is $\begin{bmatrix} d & c \\ b & a \end{bmatrix}$. The $(1,1)$ cofactor is $c_{11} = (-1)^{1+1} m_{11} = (-1)^2 d = d$. The $(1,2)$ cofactor is $c_{12} = (-1)^{1+2} m_{12} = (-1)^3 c = -c$, and so on. The matrix of cofactors is $\begin{bmatrix} d & -c \\ -b & a \end{bmatrix}$ and the adjoint of A is $\operatorname{adj} A = C^T = \begin{bmatrix} d & -b \\ -c & a \end{bmatrix}$. Look familiar?

(Hint: Start reading this section again, or look at Proposition 7.11.) ☺

9.3 Example. Let $A = \begin{bmatrix} 1 & 2 & 3 \\ 0 & -2 & 2 \\ 3 & 7 & -4 \end{bmatrix}$. The $(1,1)$ minor is the determinant of the matrix

$$\begin{bmatrix} -2 & 2 \\ 7 & -4 \end{bmatrix},$$

which is what remains when we remove row one and column one from A, so $m_{11} = -2(-4) - 2(7) = -6$. The $(1,1)$ cofactor is $c_{11} = (-1)^{1+1}m_{11} = (-1)^2(-6) = -6$. The $(2,3)$ minor is the determinant of

$$\begin{bmatrix} 1 & 2 \\ 3 & 7 \end{bmatrix},$$

which is what remains after removing row two and column three from A. Thus $m_{23} = 1$ and $c_{23} = (-1)^{2+3}m_{23} = (-1)^5(1) = -1$. The $(3,1)$ minor of A is the determinant of

$$\begin{bmatrix} 2 & 3 \\ -2 & 2 \end{bmatrix},$$

which is the matrix obtained from A by removing row three and column one. Thus $m_{31} = 10$ and $c_{31} = (-1)^{3+1}m_{31} = (-1)^4(10) = +10$.

Continuing in this way, we can find a minor and a cofactor corresponding to every position (i,j) of A. The *matrix of minors* of A is

$$M = \begin{bmatrix} -6 & -6 & 6 \\ -29 & -13 & 1 \\ 10 & 2 & -2 \end{bmatrix}$$

and the *matrix of cofactors* is $C = \begin{bmatrix} -6 & 6 & 6 \\ 29 & -13 & -1 \\ 10 & -2 & -2 \end{bmatrix}$. The transpose of this is the adjoint of A:

$$\operatorname{adj} A = C^T = \begin{bmatrix} -6 & 29 & 10 \\ 6 & -13 & -2 \\ 6 & -1 & -2 \end{bmatrix}.$$

The product of A and $\operatorname{adj} A$ is

$$A(\operatorname{adj} A) = AC^T = \begin{bmatrix} 1 & 2 & 3 \\ 0 & -2 & 2 \\ 3 & 7 & -4 \end{bmatrix}\begin{bmatrix} -6 & 29 & 10 \\ 6 & -13 & -2 \\ 6 & -1 & -2 \end{bmatrix} = \begin{bmatrix} 24 & 0 & 0 \\ 0 & 24 & 0 \\ 0 & 0 & 24 \end{bmatrix} = 24I,$$

a scalar multiple of the identity matrix, just as in the 2×2 case. Notice too that

$$(\operatorname{adj} A)A = \begin{bmatrix} -6 & 29 & 10 \\ 6 & -13 & -2 \\ 6 & -1 & -2 \end{bmatrix}\begin{bmatrix} 1 & 2 & 3 \\ 0 & -2 & 2 \\ 3 & 7 & -4 \end{bmatrix} = \begin{bmatrix} 24 & 0 & 0 \\ 0 & 24 & 0 \\ 0 & 0 & 24 \end{bmatrix} = 24I.$$

This implies of course that A is invertible with $A^{-1} = \frac{1}{24}\operatorname{adj}A = \frac{1}{\det A}\operatorname{adj}A$. ☺

By the way, the $(-1)^{i+j}$ in the definition of cofactor isn't as difficult as it may seem. First, since the only powers of -1 are $+1$ and -1, any cofactor is either plus or minus the corresponding minor. Second, instead of worrying about i and j, the best way to decide plus or minus is to remember the following pattern, which shows the values of $(-1)^{i+j}$ for all positive integers i and j.

$$\begin{matrix} + & - & + & - & \cdots \\ - & + & - & + & \cdots \\ + & - & + & - & \cdots \\ - & + & - & + & \cdots \\ \vdots & \vdots & & & \end{matrix} \tag{29}$$

Each "+" means "multiply by $+1$," that is, leave the minor for that position unchanged. Each "−" means "multiply by -1," that is, change the sign of the minor. Look at how C is obtained from M above by alternately leaving alone or changing signs.

It's a fact (that I'm not going to justify) that

9.4 A square matrix A commutes with its adjoint and the product of A and its adjoint is a scalar multiple of the identity: $A(\operatorname{adj}A) = (\operatorname{adj}A)A = kI$ for some scalar k. This k is called the *determinant* of A.

9.5 $$A(\operatorname{adj}A) = (\operatorname{adj}A)A = (\det A)I.$$

For example, the determinant of the matrix A in Example 9.3 is 24 since $A(\operatorname{adj}A) = (\operatorname{adj}A)A = 24I$.

9.6 Example. Let $A = \begin{bmatrix} -1 & 2 & 0 \\ 0 & 1 & 5 \\ 2 & 3 & -4 \end{bmatrix}$. The matrix of minors is $M = \begin{bmatrix} -19 & -10 & -2 \\ -8 & 4 & -7 \\ 10 & -5 & -1 \end{bmatrix}$, the matrix of cofactors is $C = \begin{bmatrix} -19 & 10 & -2 \\ 8 & 4 & 7 \\ 10 & 5 & -1 \end{bmatrix}$ and the adjoint of A is $\operatorname{adj}A = C^T = \begin{bmatrix} -19 & 8 & 10 \\ 10 & 4 & 5 \\ -2 & 7 & -1 \end{bmatrix}$. The product of A and $\operatorname{adj}A$ is

$$A(\operatorname{adj}A) = \begin{bmatrix} -1 & 2 & 0 \\ 0 & 1 & 5 \\ 2 & 3 & -4 \end{bmatrix}\begin{bmatrix} -19 & 8 & 10 \\ 10 & 4 & 5 \\ -2 & 7 & -1 \end{bmatrix} = \begin{bmatrix} 39 & 0 & 0 \\ 0 & 39 & 0 \\ 0 & 0 & 39 \end{bmatrix} = 39I,$$

so $\det A = 39$ and A is invertible with $A^{-1} = \frac{1}{39}\operatorname{adj}A = \frac{1}{\det A}\operatorname{adj}A$. ☺

Computing Determinants

Actually, if you only want the determinant, it isn't necessary to compute the entire matrix of cofactors. Suppose

$$A = \begin{bmatrix} a_{11} & a_{12} & a_{13} \\ a_{21} & a_{22} & a_{23} \\ a_{31} & a_{32} & a_{33} \end{bmatrix}$$

is the general 3×3 matrix and

$$C = \begin{bmatrix} c_{11} & c_{12} & c_{13} \\ c_{21} & c_{22} & c_{23} \\ c_{31} & c_{32} & c_{33} \end{bmatrix}$$

is the matrix of cofactors. Then

$$\begin{aligned} A(\operatorname{adj} A) &= \begin{bmatrix} a_{11} & a_{12} & a_{13} \\ a_{21} & a_{22} & a_{23} \\ a_{31} & a_{32} & a_{33} \end{bmatrix} \begin{bmatrix} c_{11} & c_{21} & c_{31} \\ c_{12} & c_{22} & c_{32} \\ c_{13} & c_{23} & c_{33} \end{bmatrix} \\ &= \begin{bmatrix} \det A & 0 & 0 \\ 0 & \det A & 0 \\ 0 & 0 & \det A \end{bmatrix} = (\det A)I. \end{aligned}$$

The $(1,1)$ entry of $A(\operatorname{adj} A)$, namely $a_{11}c_{11} + a_{12}c_{12} + a_{13}c_{13}$, is $\det A$:

$$a_{11}c_{11} + a_{12}c_{12} + a_{13}c_{13} = \det A.$$

The $(2,2)$ and the $(3,3)$ entries of $A(\operatorname{adj} A)$ are also $\det A$:

$$\begin{aligned} a_{21}c_{21} + a_{22}c_{22} + a_{23}c_{23} &= \det A \\ a_{31}c_{31} + a_{32}c_{32} + a_{33}c_{33} &= \det A. \end{aligned}$$

Entries not on the main diagonal are 0. For example, the $(2,3)$ and $(1,2)$ entries are

$$a_{21}c_{31} + a_{22}c_{32} + a_{23}c_{33} = 0$$

and

$$a_{11}c_{21} + a_{12}c_{22} + a_{13}c_{23} = 0.$$

These calculations illustrate for a 3×3 matrix A what is true for an $n \times n$ matrix, with any n. For any i, j,

$$a_{i1}c_{j1} + a_{i2}c_{j2} + a_{i3}c_{j3} + \cdots + a_{in}c_{jn} = \begin{cases} 1 & \text{if } i = j \\ 0 & \text{if } i \neq j. \end{cases}$$

In particular (with $i = j$), $\det A$ is the dot product of any row of A with the corresponding (row of) cofactors. Had we examined the equation $(\operatorname{adj} A)A = \det A$, we would have discovered the same thing for columns.

9.7 **Laplace Expansion:** The determinant of a matrix A is the dot product of any row (or column) of A with the corresponding cofactors.

This method of computing a determinant is called the "Laplace Expansion" after the mathematician and astronomer Pierre-Simon Laplace (1749–1827).

Looking again at Example 9.3, the dot product of the first column of A and the first column of C is 24:

$$\begin{bmatrix} 1 \\ 0 \\ 3 \end{bmatrix} \cdot \begin{bmatrix} -6 \\ 29 \\ 10 \end{bmatrix} = 24 = \det A.$$

The dot product of the second row of A and the second row of C is 24:

$$\begin{bmatrix} 0 \\ -2 \\ 2 \end{bmatrix} \cdot \begin{bmatrix} 29 \\ -13 \\ -1 \end{bmatrix} = 24 = \det A.$$

Suppose we had to compute the determinant of $A = \begin{bmatrix} -1 & 0 & 2 \\ 3 & -7 & 8 \\ -4 & 4 & -5 \end{bmatrix}$. We simply take any row or column and compute the dot product with the corresponding cofactors. Given this choice, we should take advantage of the 0 in the $(1,2)$ position and choose either the first row of A or second column of A. If we choose to **expand by cofactors of the first row** (notice the language), we get

$$-1 \begin{vmatrix} -7 & 8 \\ 4 & -5 \end{vmatrix} - 0 \begin{vmatrix} 3 & 8 \\ -4 & -5 \end{vmatrix} + 2 \begin{vmatrix} 3 & -7 \\ -4 & 4 \end{vmatrix} = -1(3) + 0 + 2(-16) = -35$$

whereas, if we **expand by cofactors of the second column**, we get

$$-7 \begin{vmatrix} -1 & 2 \\ -4 & -5 \end{vmatrix} \overset{\downarrow}{-} 4 \begin{vmatrix} -1 & 2 \\ 3 & 8 \end{vmatrix} = -7(13) - 4(-14) = -35.$$

Did you notice that minus sign? We must change the sign of the 4 in the $(3,2)$ position of A because the corresponding cofactor requires multiplication of the minor by -1. There's a minus sign in position $(3,2)$ of the pattern 29. (Minus means change the sign.)

9.8 Example. The easiest way to find the determinant of $A = \begin{bmatrix} -1 & 2 & -3 & 4 \\ 0 & -2 & 3 & 0 \\ 1 & 2 & 3 & -5 \\ -2 & 9 & -1 & 2 \end{bmatrix}$ is to expand by cofactors of the second row (because there are two 0s in this row, hence two 3×3 determinants we don't have to compute). We get

$$\det A = -2 \begin{vmatrix} -1 & -3 & 4 \\ 1 & 3 & -5 \\ -2 & -1 & 2 \end{vmatrix} - 3 \begin{vmatrix} -1 & 2 & 4 \\ 1 & 2 & -5 \\ -2 & 9 & 2 \end{vmatrix}. \qquad (30)$$

We still have to find two 3×3 determinants, but this is better than four! Expanding by cofactors of the first row, the first determinant is

$$\begin{vmatrix} -1 & -3 & 4 \\ 1 & 3 & -5 \\ -2 & -1 & 2 \end{vmatrix} = -1 \begin{vmatrix} 3 & -5 \\ -1 & 2 \end{vmatrix} + 3 \begin{vmatrix} 1 & -5 \\ -2 & 2 \end{vmatrix} + 4 \begin{vmatrix} 1 & 3 \\ -2 & -1 \end{vmatrix}$$
$$= -1(1) + 3(-8) + 4(5) = -5$$

and, expanding by cofactors of the first row, the second determinant is

$$\begin{vmatrix} -1 & 2 & 4 \\ 1 & 2 & -5 \\ -2 & 9 & 2 \end{vmatrix} = -1 \begin{vmatrix} 2 & -5 \\ 9 & 2 \end{vmatrix} - 2 \begin{vmatrix} 1 & -5 \\ -2 & 2 \end{vmatrix} + 4 \begin{vmatrix} 1 & 2 \\ -2 & 9 \end{vmatrix}$$
$$= -1(49) - 2(-8) + 4(13) = 19.$$

Returning to (30), we obtain $\det A = -2(-5) - 3(19) = -47$. ☺

Before continuing, we summarize a consequence of the formula

$$A(\operatorname{adj} A) = (\det A)I$$

which we have already noted several times. If $\det A \neq 0$, we can divide by $\det A$ giving I on the right side and $\frac{1}{\det A}\big(A(\operatorname{adj} A)\big)$ on the left. By scalar associativity of matrix multiplication, we can move the $\frac{1}{\det A}$ next to $\operatorname{adj} A$ giving $A\big(\frac{1}{\det A} \operatorname{adj} A\big)$ and this equals I. This says that A is invertible and $A^{-1} = \dfrac{1}{\det A} \operatorname{adj} A$. So we have shown

9.9 $\boxed{\det A \neq 0 \text{ implies } A \text{ is invertible and } A^{-1} = \dfrac{1}{\det A} \operatorname{adj} A.}$

The converse of this statement is also true too as we show in the next section: if A is invertible, then $\det A \neq 0$—see **10.13**.

True/False Questions for Week 9

Decide, with as little calculation as possible, whether each of the following statements is true or false and explain your answer whenever you say "false."

1. The $(1,2)$ cofactor of $\begin{bmatrix} -6 & 7 \\ 2 & 1 \end{bmatrix}$ is -2.

2. If the matrix of minors of a certain matrix is $M = \begin{bmatrix} 2 & -3 \\ 4 & 1 \end{bmatrix}$, then the matrix of cofactors is $C = \begin{bmatrix} 2 & 4 \\ -3 & 1 \end{bmatrix}$.

3. Adjoint is another name for the matrix of cofactors.

4. The determinant of a matrix A can be found by multiplying A by its adjoint.

5. The adjoint of $\begin{bmatrix} a & b \\ c & d \end{bmatrix}$ is $\begin{bmatrix} d & -b \\ -c & a \end{bmatrix}$.

6. Every square matrix commutes with its adjoint.

7. This week, we learned about the Laplace expansion of the determinant.

Week 9 Test Yourself

Here are a few problems with short answers that you can use to test your understanding of the concepts you have met this week.

1. Given $A = \begin{bmatrix} 2 & 3 & 4 \\ 0 & -1 & 3 \\ 4 & 7 & 5 \end{bmatrix}$,

   i. find M, the matrix of minors, find C, the matrix of cofactors, find AC^T and C^TA;

   ii. find $\det A$;

   iii. if A is invertible, find A^{-1}.

2. (a) Find the determinant of $A = \begin{bmatrix} 1 & -1 & 2 \\ 3 & 1 & 1 \\ 2 & -1 & 3 \end{bmatrix}$ with a Laplace expansion along the third row.

   (b) Find the determinant of A with a Laplace expansion down the second column.

3. Let $A = \begin{bmatrix} 1 & 0 & 1 \\ 2 & 1 & 1 \\ 1 & 2 & 1 \end{bmatrix}$ and suppose $C = \begin{bmatrix} -1 & -1 & c_{13} \\ c_{21} & 0 & -2 \\ -1 & c_{32} & 1 \end{bmatrix}$ is the matrix of cofactors of A.

   (a) Find the values of c_{21}, c_{13} and c_{32}.

   (b) Find $\det A$.

   (c) Is A invertible? Explain without further calculation.

   (d) Find A^{-1}, if this exists.

4. Can a 3×2 matrix have determinant -1? Explain.

Week 10

Properties of Determinants

It's been a while since we discussed the properties of something or other! Well, the determinant has many fascinating properties. The determinant of a triangular matrix is easy.

10.1 Property 1. The determinant of a triangular matrix is the product of its diagonal entries. In particular, the determinant of a diagonal matrix is the product of its diagonal entries.

10.2 Examples.

- $\det \begin{bmatrix} a & 0 \\ b & c \end{bmatrix} = ac$, the product of the two diagonal entries a and c

- With a Laplace expansion down the first column, $\det \begin{bmatrix} a & b & c \\ 0 & d & e \\ 0 & 0 & f \end{bmatrix} = a \det \begin{bmatrix} d & e \\ 0 & f \end{bmatrix} = adf$, the product of the diagonal entries

- $\det \begin{bmatrix} a & 0 & 0 & 0 \\ b & c & 0 & 0 \\ d & e & f & 0 \\ g & h & i & j \end{bmatrix} = a \det \begin{bmatrix} c & 0 & 0 \\ e & f & 0 \\ h & i & j \end{bmatrix} = acfj$, assuming the result for 3×3 matrices. At this point, the student should see why Property 1 holds in general. If it holds for matrices of a certain size, then it holds for matrices of the next size because if $A = \begin{bmatrix} a_{11} & 0 & \cdots & 0 \\ a_{21} & & & \\ \vdots & & B & \\ a_{k1} & & & \end{bmatrix}$, then (with a Laplace expansion along the first row) $\det A = a_{11} \det B$ so, if $\det B$ is the product of its triangular entries, so is $\det A$. This is an argument by "mathematical induction" with which some readers may be familiar.

- $\det \begin{bmatrix} -2 & 0 & 0 & 0 \\ 0 & 3 & 0 & 0 \\ 0 & 0 & 7 & 0 \\ 0 & 0 & 0 & 5 \end{bmatrix} = (-2)(3)(7)(5) = -210.$ ☺

10.3 Property 2. The determinant is linear in each row.

What in the world does this mean?

Suppose $A = \begin{bmatrix} \mathsf{a}_1 & \rightarrow \\ \mathsf{a}_2 & \rightarrow \\ \vdots & \\ \mathsf{a}_n & \rightarrow \end{bmatrix}$ is an $n \times n$ matrix with rows $\mathsf{a}_1, \ldots, \mathsf{a}_n$. Because we want to focus on rows, we will write $\det(\mathsf{a}_1, \ldots, \mathsf{a}_n)$ to mean the determinant of A. "The determinant is linear in each row" means

1. $\det(\mathsf{a}_1, \ldots, \overset{\downarrow}{\mathsf{x} + \mathsf{y}}, \ldots, \mathsf{a}_n) = \det(\mathsf{a}_1, \ldots, \overset{\downarrow}{\mathsf{x}}, \ldots, \mathsf{a}_n) + \det(\mathsf{a}_1, \ldots, \overset{\downarrow}{\mathsf{y}}, \ldots, \mathsf{a}_n)$, and
2. $\det(\mathsf{a}_1, \ldots, c\mathsf{a}_i, \ldots, \mathsf{a}_n) = c \det(\mathsf{a}_1, \ldots \mathsf{a}_i, \ldots, \mathsf{a}_n)$

for all vectors x and y and for all scalars c.

The first point says that if some row of a matrix is the sum of two rows x and y, then the determinant of the new matrix is the sum of the determinants of the matrices with rows x and y, respectively, and all other rows the same. For instance,

$$\det\begin{bmatrix} a_1+a_2 & b_1+b_2 \\ c & d \end{bmatrix} = \det\begin{bmatrix} a_1 & b_1 \\ c & d \end{bmatrix} + \det\begin{bmatrix} a_2 & b_2 \\ c & d \end{bmatrix}.$$

The second point says that if some row of A is multiplied by a scalar c, then so is the determinant. For instance, if $A = \begin{bmatrix} x & y \\ z & w \end{bmatrix}$ and row two is multiplied by c, the determinant of the new matrix,

$$\det\begin{bmatrix} x & y \\ cz & cw \end{bmatrix} = x(cw) - y(cz) = c(xw - yz) = c \det A,$$

is c times the determinant of A.

Be careful with the concept of linearity **in a row**. This does not say $\det(A+B) = \det A + \det B$ (which is not true in general). Also, in general, $\det(cA) \neq c \det A$. With $A = \begin{bmatrix} x & y \\ z & w \end{bmatrix}$, we have $cA = \begin{bmatrix} cx & cy \\ cz & cw \end{bmatrix}$, so $\det cA = (cx)(cw) - (cy)(cz) = c^2(xw - yz) = c^2 \det A \neq c \det A$.

This little calculation does, however, illustrate a nice formula for $\det cA$.

Suppose A is an $n \times n$ matrix with rows $\mathsf{a}_1, \mathsf{a}_2, \ldots, \mathsf{a}_n$. Then cA has rows $c\mathsf{a}_1, c\mathsf{a}_2, \ldots, c\mathsf{a}_n$. Linearity in each row says we can factor c from each row:

$$\det cA = \det(c\mathsf{a}_1, c\mathsf{a}_2, \ldots, c\mathsf{a}_n) = \underbrace{(c \cdot c \cdot \cdots \cdot c)}_{n \text{ times}} \det(\mathsf{a}_1, \mathsf{a}_2, \ldots, \mathsf{a}_n) = c^n \det A.$$

10.4 Property 3. If A is an $n \times n$ matrix and c is a scalar, $\det cA = c^n \det A$.

10.5 Example. If A is a 5×5 matrix and $\det A = 3$, then $\det(-2A) = (-2)^5 \det A = (-2)^5 3 = -32(3) = -96$. :)

10.6 Property 4. The determinant of a matrix with a row of 0s is 0. This is quite easy to see—if one row of a matrix A consists entirely of 0s, the Laplace expansion along that row gives 0 for the determinant.

10.7 Property 5. The determinant of a matrix with two equal rows is 0.

This is clear for 2×2 matrices because $\det \begin{bmatrix} a & b \\ a & b \end{bmatrix} = ab - ab = 0$ and an argument by mathematical induction finishes the job in general. Specifically, suppose an $n \times n$ matrix, $n \geq 3$, has two equal rows. Using a Laplace expansion along any row different from the two equal rows produces a linear combination of determinants, each determinant that of an $(n-1)\times(n-1)$ matrix with two equal rows. So, if each such determinant is 0, so is the the $n\times n$ determinant.

10.8 **Property 6.** If one row of a matrix is a scalar multiple of another, the determinant is 0.

Suppose A is a matrix with two rows of the form x and $c\mathsf{x}$. Using linearity to factor c from row $c\mathsf{x}$ gives a matrix with two equal rows,

$$\det(\mathsf{a}_1,\ldots,\mathsf{x},\ldots,c\mathsf{x},\ldots,\mathsf{a}_n) = c\det(\mathsf{a}_1,\ldots,\mathsf{x},\ldots,\mathsf{x},\ldots,\mathsf{a}_n),$$

which is 0 by **10.7**.

10.9 **Property 7.** The determinant of a matrix changes sign if two rows are interchanged.

This follows quickly from linearity in each row and **10.7**. We wish to show that if x and y are two rows of a matrix, say rows i and j, then

$$\det(\mathsf{a}_1,\ldots,\mathsf{y},\ldots,\mathsf{x},\ldots,\mathsf{a}_n) = -\det(\mathsf{a}_1,\ldots,\mathsf{x},\ldots,\mathsf{y},\ldots,\mathsf{a}_n).$$

To see this, note that if we make a matrix with $\mathsf{x}+\mathsf{y}$ in each of rows i and j, then the determinant is 0.

$$\det(\mathsf{a}_1,\ldots,\mathsf{x}+\mathsf{y},\ldots,\mathsf{x}+\mathsf{y},\ldots,\mathsf{a}_n) = 0.$$

Linearity in row i, however, says that this determinant is

$$\det(\mathsf{a}_1,\ldots,\mathsf{x},\ldots,\mathsf{x}+\mathsf{y},\ldots,\mathsf{a}_n) + \det(\mathsf{a}_1,\ldots,\mathsf{y},\ldots,\mathsf{x}+\mathsf{y},\ldots,\mathsf{a}_n)$$

and linearity in row j allows us to write each of these determinants as the sum of two more determinants. Altogether, we get

$$\det(\mathsf{a}_1,\ldots,\mathsf{x},\ldots,\mathsf{x},\ldots,\mathsf{a}_n) + \det(\mathsf{a}_1,\ldots,\mathsf{x},\ldots,\mathsf{y},\ldots,\mathsf{a}_n) \\ + \det(\mathsf{a}_1,\ldots,\mathsf{y},\ldots,\mathsf{x},\ldots,\mathsf{a}_n) + \det(\mathsf{a}_1,\ldots,\mathsf{y},\ldots,\mathsf{y},\ldots,\mathsf{a}_n) = 0.$$

The first and last of these four determinants are 0 because they have two equal rows, so

$$\det(\mathsf{a}_1,\ldots,\mathsf{x},\ldots,\mathsf{y},\ldots,\mathsf{a}_n) + \det(\mathsf{a}_1,\ldots,\mathsf{y},\ldots,\mathsf{x},\ldots,\mathsf{a}_n) = 0$$

which gives what we wanted.

Recall that a permutation matrix is a matrix whose rows (or columns) are those of the identity matrix, in some order. Now a rearrangement of rows can be achieved by successively interchanging pairs of rows and each such interchange changes the sign of the determinant. Since $\det I = 1$,

10.10 The determinant of any permutation matrix is ± 1.

10.11 Example. The matrix $P = \begin{bmatrix} 0 & 1 & 0 & 0 \\ 0 & 0 & 1 & 0 \\ 0 & 0 & 0 & 1 \\ 1 & 0 & 0 & 0 \end{bmatrix}$ is the permutation matrix whose rows are the rows of I_4, the 4×4 identity matrix, in order 2, 3, 4, 1. We can compute $\det P$ by interchanging rows until we get I_4, as follows:.

$$\det P = \begin{vmatrix} 0 & 1 & 0 & 0 \\ 0 & 0 & 1 & 0 \\ 0 & 0 & 0 & 1 \\ 1 & 0 & 0 & 0 \end{vmatrix} = - \begin{vmatrix} 1 & 0 & 0 & 0 \\ 0 & 0 & 1 & 0 \\ 0 & 0 & 0 & 1 \\ 0 & 1 & 0 & 0 \end{vmatrix} \quad \text{interchanging rows one and four}$$

$$= + \begin{vmatrix} 1 & 0 & 0 & 0 \\ 0 & 1 & 0 & 0 \\ 0 & 0 & 0 & 1 \\ 0 & 0 & 1 & 0 \end{vmatrix} \quad \text{interchanging rows two and four}$$

$$= - \begin{vmatrix} 1 & 0 & 0 & 0 \\ 0 & 1 & 0 & 0 \\ 0 & 0 & 1 & 0 \\ 0 & 0 & 0 & 1 \end{vmatrix} \quad \text{interchanging rows three and four}$$

$$= -1.$$

☝

10.12 Property 8. The determinant is *multiplicative*: $\det AB = \det A \det B$ for any two square matrices of the same size. For example, if $A = \begin{bmatrix} -1 & 2 \\ 4 & 5 \end{bmatrix}$ and $B = \begin{bmatrix} 0 & 3 \\ 4 & -2 \end{bmatrix}$, then $AB = \begin{bmatrix} 8 & -7 \\ 20 & 2 \end{bmatrix}$ and $\det AB = 156 = (-13)(-12) = \det A \det B$.

One proof of Property 8 which I find particularly instructive involves three preliminary steps.

Step 1. Show that $\det AB = (\det A)(\det B)$ if A is singular (that is, not invertible).

It remains to get the same result when A is invertible.

Step 2. If E is an elementary matrix, show that $\det(EB) = (\det E)(\det B)$.

Step 3. If $E_1, E_2, \ldots, E_k$ are elementary matrices, then

$$\det(E_1 E_2 \cdots E_k B) = (\det E_1)(\det E_2) \cdots (\det E_k)(\det B).$$

If we set $B = I$ in the third step, we get $\det(E_1 E_2 \cdots E_k) = (\det E_1)(\det E_2) \cdots (\det E_k)$ for any elementary matrices $E_1, E_2, \ldots, E_k$. Now the fact that an invertible matrix is the product of elementary matrices (Theorem 8.22) completes the argument.

The multiplicative nature of the determinant is extremely important, and useful.

10.13 Property 9. If A is invertible, then $\det A \neq 0$ and $\det A^{-1} = \dfrac{1}{\det A}$.

This follows quickly from Property 8. We know that $AA^{-1} = I$, the identity matrix, and that $\det I = 1$. So $\det AA^{-1} = \det A \det A^{-1} = 1$. If the product of two numbers is not 0, neither number can be 0. Here, this means $\det A \neq 0$ and, moreover, $\det A^{-1} = \dfrac{1}{\det A}$.

Last week, we showed that if $\det A \neq 0$, then A is invertible—see **9.9**. Now we have the result in the other direction too and hence a perfect generalization to matrices of any size of what we saw for 2×2 matrices in Proposition 7.11.

10.14 Theorem. *If A is a square matrix, then A is invertible if and only if* $\det A \neq 0$, *in which case* $A^{-1} = \dfrac{1}{\det A} \operatorname{adj} A$.

10.15 Problem. Determine whether or not $A = \begin{bmatrix} -1 & 2 & 2 \\ 3 & 0 & -1 \\ 1 & 4 & 3 \end{bmatrix}$ is invertible.

Solution. Expanding by cofactors of the second row, we have $\det A = -3(6-8) - (-1)(-4-2) = -3(-2) + (-6) = 0$. So A is not invertible.

Here are some more applications of Property 9.

10.16 Proposition. *If P is a permutation matrix,* $\det P^{-1} = \det P$.

Proof. As noted in **10.10**, $\det P = \pm 1$. Thus

$$\det P^{-1} = \frac{1}{\det P} = \begin{cases} +1 & \text{if } \det P = 1 \\ -1 & \text{if } \det P = -1 \end{cases}$$

so $\det P^{-1} = \det P$ in either case. ∎

We also have an easy test for linear independence.

10.17 Proposition. *If A is a square matrix with nonzero determinant, then the columns of A are linearly independent.*

Proof. The columns of a matrix are linearly independent if and only if the only solution to $A\mathsf{x} = \mathsf{0}$ is the trivial one, $\mathsf{x} = \mathsf{0}$—see **7.2**. And that's just what we have here because if $\det A \neq 0$, then A is invertible, so $A\mathsf{x} = \mathsf{0}$ implies $A^{-1}A\mathsf{x} = \mathsf{0}$ which is $\mathsf{x} = \mathsf{0}$. ∎

10.18 Example. To determine whether or not the three vectors $\mathsf{x}_1 = \begin{bmatrix} 1 \\ 2 \\ 3 \end{bmatrix}$, $\mathsf{x}_2 = \begin{bmatrix} -2 \\ 0 \\ 5 \end{bmatrix}$, and $\mathsf{x}_3 = \begin{bmatrix} 1 \\ -3 \\ 7 \end{bmatrix}$ are linearly independent, we form the matrix $A = \begin{bmatrix} 1 & -2 & 1 \\ 2 & 0 & -3 \\ 3 & 5 & 7 \end{bmatrix}$ whose columns are the given vectors and compute (with a Laplace expansion along the second row):

$$\det A = -2 \begin{vmatrix} -2 & 1 \\ 5 & 7 \end{vmatrix} - (-3) \begin{vmatrix} 1 & -2 \\ 3 & 5 \end{vmatrix} = -2(-19) + 3(11) = 71.$$

Since $\det A \neq 0$, the columns of A are linearly independent. ☺

10.19 Remark. The *converse* of Proposition 10.17 is also true: If the columns of a matrix A are linearly independent, then $\det A \neq 0$ (and so A is invertible). This is easily seen within the sequel to this course, Math 2051.

There is a commonly used term for "not invertible."

10.20 Definition. A square matrix is *singular* if it is not invertible.

So Theorem 10.14 says that a matrix is singular if and only if its determinant is 0.

10.21 Problem. For what value(s) of c is the matrix $A = \begin{bmatrix} 1 & 0 & -c \\ -1 & 3 & 1 \\ 0 & 2c & -4 \end{bmatrix}$ singular?

Solution. The third row is twice $[0 \; c \; -4]$ so, using linearity of the determinant in the third row, we have $\det A = 2\begin{vmatrix} 1 & 0 & -c \\ -1 & 3 & 1 \\ 0 & c & -2 \end{vmatrix}$ and then, expanding by cofactors of the first row

$$\det A = 2\left\{\begin{vmatrix} 3 & 1 \\ c & -2 \end{vmatrix} - c\begin{vmatrix} -1 & 3 \\ 0 & c \end{vmatrix}\right\} = 2[(-6-c) - c(-c)]$$
$$= 2(c^2 - c - 6) = 2(c-3)(c+2).$$

The matrix is singular if and only if $\det A = 0$, and this is the case if and only if $c = 3$ or $c = -2$.

Like Property 8, our final property of determinants is quite amazing.

10.22 Property 10. $\det A^T = \det A$.

Here's a way to see this. If P is a permutation matrix, then $P^T = P^{-1}$ and $\det P^{-1} = \det P$, so $\det P^T = \det P$, the desired result for permutation matrices. Now remember that $A = PLU$ can be factored as the product of a permutation matrix P (perhaps $P = I$), a lower triangular matrix L with 1s on the diagonal and a row echelon matrix U—see **8.20**. Thus $\det L = 1$ and $\det A = \det PLU = (\det P)(\det L)(\det U) = (\det P)(\det U)$. On the other hand, $A = PLU$ implies $A^T = (PLU)^T = U^T L^T P^T$. We have already shown that $\det P^T = \det P$. Since L^T is triangular with 1s on the diagonal, $\det L^T = 1$, and since U^T is triangular with the same diagonal as U, $\det U^T = \det U$. Thus $\det A^T = \det U^T \det L^T \det P^T = (\det U)(\det P) = (\det P)(\det U)$, giving $\det A^T = \det A$.

The fact that a matrix and its transpose have the same determinant implies that any property of the determinant that we have stated about rows, holds equally for columns. For example, if A is a matrix with two equal columns, then A^T is a matrix with two equal rows, so $\det A^T = 0$. Since $\det A^T = \det A$, we have $\det A = 0$ too.

The next example and problem are representative of questions frequently asked on exams.

10.23 Example. Suppose A and B are square $n \times n$ matrices with $\det A = 2$ and $\det B = 5$. What is $\det A^3B^{-1}A^T$? By Property 10.12, $\det A^3 = \det AAA = (\det A)(\det A)(\det A) = (\det A)^3 = 2^3 = 8$. By Property 10.13, $\det B^{-1} = \dfrac{1}{\det B} = \frac{1}{5}$. By Property 10.22, $\det A^T = \det A = 2$. A final application of 10.12 gives $\det A^3B^{-1}A^T = 8(\frac{1}{5})(2) = \frac{16}{5}$. ☺

10.24 Problem. Suppose $A = \begin{bmatrix} a & b & c \\ d & e & f \\ g & h & i \end{bmatrix}$ is a 3×3 matrix with $\det A = 2$.

Find the determinant of $\begin{bmatrix} -g & 3d+5a & 7d \\ -h & 3e+5b & 7e \\ -i & 3f+5c & 7f \end{bmatrix}$.

Solution. We use the fact that the all properties of determinants stated for rows hold equally for columns. Factoring 7 from the third column,

$$\begin{vmatrix} -g & 3d+5a & 7d \\ -h & 3e+5b & 7e \\ -i & 3f+5c & 7f \end{vmatrix} = 7\begin{vmatrix} -g & 3d+5a & d \\ -h & 3e+5b & e \\ -i & 3f+5c & f \end{vmatrix}$$

$$= -7\begin{vmatrix} g & 3d+5a & d \\ h & 3e+5b & e \\ i & 3f+5c & f \end{vmatrix} \qquad \text{factoring } -1 \text{ from column one}$$

$$= -7\left\{\begin{vmatrix} g & 3d & d \\ h & 3e & e \\ i & 3f & f \end{vmatrix} + \begin{vmatrix} g & 5a & d \\ h & 5b & e \\ i & 5c & f \end{vmatrix}\right\} \qquad \text{by linearity of the determinant in column two}$$

$$= -7\begin{vmatrix} g & 5a & d \\ h & 5b & e \\ i & 5c & f \end{vmatrix} \qquad \text{the first determinant is 0 because one column is a multiple of another}$$

$$= -35\begin{vmatrix} g & a & d \\ h & b & e \\ i & c & f \end{vmatrix} \qquad \text{factoring 5 from the second column}$$

$$= -35\begin{vmatrix} g & h & i \\ a & b & c \\ d & e & f \end{vmatrix} \qquad \text{taking the transpose}$$

$$= -35\begin{vmatrix} a & b & c \\ g & h & i \\ d & e & f \end{vmatrix} \qquad R1 \leftrightarrow R2$$

$$= -35\begin{vmatrix} a & b & c \\ d & e & f \\ g & h & i \end{vmatrix} \qquad R2 \leftrightarrow R3.$$

The desired determinant is $-35(\det A) = -70$. ♤

Using Elementary Row Operations to find Determinants

There is another way to find the determinant of a matrix which is often simpler than the method of cofactors we described last week. I know you remember the three elementary row operations:

1. Interchange two rows.
2. Multiply a row by any nonzero scalar.
3. Replace a row by that row less a multiple of another row.

The effect of the first two operations on the determinant has been noted, in Properties 10.9 and 10.3.

- The determinant changes sign if two rows are interchanged.
- If a certain row of a matrix contains the factor c, then c factors out of the determinant.

$$\det(\mathrm{a}_1, \ldots, c\mathrm{x}, \ldots, \mathrm{a}_n) = c \, \det(\mathrm{a}_1, \ldots, \mathrm{x}, \ldots, \mathrm{a}_n) \, .$$

What happens to the determinant when we apply the third elementary row operation? Suppose A has rows $\mathrm{a}_1, \mathrm{a}_2, \ldots, \mathrm{a}_n$ and we apply the operation $Rj \to Rj - c(Ri)$ to get a new matrix B. Since the determinant is linear in row j,

$$\det B = \det(\mathrm{a}_1, \ldots, \overset{i}{\mathrm{a}_i}, \ldots \mathrm{a}_j \overset{j}{-} c\mathrm{a}_i, \ldots, \mathrm{a}_n)$$

$$= \det(\mathrm{a}_1, \ldots, \overset{i}{\mathrm{a}_i}, \ldots, \overset{j}{\mathrm{a}_j}, \ldots \mathrm{a}_n) + \det(\mathrm{a}_1, \ldots, \overset{i}{\mathrm{a}_i}, \ldots, \overset{j}{-c\mathrm{a}_i}, \ldots, \mathrm{a}_n).$$

The second determinant here is 0 because row j is a multiple of row i. Thus

$$\det B = \det(\mathrm{a}_1, \ldots, \mathrm{a}_i, \ldots, \mathrm{a}_j, \ldots \mathrm{a}_n) = \det A.$$

The determinant has not changed!

We summarize.

Matrix operation	**Effect on determinant**
interchange two rows	multiply by -1
factor $c \neq 0$ from a row	factor c from the determinant
replace a row by that row less a multiple of another row	no change

This suggests a new way to compute the determinant of a matrix.

10.25

> **To find the determinant of a matrix A:**
>
> 1. Reduce A to an upper triangular matrix using the elementary row operations, making note of the effect of each operation on the determinant.
> 2. Use the fact that the determinant of a triangular matrix is the product of its diagonal entries.

10.26 Problem. Find the determinant of $\begin{bmatrix} 1 & 2 & 3 & 4 \\ 5 & 6 & 7 & 8 \\ 9 & 10 & 11 & 12 \\ 13 & 14 & 15 & 16 \end{bmatrix}$.

Solution. We start Gaussian elimination, getting two 0s in the first column by means of the third elementary row operation (which does not change the determinant), $A \to \begin{bmatrix} 1 & 2 & 3 & 4 \\ 0 & -4 & -8 & -12 \\ 0 & -8 & -16 & -24 \\ * & * & * & * \end{bmatrix}$, and get a matrix in which one row is a multiple of another. The determinant of such a matrix is 0, so $\det A = 0$ too. ♤

10.27 Example. $\begin{vmatrix} 2 & -1 & 4 & 0 \\ 2 & 1 & -1 & 5 \\ -2 & -2 & 17 & -3 \\ 4 & 6 & 1 & 9 \end{vmatrix} = \begin{vmatrix} 2 & -1 & 4 & 0 \\ 0 & 2 & -5 & 5 \\ 0 & -3 & 21 & -3 \\ 0 & 8 & -7 & 9 \end{vmatrix}$

because the third elementary row operation does not change the determinant

$$= 3\begin{vmatrix} 2 & -1 & 4 & 0 \\ 0 & 2 & -5 & 5 \\ 0 & -1 & 7 & -1 \\ 0 & 8 & -7 & 9 \end{vmatrix} \quad \text{factoring a 3 from row three}$$

$$= -3\begin{vmatrix} 2 & -1 & 4 & 0 \\ 0 & -1 & 7 & -1 \\ 0 & 2 & -5 & 5 \\ 0 & 8 & -7 & 9 \end{vmatrix} \quad \text{interchanging rows two and three}$$

$$= -3\begin{vmatrix} 2 & -1 & 4 & 0 \\ 0 & -1 & 7 & -1 \\ 0 & 0 & 9 & 3 \\ 0 & 0 & 49 & 1 \end{vmatrix} \quad \text{the third elementary row operation does change the determinant}$$

At this point, we switch to a Laplace expansion down the first column, so the determinant becomes $-3(2)\begin{vmatrix} -1 & 7 & -1 \\ 0 & 9 & 3 \\ 0 & 49 & 1 \end{vmatrix}$. Then, with another expansion down the first column, we have $-3(2)(-1)\begin{vmatrix} 9 & 3 \\ 49 & 1 \end{vmatrix} = 6(9 - 147) = 6(-138) = -828$. ☺

10.28 Problem. Here is a problem we met before solved again by different means.

Suppose $A = \begin{bmatrix} a & b & c \\ d & e & f \\ g & h & i \end{bmatrix}$ has $\det A = 2$. Find the determinant of $\begin{bmatrix} -g & 3d+5a & 7d \\ -h & 3e+5b & 7e \\ -i & 3f+5c & 7f \end{bmatrix}$.

Solution. We use the fact that $\det X = \det X^T$ and then use elementary row operations to move the new matrix towards A.

$$\begin{vmatrix} -g & 3d+5a & 7d \\ -h & 3e+5b & 7e \\ -i & 3f+5c & 7f \end{vmatrix} = \begin{vmatrix} -g & -h & -i \\ 3d+5a & 3e+5b & 3f+5c \\ 7d & 7e & 7f \end{vmatrix}$$

$$= -7 \begin{vmatrix} g & h & i \\ 3d+5a & 3e+5b & 3f+5c \\ d & e & f \end{vmatrix} \qquad \text{factoring } -1 \text{ from row one and 7 from row 3}$$

$$= -7 \begin{vmatrix} g & h & i \\ 5a & 5b & 5c \\ d & e & f \end{vmatrix} \qquad R2 \to R2 - 3(R3)$$

$$= -35 \begin{vmatrix} g & h & i \\ a & b & c \\ d & e & f \end{vmatrix} \qquad \text{factoring 5 from row two}$$

$$= -35 \begin{vmatrix} a & b & c \\ g & h & i \\ d & e & f \end{vmatrix} \qquad R1 \longleftrightarrow R2$$

$$= -35 \begin{vmatrix} a & b & c \\ d & e & f \\ g & h & i \end{vmatrix} \qquad R2 \longleftrightarrow R3$$

The desired determinant is $-35 \det A = -70$, as before.

Comparing Methods

As we have seen, there are a number of ways to compute a determinant. Which is best depends on the matrix and who is doing the calculation. Frankly, my preference is often to use a computer! If no computer is handy, I find a 2×2 determinant in my head, a 3×3 determinant by the method of cofactors (because I can do 2×2 determinants in my head), and bigger determinants by reducing to triangular form. There are always exceptions to these general rules, of course, depending on the nature of the matrix.

True/False Questions for Week 10

Decide, with as little calculation as possible, whether each of the following statements is true or false and explain your answer whenever you say "false."

1. If A is a 4×4 matrix with determinant 3, then $\det(-2A) = -6$.
2. The columns of $A = \begin{bmatrix} -2 & 0 & 0 \\ 1 & 3 & 0 \\ 1 & 5 & -4 \end{bmatrix}$ are linearly independent.

3. $\det\begin{bmatrix} a+e & b+f & x+p \\ c+g & d+h & y+q \\ r & s & t \end{bmatrix} = \det\begin{bmatrix} a+e & b+f & x+p \\ c & d & y \\ r & s & t \end{bmatrix} + \det\begin{bmatrix} a+e & b+f & x+p \\ g & h & q \\ r & s & t \end{bmatrix}$.

4. The determinant of a permutation matrix cannot be 0.

5. Given square matrices A and B, we have $\det(A+B) = \det A + \det B$.

6. If A and P are 5×5 matrices with $\det A = 7$ and $\det P = 21$, Then $\det A^{-1}P = 3$.

7. If A is a square matrix and $\det A = 0$, then A must have a row of 0s.

8. If A and B are square matrices of the same size, then $\det AB = \det BA$.

9. If we divide a row of matrix by 2, the determinant is multiplied by 2.

10. $\begin{vmatrix} 2 & 4 & 1 \\ -1 & 0 & 3 \\ 7 & 1 & -3 \end{vmatrix} = \begin{vmatrix} 0 & 4 & 7 \\ -1 & 0 & 3 \\ 7 & 1 & -3 \end{vmatrix}$

11. $\begin{vmatrix} 2 & 4 & 1 \\ -1 & -2 & -3 \\ 2 & 4 & 6 \end{vmatrix} = -2\begin{vmatrix} 2 & 4 & 1 \\ 1 & 2 & 3 \\ 1 & 2 & 3 \end{vmatrix}$

12. $\begin{vmatrix} a & b & c \\ d+x & e+y & f+z \\ g & h & i \end{vmatrix} = \begin{vmatrix} a & b & c \\ d & e & f \\ g & h & i \end{vmatrix} + \begin{vmatrix} a & b & c \\ x & y & z \\ g & h & i \end{vmatrix}$.

13. $\begin{vmatrix} a & b & c \\ d+x & e+y & f+z \\ x & y & z \end{vmatrix} = \begin{vmatrix} a & b & c \\ d & e & f \\ x & y & z \end{vmatrix}$.

14. A permutation matrix has determinant 1.

15. A 3×3 matrix is invertible if and only if it not the 0 matrix.

16. If A is not singular, then A^{-1} is not singular.

Week 10 Test Yourself

Here are a few problems with short answers that you can use to test your understanding of the concepts you have met this week.

1. Given that $\begin{vmatrix} -1 & 7 & 3 & 4 \\ 0 & -5 & -3 & 2 \\ 6 & 1 & -1 & 3 \\ 2 & 2 & 0 & 1 \end{vmatrix} = 118$, find (a) $\begin{vmatrix} -1 & 7 & 3 & 4 \\ 0 & -5 & -3 & 2 \\ 6 & 1 & -1 & 3 \\ -1 & 7 & 3 & 4 \end{vmatrix}$ and (b) $\begin{vmatrix} -1 & 7 & 3 & 4 \\ 0 & -5 & -3 & 2 \\ 12 & 2 & -2 & 6 \\ 2 & 2 & 0 & 1 \end{vmatrix}$.

2. Let $A = \begin{bmatrix} 1 & 7 & 0 & 17 & 9 \\ 0 & 2 & -35 & 10 & 15 \\ 0 & 0 & 3 & -5 & 11 \\ 0 & 0 & 0 & 2 & 77 \\ 0 & 0 & 0 & 0 & 5 \end{bmatrix}$. Find $\det A$, $\det A^{-1}$, and $\det A^2$.

3. What does it mean to say that the determinant is a "multiplicative" function?

4. Without calculation, explain why the columns of $A = \begin{bmatrix} 1 & 2 & 3 & 4 \\ 0 & -1 & 1 & 2 \\ 0 & 0 & 3 & 4 \\ 0 & 0 & 0 & 5 \end{bmatrix}$ are linearly independent.

5. What is a "singular" matrix?

6. For what values of x are the given matrices singular?

   (a) $\begin{bmatrix} x & 1 \\ -2 & x+3 \end{bmatrix}$ (b) $\begin{bmatrix} 1 & 2 & x \\ 0 & -3 & 0 \\ 4 & x & 7 \end{bmatrix}$

7. (a) Find the determinant of $A = \begin{bmatrix} -1 & -1 & 1 & 0 \\ 2 & 1 & 1 & 3 \\ 0 & 1 & 1 & 2 \\ 1 & 3 & -1 & 2 \end{bmatrix}$ by reducing to an upper triangular matrix.

   (b) Are the columns of A linearly independent? Explain.

Week 11

UNIT FOUR: THE EQUATION $A\mathbf{x} = \lambda\mathbf{x}$

Eigenvalues and Eigenvectors

This unit highlights the equation $A\mathsf{x} = \lambda\mathsf{x}$, where A is a (necessarily square) matrix and x is a vector. By tradition, the Greek symbol λ, pronounced "lambda," is used to denote a scalar in this special equation, which says that multiplication by A transforms x into a multiple of itself.

11.1 **Definitions.** An *eigenvector* of a square matrix A is a nonzero vector x such that $A\mathsf{x} = \lambda\mathsf{x}$ for some scalar λ. The scalar λ is called an *eigenvalue of A corresponding to* x. The set of all solutions to $A\mathsf{x} = \lambda\mathsf{x}$ is called the *eigenspace of A corresponding to* λ.

11.2 **Remark.** The equation $A\mathsf{x} = \lambda\mathsf{x}$ is not very interesting if $\mathsf{x} = 0$ since, in this case, it holds for any scalar λ. This is why eigenvectors are required to be **nonzero** vectors. On the other hand, the eigenspace of A does include the zero vector. The eigenspace of A corresponding to λ consists of all eigenvectors corresponding to λ (the nonzero solutions to $A\mathsf{x} = \lambda\mathsf{x}$) together with the zero vector.

11.3 **Example.** Let $A = \begin{bmatrix} 1 & 4 \\ 2 & 3 \end{bmatrix}$ and $\mathsf{x} = \begin{bmatrix} 1 \\ 1 \end{bmatrix}$. Then x is an eigenvector of A corresponding to $\lambda = 5$ because

$$A\mathsf{x} = \begin{bmatrix} 1 & 4 \\ 2 & 3 \end{bmatrix}\begin{bmatrix} 1 \\ 1 \end{bmatrix} = \begin{bmatrix} 5 \\ 5 \end{bmatrix} = 5\mathsf{x}.$$

The eigenspace of A corresponding to $\lambda = 5$ is the set of solutions to $A\mathsf{x} = 5\mathsf{x}$. Rewriting this equation in the form $(A - 5I)\mathsf{x} = 0$, we must solve the homogeneous system with coefficient matrix $A - 5I$. Gaussian elimination proceeds $A - 5I = \begin{bmatrix} -4 & 4 \\ 2 & -2 \end{bmatrix} \to \begin{bmatrix} 1 & -1 \\ 0 & 0 \end{bmatrix}$. So if $\mathsf{x} = \begin{bmatrix} x_1 \\ x_2 \end{bmatrix}$ is an eigenvector, then $x_2 = t$ is free and $x_1 - x_2 = 0$. Thus $x_1 = x_2 = t$ and $\mathsf{x} = \begin{bmatrix} t \\ t \end{bmatrix} = t\begin{bmatrix} 1 \\ 1 \end{bmatrix}$. The eigenspace corresponding to $\lambda = 5$ is the set of all multiples of $\begin{bmatrix} 1 \\ 1 \end{bmatrix}$. The nonzero multiples are the eigenvectors corresponding to 5. ☺

11.4 Example. The vector $\mathsf{x} = \begin{bmatrix} 3 \\ -1 \\ -1 \end{bmatrix}$ is an eigenvector of $A = \begin{bmatrix} 5 & 8 & 16 \\ 4 & 1 & 8 \\ -4 & -4 & -11 \end{bmatrix}$ with $\lambda = -3$ the corresponding eigenvector because

$$A\mathsf{x} = \begin{bmatrix} 5 & 8 & 16 \\ 4 & 1 & 8 \\ -4 & -4 & -11 \end{bmatrix} \begin{bmatrix} 3 \\ -1 \\ -1 \end{bmatrix} = \begin{bmatrix} -9 \\ 3 \\ 3 \end{bmatrix} = -3\mathsf{x}.$$

To find the eigenspace of $\lambda = -3$, we must find all the vectors x that satisfy $A\mathsf{x} = -3\mathsf{x}$. As in Example 11.3, we rewrite this equation as $(A + 3I)\mathsf{x} = \mathsf{0}$, where I is the 3×3 identity matrix, so that the desired eigenspace is the set of solutions to the homogeneous system with coefficient matrix $A + 3I$. Gaussian elimination proceeds

$$A + 3I = \begin{bmatrix} 8 & 8 & 16 \\ 4 & 4 & 8 \\ -4 & -4 & -8 \end{bmatrix} \to \begin{bmatrix} 1 & 1 & 2 \\ 4 & 4 & 8 \\ -4 & -4 & -8 \end{bmatrix} \to \begin{bmatrix} 1 & 1 & 2 \\ 0 & 0 & 0 \\ 0 & 0 & 0 \end{bmatrix}.$$

With $\mathsf{x} = \begin{bmatrix} x_1 \\ x_2 \\ x_3 \end{bmatrix}$, we have free variables $x_2 = t$ and $x_3 = s$, and $x_1 = -x_2 - 2x_3 = -t - 2s$. So $\mathsf{x} = \begin{bmatrix} -t - 2s \\ t \\ s \end{bmatrix} = t\begin{bmatrix} -1 \\ 1 \\ 0 \end{bmatrix} + s\begin{bmatrix} -2 \\ 0 \\ 1 \end{bmatrix}$. The eigenspace of A corresponding to $\lambda = -3$ consists of all linear combinations of $\begin{bmatrix} -1 \\ 1 \\ 0 \end{bmatrix}$ and $\begin{bmatrix} -2 \\ 0 \\ 1 \end{bmatrix}$, which is a plane in Euclidean 3-space—see **1.23**. The nonzero vectors in this plane are the eigenvectors corresponding to -3. ☺

How to Find Eigenvalues and Eigenvectors

Examples 11.3 and 11.4 illustrate a general method of finding the eigenvalues and eigenspaces of a matrix. The key idea is that if $A\mathsf{x} = \lambda\mathsf{x}$ with $\mathsf{x} \neq \mathsf{0}$, then $A\mathsf{x} - \lambda\mathsf{x} = \mathsf{0}$, so $(A - \lambda I)\mathsf{x} = \mathsf{0}$. Since $\mathsf{x} \neq \mathsf{0}$, the matrix $A - \lambda I$ cannot be invertible, so its determinant must be 0. This determinant, $\det(A - \lambda I)$, is a polynomial in the variable λ called the *characteristic polynomial* of A. Its roots, which are the solutions to $\det(A - \lambda I) = 0$, are the eigenvalues of A.

11.5

To find the eigenvalues and eigenvectors of a matrix A:

1. Compute the characteristic polynomial $\det(A - \lambda I)$.
2. Set $\det(A - \lambda I) = 0$. The roots of this polynomial are the eigenvalues.
3. For each eigenvalue λ, solve the homogeneous system $(A - \lambda I)\mathsf{x} = \mathsf{0}$. The solutions to this system are the vectors of the eigenspace corresponding to λ.

11.6 Example. The characteristic polynomial of the matrix $A = \begin{bmatrix} 1 & 4 \\ 2 & 3 \end{bmatrix}$ is

$$\det(A - \lambda I) = \det \begin{bmatrix} 1-\lambda & 4 \\ 2 & 3-\lambda \end{bmatrix} = (1-\lambda)(3-\lambda) - 8 = \lambda^2 - 4\lambda - 5. \tag{31}$$

This is a polynomial of degree 2 in λ. It factors, $\lambda^2 - 4\lambda - 5 = (\lambda - 5)(\lambda + 1)$, and its roots, $\lambda = 5$, $\lambda = -1$, are the eigenvalues of A. In Example 11.3, we discovered that the eigenspace corresponding to $\lambda = 5$ consists of all scalar multiples of $\begin{bmatrix} 1 \\ 1 \end{bmatrix}$. To find the eigenspace corresponding to $\lambda = -1$, we solve the homogeneous system $(A + 1I)\mathsf{x} = 0$. The coefficient matrix is $A - \lambda I$ with $\lambda = -1$. So we put $\lambda = -1$ in the matrix $A - \lambda I$ which appears in (31). This gives $\begin{bmatrix} 2 & 4 \\ 2 & 4 \end{bmatrix}$. Here is Gaussian elimination:

$$\begin{bmatrix} 2 & 4 \\ 2 & 4 \end{bmatrix} \to \begin{bmatrix} 1 & 2 \\ 2 & 4 \end{bmatrix} \to \begin{bmatrix} 1 & 2 \\ 0 & 0 \end{bmatrix}.$$

Variable $x_2 = t$ is free and $x_1 = -2x_2 = -2t$. The eigenspace corresponding to $\lambda = -1$ is the set of vectors of the form

$$\mathsf{x} = \begin{bmatrix} x_1 \\ x_2 \end{bmatrix} = \begin{bmatrix} -2t \\ t \end{bmatrix} = t \begin{bmatrix} -2 \\ 1 \end{bmatrix}.$$

To check that our calculations are correct, we compute

$$A\mathsf{x} = \begin{bmatrix} 1 & 4 \\ 2 & 3 \end{bmatrix} \begin{bmatrix} -2 \\ 1 \end{bmatrix} = \begin{bmatrix} 2 \\ -1 \end{bmatrix} = (-1) \begin{bmatrix} -2 \\ 1 \end{bmatrix} = (-1)\mathsf{x},$$

so $\mathsf{x} = \begin{bmatrix} -2 \\ 1 \end{bmatrix}$ is indeed an eigenvector corresponding to $\lambda = -1$. ☺

11.7 Example. The characteristic polynomial of $A = \begin{bmatrix} 5 & 8 & 16 \\ 4 & 1 & 8 \\ -4 & -4 & -11 \end{bmatrix}$ is

$$A - \lambda I = \begin{bmatrix} 5 & 8 & 16 \\ 4 & 1 & 8 \\ -4 & -4 & -11 \end{bmatrix} - \begin{bmatrix} \lambda & 0 & 0 \\ 0 & \lambda & 0 \\ 0 & 0 & \lambda \end{bmatrix} = \begin{bmatrix} 5-\lambda & 8 & 16 \\ 4 & 1-\lambda & 8 \\ -4 & -4 & -11-\lambda \end{bmatrix}. \tag{32}$$

Using a Laplace expansion along the first row,

$$\begin{aligned} &\det(A - \lambda I) \\ &= (5-\lambda) \begin{vmatrix} 1-\lambda & 8 \\ -4 & -11-\lambda \end{vmatrix} - 8 \begin{vmatrix} 4 & 8 \\ -4 & -11-\lambda \end{vmatrix} + 16 \begin{vmatrix} 4 & 1-\lambda \\ -4 & -4 \end{vmatrix} \\ &= (5-\lambda)[(1-\lambda)(-11-\lambda) + 32] - 8[4(-11-\lambda) + 32] + 16[-16 + 4(1-\lambda)] \\ &= (5-\lambda)(21 + 10\lambda + \lambda^2) - 8(-12 - 4\lambda) + 16(-12 - 4\lambda) \\ &= -\lambda^3 - 5\lambda^2 - 3\lambda + 9 = -(\lambda^3 + 5\lambda^2 + 3\lambda - 9). \end{aligned}$$

$$
\begin{array}{r|rrrrrrr}
 & \lambda^2 & + & 2\lambda & - & 3 & & \\
\hline
\lambda+3 & \lambda^3 & + & 5\lambda^2 & + & 3\lambda & - & 9 \\
 & \lambda^3 & + & 3\lambda^2 & & & & \\
\cline{2-4}
 & & & 2\lambda^2 & + & 3\lambda & - & 9 \\
 & & & 2\lambda^2 & + & 6\lambda & & \\
\cline{4-6}
 & & & & & -3\lambda & - & 9 \\
 & & & & & -3\lambda & - & 9 \\
\cline{6-8}
 & & & & & & & 0
\end{array}
$$

Figure 13: Long Division of Polynomials

In Example 11.4, we noted that $\lambda = -3$ is an eigenvalue. Thus $\lambda = -3$ is one root of this polynomial and we can find the others by long division—see Figure 13. We obtain

$$\lambda^3 + 5\lambda^2 + 3\lambda - 9 = (\lambda + 3)(\lambda^2 + 2\lambda - 3) = (\lambda + 3)(\lambda + 3)(\lambda - 1) = (\lambda + 3)^2(\lambda - 1).$$

So the eigenvalues are $\lambda = 1$ and $\lambda = -3$. We found the eigenspace corresponding to $\lambda = -3$ previously. The eigenspace corresponding to $\lambda = 1$ is the set of solutions to $(A - \lambda I)\mathsf{x} = 0$ with $\lambda = 1$. The coefficient matrix of this homogeneous system is just the matrix in (32) with $\lambda = 1$, namely,

$$\begin{bmatrix} 4 & 8 & 16 \\ 4 & 0 & 8 \\ -4 & -4 & -12 \end{bmatrix}.$$

Gaussian elimination proceeds

$$\begin{bmatrix} 4 & 8 & 16 \\ 4 & 0 & 8 \\ -4 & -4 & -12 \end{bmatrix} \to \begin{bmatrix} 1 & 2 & 4 \\ 4 & 0 & 8 \\ -4 & -4 & -12 \end{bmatrix} \to \begin{bmatrix} 1 & 2 & 4 \\ 0 & -8 & -8 \\ 0 & 4 & 4 \end{bmatrix}$$
$$\to \begin{bmatrix} 1 & 2 & 4 \\ 0 & 1 & 1 \\ 0 & 4 & 4 \end{bmatrix} \to \begin{bmatrix} 1 & 2 & 4 \\ 0 & 1 & 1 \\ 0 & 0 & 0 \end{bmatrix}.$$

With $\mathsf{x} = \begin{bmatrix} x_1 \\ x_2 \\ x_3 \end{bmatrix}$, $x_3 = t$ is a free variable. Then $x_2 = -x_3 = -t$ and $x_1 = -2x_2 - 4x_3 = 2t - 4t = -2t$. The eigenspace corresponding to $\lambda = 1$ is the set of vectors of the form

$$\mathsf{x} = \begin{bmatrix} x_1 \\ x_2 \\ x_3 \end{bmatrix} = \begin{bmatrix} -2t \\ -t \\ t \end{bmatrix} = t \begin{bmatrix} -2 \\ -1 \\ 1 \end{bmatrix}.$$

To check that $\mathsf{x} = \begin{bmatrix} -2 \\ -1 \\ 1 \end{bmatrix}$ really is an eigenvector corresponding to $\lambda = 1$, we compute

$$A\mathsf{x} = \begin{bmatrix} 5 & 8 & 16 \\ 4 & 1 & 8 \\ -4 & -4 & -11 \end{bmatrix} \begin{bmatrix} -2 \\ -1 \\ 1 \end{bmatrix} = \begin{bmatrix} -2 \\ -1 \\ 1 \end{bmatrix} = 1 \begin{bmatrix} -2 \\ -1 \\ 1 \end{bmatrix} = 1\mathsf{x}.$$

11.8 Remark. Finding the eigenvalues and eigenvectors of a matrix can involve a lot of calculation. Without the assistance of a computer or sophisticated calculator, errors can creep into the work of even the most careful person. Fortunately, there are ways to check our work as it progresses. In Example 11.7, for instance, we found that the characteristic polynomial was $-(\lambda+3)^2(\lambda-1)$ and concluded that $\lambda = 1$ was an eigenvalue of A. If this really is the case, then the homogeneous system $(A-\lambda I)\mathsf{x} = \mathbf{0}$ **must** have nonzero solutions when $\lambda = 1$. It was reassuring to find that $A - I$ had row echelon form $\begin{bmatrix} 1 & 2 & 4 \\ 0 & 1 & 1 \\ 0 & 0 & 0 \end{bmatrix}$, guaranteeing nonzero solutions.

11.9 Remark. This section might suggest that all matrices have eigenvalues whereas this is far from the case. The characteristic polynomial of $A = \begin{bmatrix} 0 & 1 \\ -1 & 0 \end{bmatrix}$, for example, is

$$\det(A - \lambda I) = \begin{vmatrix} -\lambda & 1 \\ -1 & -\lambda \end{vmatrix} = \lambda^2 + 1.$$

This polynomial has no **real** roots, so A has no **real** eigenvalues. In this course, however, Math 2050, all numbers are assumed to be real, not imaginary.

Similarity and Diagonalization

Now, finally, at the end of a long semester—hasn't it been fun?—we focus attention on diagonal matrices. Remember that a diagonal matrix is a matrix like $A = \begin{bmatrix} 2 & 0 \\ 0 & 7 \end{bmatrix}$ whose only nonzero entries lie on the (main) diagonal. Diagonal matrices are nice!

- Easy determinant: $\det A = 2(7) = 14$.
- Easy to decide invertibility.
- Easy to invert: $A^{-1} = \begin{bmatrix} \frac{1}{2} & 0 \\ 0 & \frac{1}{7} \end{bmatrix}$.
- Easy to multiply: $\begin{bmatrix} 2 & 0 \\ 0 & 7 \end{bmatrix} \begin{bmatrix} -3 & 0 \\ 0 & 2 \end{bmatrix} = \begin{bmatrix} -6 & 0 \\ 0 & 14 \end{bmatrix}$.
- Easy to find powers: $A^{100} = \begin{bmatrix} 2^{100} & 0 \\ 0 & 7^{100} \end{bmatrix}$.
- Easy to find eigenvalues: $\det(A - \lambda I) = \det \begin{bmatrix} 2-\lambda & 0 \\ 0 & 7-\lambda \end{bmatrix} = (2-\lambda)(7-\lambda)$, so the eigenvalues are the diagonal entries, 2 and 7.

If diagonal is good, the next best thing is "similar to diagonal."

11.10 Definitions. Matrices A and B are *similar* if $P^{-1}AP = B$ for some invertible matrix P. A matrix A is *diagonalizable* if it is similar to a diagonal matrix—that is, if there exists an invertible matrix P and a diagonal matrix D such that $P^{-1}AP = D$.

Let's think about similarity first.

11.11 Example. The two matrices $A = \begin{bmatrix} 1 & 2 \\ 3 & 4 \end{bmatrix}$ and $B = \begin{bmatrix} -55 & -97 \\ 34 & 60 \end{bmatrix}$ are similar because if $P = \begin{bmatrix} 3 & 5 \\ 1 & 2 \end{bmatrix}$, then $P^{-1} = \begin{bmatrix} 2 & -5 \\ -1 & 3 \end{bmatrix}$ and

$$P^{-1}AP = \begin{bmatrix} 2 & -5 \\ -1 & 3 \end{bmatrix}\begin{bmatrix} 1 & 2 \\ 3 & 4 \end{bmatrix}\begin{bmatrix} 3 & 5 \\ 1 & 2 \end{bmatrix} = \begin{bmatrix} -55 & -97 \\ 34 & 60 \end{bmatrix} = B.$$

:-)

As this example shows, similar matrices need not look much alike! They are, however, "similar" in lots of ways.

- Similar matrices have the same determinant since

$$\det(P^{-1}AP) = \det(P^{-1})\det A \det P = \frac{1}{\det P}\det A \det P = \det A.$$

- Similar matrices have the same characteristic polynomial because

$$P^{-1}AP - \lambda I = P^{-1}(A - \lambda I)P$$

  and the determinant of $P^{-1}(A - \lambda I)P$ is $\det(A - \lambda I)$.

- Similar matrices have the same eigenvalues because the eigenvalues are the roots of the characteristic polynomials and we just showed that these were the same.

  It is not a coincidence that the sum of the diagonal entries of the matrices A and B in Example 11.11 is 5 in each case.

- Similar matrices even have the same trace. (The *trace* of a matrix is the sum of its diagonal entries.)

Diagonalizable means "similar to diagonal."

11.12 Example. Let $A = \begin{bmatrix} 1 & 4 \\ 2 & 3 \end{bmatrix}$ and $P = \begin{bmatrix} -2 & 1 \\ 1 & 1 \end{bmatrix}$. Then P is invertible, $P^{-1} = \frac{1}{3}\begin{bmatrix} -1 & 1 \\ 1 & 2 \end{bmatrix}$ and

$$P^{-1}AP = \frac{1}{3}\begin{bmatrix} -1 & 1 \\ 1 & 2 \end{bmatrix}\begin{bmatrix} 1 & 4 \\ 2 & 3 \end{bmatrix}\begin{bmatrix} -2 & 1 \\ 1 & 1 \end{bmatrix} = \begin{bmatrix} -1 & 0 \\ 0 & 5 \end{bmatrix} = D$$

is a diagonal matrix, so A is diagonalizable. :-)

When we said that similar to diagonal is the next best thing to diagonal notice, for instance, how easy it is to compute the characteristic polynomial and to find the eigenvalues of a diagonalizable matrix. In the last example, since A is similar to D, these two matrices have the same characteristic polynomial. Since D is diagonal, its characteristic polynomial is $(-1-\lambda)(5-\lambda)$ and its eigenvalues are -1 and 5. So this polynomial is the characteristic polynomial of A and the eigenvalues of A are the same as those of D, -1 and 5.

Look how easy it is to compute powers of a diagonalizable matrix. If $P^{-1}AP = D$ is diagonal, then $A = PDP^{-1}$, so

$$\begin{aligned} A^{100} &= (PDP^{-1})(PDP^{-1})(PDP^{-1})(PDP^{-1})\cdots(PDP^{-1}) \\ &= PDP^{-1}PDP^{-1}PDP^{-1}PDP^{-1}\cdots PDP^{-1} = PD^{100}P^{-1} \end{aligned}$$

because all $P^{-1}P = I$ is the identity. For example, if $A = \begin{bmatrix} 1 & 4 \\ 2 & 3 \end{bmatrix}$, then

$$\begin{aligned} A^{100} &= PD^{100}P^{-1} \\ &= \frac{1}{3}\begin{bmatrix} -2 & 1 \\ 1 & 1 \end{bmatrix}\begin{bmatrix} -1 & 0 \\ 0 & 5^{100} \end{bmatrix}\begin{bmatrix} -1 & 1 \\ 1 & 2 \end{bmatrix} = \frac{1}{3}\begin{bmatrix} -2+5^{100} & 2+2(5^{100}) \\ 1+5^{100} & -1+2(5^{100}) \end{bmatrix}. \end{aligned}$$

Isn't this easier than working directly with A?

$$A^2 = \begin{bmatrix} 9 & 16 \\ 8 & 17 \end{bmatrix}, \quad A^3 = \begin{bmatrix} 41 & 84 \\ 42 & 83 \end{bmatrix}, \quad A^4 = \begin{bmatrix} 209 & 416 \\ 208 & 417 \end{bmatrix}, \ldots$$

The Equation $P^{-1}AP = D$

Let's investigate the equation $P^{-1}AP = D$, which we rewrite as $AP = PD$. Suppose P has columns $\mathsf{x}_1, \ldots, \mathsf{x}_n$. The rules for matrix multiplication give

$$AP = A\begin{bmatrix} \mathsf{x}_1 & \mathsf{x}_2 & \cdots & \mathsf{x}_n \\ \downarrow & \downarrow & & \downarrow \\ & & & \end{bmatrix} = \begin{bmatrix} A\mathsf{x}_1 & A\mathsf{x}_2 & \cdots & A\mathsf{x}_n \\ \downarrow & \downarrow & & \downarrow \\ & & & \end{bmatrix}.$$

On the other hand, what is PD? If the diagonal entries of D are $\lambda_1, \lambda_2, \ldots, \lambda_n$, then

$$PD = \begin{bmatrix} \mathsf{x}_1 & \mathsf{x}_2 & \cdots & \mathsf{x}_n \\ \downarrow & \downarrow & & \downarrow \\ & & & \\ & & & \end{bmatrix}\begin{bmatrix} \lambda_1 & 0 & \cdots & 0 \\ 0 & \lambda_2 & \cdots & 0 \\ \vdots & & \ddots & \\ 0 & \cdots & & \lambda_n \end{bmatrix} = \begin{bmatrix} \lambda_1\mathsf{x}_1 & \lambda_2\mathsf{x}_2 & \ldots & \lambda_n\mathsf{x}_n \\ \downarrow & \downarrow & & \downarrow \\ & & & \end{bmatrix}$$

as we showed in Example 5.25. Equating corresponding columns of the equal matrix AP and PD, we see that $A\mathsf{x}_1 = \lambda_1\mathsf{x}_1$, $A\mathsf{x}_2 = \lambda_2\mathsf{x}_2$, ..., $A\mathsf{x}_n = \lambda_n\mathsf{x}_n$. In other words, the columns of P are eigenvectors corresponding to eigenvalues which are the diagonal entries of D.

So we have shown that if A is diagonalizable and $P^{-1}AP = D$, then the columns of P are eigenvectors corresponding to eigenvalues which are the diagonal entries of D, the order of the eigenvectors in P corresponding to the order of the diagonal entries of D.

11.13 Example. Let's look at Example 11.12 again. There we showed that the matrix $A = \begin{bmatrix} 1 & 4 \\ 2 & 3 \end{bmatrix}$ is diagonalizable. Specifically, with $P = \begin{bmatrix} -2 & 1 \\ 1 & 1 \end{bmatrix}$, we have $P^{-1}AP = D = \begin{bmatrix} -1 & 0 \\ 0 & 5 \end{bmatrix} = D$. The first column of P, $\begin{bmatrix} -2 \\ 1 \end{bmatrix}$ is an eigenvector of A with corresponding eigenvalue -1, the first diagonal entry of D. The second column of P, $\begin{bmatrix} 1 \\ 1 \end{bmatrix}$, is an eigenvector of A with corresponding eigenvalue 5, the second diagonal entry of D. These facts agree with what we found in Example 11.6, where this matrix first appeared.

Suppose we let $Q = \begin{bmatrix} 3 & 4 \\ 3 & -2 \end{bmatrix}$. Notice that the columns of Q are eigenvectors of A corresponding to the eigenvalues 5 and -1, respectively. It follows that $Q^{-1}AQ = \begin{bmatrix} 5 & 0 \\ 0 & -1 \end{bmatrix}$. ⌣̈

We have been trying to show that a matrix is diagonalizable if we can find enough eigenvectors to build an **invertible** matrix P.

11.14 Example. Let $A = \begin{bmatrix} 2 & 5 \\ 0 & 2 \end{bmatrix}$. The characteristic polynomial of A is $\det \begin{bmatrix} 2-\lambda & 5 \\ 0 & 2-\lambda \end{bmatrix} = (2-\lambda)^2$, so the only eigenvalue of A is $\lambda = 2$. To find the corresponding eigenspace, we solve the homogeneous system $(A - \lambda I)\mathrm{x} = 0$ for $\mathrm{x} = \begin{bmatrix} x_1 \\ x_2 \end{bmatrix}$ with $\lambda = 2$. Gaussian elimination proceeds $A - \lambda I = \begin{bmatrix} 0 & 5 \\ 0 & 0 \end{bmatrix}$, so $x_1 = t$ is free and $x_2 = 0$. The eigenspace corresponding to $\lambda = 2$ consists of vectors of the form $\begin{bmatrix} t \\ 0 \end{bmatrix} = t\begin{bmatrix} 1 \\ 0 \end{bmatrix}$. There is no way to build an invertible matrix P of eigenvectors since the only eigenvectors are multiples of $\begin{bmatrix} 1 \\ 0 \end{bmatrix}$ and a matrix of the form $\begin{bmatrix} a & b \\ 0 & 0 \end{bmatrix}$ is not invertible. ⌣̈

Suppose the columns of P are eigenvectors and P is invertible. Then the eigenvectors in the columns must be linearly independent since $P\mathrm{x} = 0$ implies $\mathrm{x} = 0$ (multiplying by P^{-1}). This argument can also be reversed. If the columns of a matrix P are linearly independent, then P is invertible—see Remark 10.19.

All this goes to show that if an $n \times n$ matrix A is diagonalizable, then there is a set of n linearly independent eigenvectors (to build the necessary P). We will spare you the details, but the converse is true too.

11.15 Theorem. *An $n \times n$ matrix is diagonalizable if and only if it has n linearly independent eigenvectors. Furthermore, $P^{-1}AP = D$ if and only if the columns of P are linearly independent eigenvectors of A and the diagonal entries of D are eigenvalues that correspond in order to the columns of P.*

11.16 Example. Let $A = \begin{bmatrix} 5 & 8 & 16 \\ 4 & 1 & 8 \\ -4 & -4 & -11 \end{bmatrix}$. In Example 11.7, we recorded the fact that A has eigenvalues $\lambda = -3$ and $\lambda = 1$. For $\lambda = 1$, we found that $\mathsf{x}_1 = \begin{bmatrix} -2 \\ -1 \\ 1 \end{bmatrix}$. In Example 11.4, we discovered that $\mathsf{x}_2 = \begin{bmatrix} -1 \\ 1 \\ 0 \end{bmatrix}$ and $\mathsf{x}_3 = \begin{bmatrix} -2 \\ 0 \\ 1 \end{bmatrix}$ are eigenvectors for $\lambda = -3$. We leave it to the reader to verify that x_1, x_2, and x_3 are linearly independent. (In addition to the "obvious" way—use the definition—another method was given in Example 10.18.) Thus the matrix $P = \begin{bmatrix} -2 & -1 & -2 \\ -1 & 1 & 0 \\ 1 & 0 & 1 \end{bmatrix}$, whose columns are $\mathsf{x}_1, \mathsf{x}_2, \mathsf{x}_3$, is invertible and $P^{-1}AP = D = \begin{bmatrix} 1 & 0 & 0 \\ 0 & -3 & 0 \\ 0 & 0 & -3 \end{bmatrix}$, the diagonal matrix whose diagonal entries are the eigenvalues of A, in the order determined by the columns of P. Since the first column of P is an eigenvector corresponding to $\lambda = 1$, the first diagonal entry of D is 1; since the last two columns of P are eigenvectors corresponding to $\lambda = -3$, the last two diagonal entries of D are -3. ☺

11.17 Remark. The *diagonalizing matrix* P and the diagonal matrix D are far from unique. You can get a variety of Ds by rearranging the columns of P, and even for a fixed D, there are many Ps that work. For example, you can always replace a column of P by a nonzero multiple of this column. (It is not hard to show that if x is an eigenvector corresponding to λ, then so is $c\mathsf{x}$ for any nonzero scalar c.) With $A = \begin{bmatrix} 5 & 8 & 16 \\ 4 & 1 & 8 \\ -4 & -4 & -11 \end{bmatrix}$ as in Example 11.16, if we let $Q = \begin{bmatrix} 3 & -4 & 6 \\ -3 & -2 & 0 \\ 0 & 2 & -3 \end{bmatrix}$ and note that the columns of this matrix are eigenvectors of A corresponding to the eigenvalues -3, 1 and -3, in this order, then $Q^{-1}AQ = \begin{bmatrix} -3 & 0 & 0 \\ 0 & 1 & 0 \\ 0 & 0 & -3 \end{bmatrix}$.

11.18 Example. Let $A = \begin{bmatrix} 2 & 1 & -12 \\ 0 & 1 & 11 \\ 1 & 1 & 4 \end{bmatrix}$.

Expanding $\det(A - \lambda I)$ by cofactors down the first column, we obtain the characteristic polynomial of A:

$$\begin{vmatrix} 2-\lambda & 1 & -12 \\ 0 & 1-\lambda & 11 \\ 1 & 1 & 4-\lambda \end{vmatrix} = (2-\lambda)[(1-\lambda)(4-\lambda)-11]+[11+12(1-\lambda)]$$
$$= -\lambda^3+7\lambda^2-15\lambda+9 = -(\lambda-1)(\lambda-3)^2.$$

The eigenvalues of A are $\lambda = 1$ and $\lambda = 3$. To find the eigenvectors $\mathsf{x} = \begin{bmatrix} x_1 \\ x_2 \\ x_3 \end{bmatrix}$ for $\lambda = 1$ we must solve $(A - \lambda I)\mathsf{x} = \mathsf{0}$ with $\lambda = 1$. Gaussian elimination proceeds

$$\begin{bmatrix} 2-\lambda & 1 & -12 \\ 0 & 1-\lambda & 11 \\ 1 & 1 & 4-\lambda \end{bmatrix} = \begin{bmatrix} 1 & 1 & -12 \\ 0 & 0 & 11 \\ 1 & 1 & 3 \end{bmatrix} \to \begin{bmatrix} 1 & 1 & -12 \\ 0 & 0 & 1 \\ 0 & 0 & 15 \end{bmatrix} \to \begin{bmatrix} 1 & 1 & -12 \\ 0 & 0 & 1 \\ 0 & 0 & 0 \end{bmatrix},$$

so $x_2 = t$ is free, $x_3 = 0$, and $x_1 + x_2 - 12x_3 = 0$, so $x_1 = -x_2 + 12x_3 = -t$. Eigenvectors are of the form

$$\begin{bmatrix} x_1 \\ x_2 \\ x_3 \end{bmatrix} = \begin{bmatrix} -t \\ t \\ 0 \end{bmatrix} = t\begin{bmatrix} -1 \\ 1 \\ 0 \end{bmatrix}.$$

To find the eigenvectors for $\lambda = 3$, we set $\lambda = 3$ in $A - \lambda I$ and reduce to row echelon form:

$$\begin{bmatrix} 2-\lambda & 1 & -12 \\ 0 & 1-\lambda & 11 \\ 1 & 1 & 4-\lambda \end{bmatrix} = \begin{bmatrix} -1 & 1 & -12 \\ 0 & -2 & 11 \\ 1 & 1 & 1 \end{bmatrix}$$
$$\to \begin{bmatrix} 1 & -1 & 12 \\ 1 & 1 & 1 \\ 0 & -2 & 11 \end{bmatrix} \to \begin{bmatrix} 1 & -1 & 12 \\ 0 & 2 & -11 \\ 0 & -2 & 11 \end{bmatrix} \to \begin{bmatrix} 1 & -1 & 12 \\ 0 & 1 & -\frac{11}{2} \\ 0 & 0 & 0 \end{bmatrix}$$

so $x_3 = t$ is free, $x_2 = \frac{11}{2}x_3 = \frac{11}{2}t$, and $x_1 - x_2 + 12x_3 = 0$, so $x_1 = x_2 - 12x_3 = \frac{11}{2}t - 12t = -\frac{13t}{2}$. Eigenvectors are of the form

$$\begin{bmatrix} x_1 \\ x_2 \\ x_3 \end{bmatrix} = \begin{bmatrix} -\frac{13t}{2} \\ \frac{11t}{2} \\ t \end{bmatrix} = t\begin{bmatrix} -\frac{13}{2} \\ \frac{11}{2} \\ 1 \end{bmatrix}.$$

So the only eigenvalues of A are multiples of $\begin{bmatrix} -1 \\ 1 \\ 0 \end{bmatrix}$ and $\begin{bmatrix} -13 \\ 11 \\ 2 \end{bmatrix}$. We need three linearly independent eigenvectors to build a diagonalizing matrix P, but we can find only two: A is not diagonalizable. ☺

There is one instance when we are assured that the condition of Theorem 11.15 holds and hence that a matrix is diagonalizable.

11.19 Theorem. *Eigenvectors corresponding to different eigenvalues are linearly independent. So if A is an $n \times n$ matrix with n different eigenvalues, then A is diagonalizable.*

To illustrate how useful this theorem is, the minute we discovered that the matrix A in Example 11.6 had different eigenvalues 5 and -1, we could have concluded that A was diagonalizable (without computing eigenvectors). Here is another example.

11.20 Example. The characteristic polynomial of $A = \begin{bmatrix} 1 & 2 \\ 1 & 1 \end{bmatrix}$ is $\begin{vmatrix} 1-\lambda & 2 \\ 1 & 1-\lambda \end{vmatrix} = (1-\lambda)^2 - 2$. The roots of this polynomial satisfy $(1-\lambda)^2 = 2$, so $|1-\lambda| = \sqrt{2}$ and $\lambda = 1 \pm \sqrt{2}$. These are distinct, so A is diagonalizable by Theorem 11.19; in fact, A is similar to $\begin{bmatrix} 1+\sqrt{2} & 0 \\ 0 & 1-\sqrt{2} \end{bmatrix}$. ☺

11.21 Remark. Theorem 11.19 doesn't explain why the matrix A in Example 11.16 is diagonalizable. This 3×3 matrix has just two eigenvalues, $\lambda = -3$ and $\lambda = 1$. For $\lambda = 1$, we found an eigenvector x_1. For $\lambda = -3$, we found two eigenvectors x_2 and x_3. Theorem 11.19 guarantees that x_1 and x_2 are linearly independent since they correspond to different eigenvalues and, for the same reason, it says that x_1 and x_3 are linearly independent. It doesn't say, however, that the **three** vectors $\mathsf{x}_1, \mathsf{x}_2, \mathsf{x}_3$ are linearly independent.

11.22 Remark. Despite all the examples of diagonalizable matrices we have provided, not all matrices are diagonalizable by any means. Suppose we try to find the eigenvalues of $A = \begin{bmatrix} 0 & 1 \\ 0 & 0 \end{bmatrix}$. We have

$$\det(A - \lambda I) = \begin{vmatrix} -\lambda & 1 \\ 0 & -\lambda \end{vmatrix} = \lambda^2,$$

so $\lambda = 0$ is the only eigenvalue. If A were diagonalizable, the diagonal matrix D, having eigenvalues of A on its diagonal, would be the zero matrix. Since the equation $P^{-1}AP = 0$ implies $A = 0$, A can't be diagonalizable.

True/False Questions for Week 11

Decide, with as little calculation as possible, whether each of the following statements is true or false and explain your answer whenever you say "false."

1. 0 is an admissible eigenvalue for a matrix.
2. 0 is an admissible eigenvector for a matrix.
3. An eigenspace of a matrix consists entirely of eigenvectors.
4. If $A\mathsf{x} = \lambda\mathsf{x}$, then $(A - \lambda)\mathsf{x} = 0$.
5. $\lambda = 1$ is an eigenvalue of $\begin{bmatrix} 1 & 2 \\ 0 & 1 \end{bmatrix}$.

6. $\lambda = 1$ is the only eigenvalue of $\begin{bmatrix} 1 & 2 \\ 0 & 1 \end{bmatrix}$.

7. $\begin{bmatrix} 2 \\ 1 \end{bmatrix}$ is an eigenvector of $\begin{bmatrix} 1 & 0 \\ 2 & -3 \end{bmatrix}$.

8. The eigenvectors of a matrix are the roots of its characteristic polynomial.

9. Every 2×2 matrix has a real eigenvalue.

10. The characteristic polynomial of $\begin{bmatrix} 1 & 2 & 3 \\ 0 & 4 & 5 \\ 0 & 0 & 6 \end{bmatrix}$ is $(1 - \lambda)(4 - \lambda)(6 - \lambda)$.

11. If $A = \begin{bmatrix} -1 & 0 \\ 0 & 2 \end{bmatrix}$, then $\lambda = +1$ is an eigenvalue of A^{10}.

12. The matrices $\begin{bmatrix} 2 & 0 & 0 \\ 0 & 3 & 0 \\ 0 & 0 & 5 \end{bmatrix}$ and $\begin{bmatrix} 1 & 1 & 3 \\ 0 & 6 & -1 \\ 0 & 0 & 5 \end{bmatrix}$ are similar.

13. The matrices $\begin{bmatrix} 2 & 20 \\ 9 & -1 \end{bmatrix}$ and $\begin{bmatrix} 3 & 30 \\ 27 & 1 \end{bmatrix}$ are similar.

14. Similar matrices are diagonal.

15. Diagonal matrices are similar.

16. The matrices $\begin{bmatrix} 2 & 5 \\ -2 & 3 \end{bmatrix}$ and $\begin{bmatrix} 4 & -1 \\ 3 & 2 \end{bmatrix}$ are similar.

17. The matrix $A = \begin{bmatrix} -1 & 0 \\ 0 & 5 \end{bmatrix}$ is diagonalizable.

18. If a 2×2 matrix has just one eigenvalue, it is not diagonalizable.

19. If a 3×3 matrix has 3 different eigenvalues, then it is diagonalizable.

20. If a 3×3 matrix is diagonalizable, then it must have 3 different eigenvalues.

21. If A is a symmetric matrix, then the eigenvalues of A are its diagonal entries.

Week 11 Test Yourself

Here are a few problems with short answers that you can use to test your understanding of the concepts you have met this week.

1. Determine whether or not each of the vectors

$$\mathsf{v}_1 = \begin{bmatrix} 0 \\ 0 \\ 0 \end{bmatrix}, \mathsf{v}_2 = \begin{bmatrix} 1 \\ 0 \\ 0 \end{bmatrix}, \mathsf{v}_3 = \begin{bmatrix} 0 \\ 2 \\ 2 \end{bmatrix}, \mathsf{v}_4 = \begin{bmatrix} 1 \\ 0 \\ -1 \end{bmatrix}, \mathsf{v}_5 = \begin{bmatrix} 3 \\ 3 \\ 3 \end{bmatrix}, \mathsf{v}_6 = \begin{bmatrix} 1 \\ 0 \\ 1 \end{bmatrix}$$

are eigenvectors of $A = \begin{bmatrix} 5 & -7 & 7 \\ 4 & -3 & 4 \\ 4 & -1 & 2 \end{bmatrix}$. Justify your answers and when you say "yes," state the corresponding eigenvalue.

2. Determine which of the scalars $-2, -1, 1, 3, 4$, if any, are eigenvalues of $A = \begin{bmatrix} 5 & -2 \\ 4 & -1 \end{bmatrix}$.

3. (a) Show that $\begin{bmatrix} -3 \\ 3 \end{bmatrix}$ is an eigenvector of $A = \begin{bmatrix} 5 & 1 \\ -1 & 3 \end{bmatrix}$ and state the corresponding eigenvalue.

   (b) What is the characteristic polynomial of A?

   (c) What are the eigenvalues of A?

4. Find the characteristic polynomial, the (real) eigenvalues, and the corresponding eigenspaces of $A = \begin{bmatrix} 4 & 2 & 2 \\ 4 & 2 & -4 \\ -2 & 0 & -4 \end{bmatrix}$.

5. Show that the only matrix similar to the $n \times n$ identity matrix I is I itself.

6. What is the "trace" of a matrix?

7. Suppose A and B are similar. Prove that A and B have the same characteristic polynomial.

8. (a) Find all eigenvalues and eigenvectors of $A = \begin{bmatrix} 0 & 1 \\ 0 & 0 \end{bmatrix}$.

   (b) Explain why the results of part (a) show that A is not diagonalizable.

9. Explain one part of Theorem 11.15, namely, if an $n \times n$ matrix A is diagonalizable, then it has n linearly independent eigenvectors.

10. Determine whether or not each of the matrices A given below is diagonalizable. If it is, find an invertible matrix P and a diagonal matrix D such that $P^{-1}AP = D$. If it isn't, explain why.

    (a) $A = \begin{bmatrix} 1 & 0 \\ 2 & 3 \end{bmatrix}$ (b) $A = \begin{bmatrix} -1 & 3 & 0 \\ 0 & 2 & 0 \\ 2 & 1 & -1 \end{bmatrix}$

11. Given that a matrix A is similar to the matrix $\begin{bmatrix} -1 & 2 & 3 & 4 \\ 0 & 1 & 1 & 2 \\ 0 & 0 & -3 & 5 \\ 0 & 0 & 0 & -3 \end{bmatrix}$, what's the characteristic polynomial of A? Explain.

12. If a matrix A is similar to $5I$, I an identity matrix, what can you conclude about A? Explain.

13. What are the eigenvalues and the corresponding eigenspaces of the matrix $A = \begin{bmatrix} 2 & 0 & 0 & 0 \\ 0 & 2 & 0 & 0 \\ 0 & 0 & 2 & 0 \\ 0 & 0 & 0 & 2 \end{bmatrix}$?

14. (a) Find an easy reason that explains why the matrix $A = \begin{bmatrix} 5 & 3 \\ 2 & 4 \end{bmatrix}$ is diagonalizable.

    (b) Find two matrices to which A is similar.

APPLICATIONS

Electric Circuits

In this brief section, we illustrate one situation where systems of linear equations naturally arise. We consider electrical networks comprised of batteries and resistors connected with wires. Each battery has a positive terminal and a negative terminal causing current to flow around the circuit out of the positive terminal and into the negative terminal. We are all familiar with nine volt batteries and 1.5 volt AA batteries. "Voltage" is a measure of the power of a battery. Each resistor, as the name implies, consumes voltage (one refers to a "voltage drop") and affects the current in the circuit. According to Ohm's Law, the connection between voltage E, current I (measured in amperes, amps for short) and resistance (measured in ohms) is given by this formula:

$$\begin{array}{ccccc} V & = & I & & R \\ \text{volts} & = & \text{amps} & \times & \text{ohms.} \end{array}$$

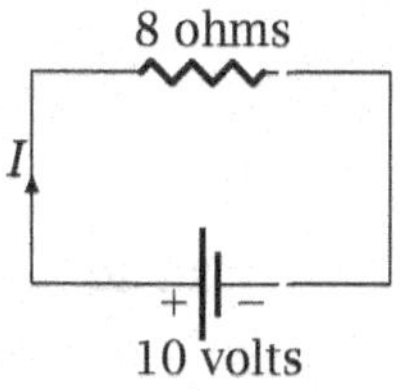

There are two ways to view this formula. In a simple circuit such as the one shown, it tells us the size of the current flowing around the circuit: $I = \frac{E}{R} = \frac{10}{8} = 1.25$ amps. Alternatively, it tells us the size of the voltage drop at the resistor: $V = IR = 1.25(8) = 10$. (It is standard to use the symbol —+||−— for battery and —∧∧∧— for resistor.)

A basic principle due to Kirchoff says that the sum of the voltage drops at the resistors on a circuit must equal the total voltage provided by the batteries. For the circuit on the left of Figure 14, the voltage drops are $4I$ at one resistor and $6I$ at the other for a total of $10I$. The total voltage provided by the batteries is $3 + 12 = 15$. We have $10I = 15$, so the current in the circuit is $I = \frac{15}{10} = 1.5$ amps.

Notice that on the left, the current I flows counterclockwise around the circuit, from the positive terminal to the negative terminal in each battery. The situation is different for the circuit on the right. Current leaves the positive terminal of the top battery, but it enters the positive terminal of the lower battery. For this reason, the lower battery contributes -3 volts to the total voltage of the circuit, which is then $12 - 3 = 9$ volts. This time, we get $10I = 9$, so $I = 0.9$ amps. For this reason, we state Kirchoff's first law like this:

12.1 Kirchoff's Circuit Law: The sum of the voltage drops around a circuit equals the algebraic

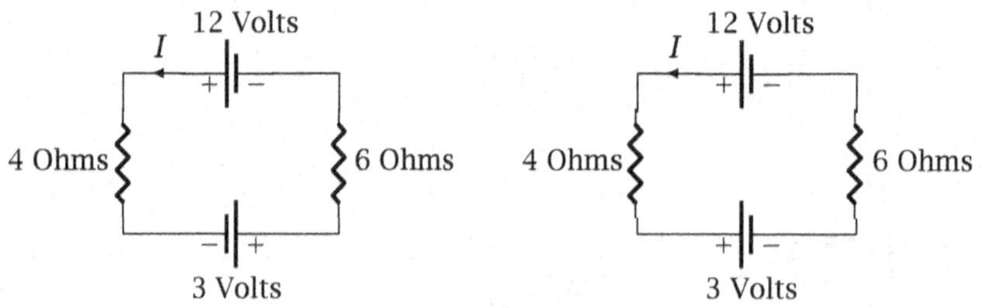

Figure 14

sum of the voltages provided by the batteries on the circuit.

"Algebraic" here means we take the direction of the current into account.

The reason for introducing electrical circuits in a chapter devoted to systems of linear equations becomes evident when we consider circuits that contain two or more "subcircuits." The circuit depicted in Figure 15 contains two subcircuits which are joined at A and B. The points A and B are called *nodes*, these being places where one subcircuit meets another. At A, there are two entering currents of I_2 and I_3 amps and a single departing current of I_1 amps. At B, the entering current of I_1 amps splits into currents of I_2 and I_3 amps. A second law of Kirchoff is needed.

12.2 **Kirchoff's Node Law:** The sum of the currents flowing into any node is the sum of the currents flowing out of that node.

Thus

$$\begin{aligned} \text{At } A: &\quad I_2 + I_3 = I_1 \\ \text{At } B: &\quad I_1 = I_2 + I_3. \end{aligned}$$

These equations both say $I_1 - I_2 - I_3 = 0$. In addition, Kirchoff's Circuit Law applied to the subcircuit on the left gives

$$6I_1 + 2I_2 = 4$$

since there are voltage drops of $6I_1$ and $2I_2$ at the two resistors on this circuit ($V = IR$) and the total voltage supplied by the battery is 4 volts. We must be careful with the subcircuit on the right. Do we wish to follow this circuit clockwise or counterclockwise? Suppose we follow it counterclockwise, in the direction of the arrow for I_3. Then there is a $4I_3$ voltage drop at the 4 Ohm resistor and a $-2I_2$ voltage drop at the 2 Ohm resistor because the current passing through this resistor, in the counterclockwise direction, is $-I_2$. Kirchoff's Circuit Law gives

$$-2I_2 + 4I_3 = 10$$

To find the three currents, we must solve the system

$$\begin{aligned} I_1 - I_2 - I_3 &= 0 \\ 6I_1 + 2I_2 \phantom{{}+4I_3} &= 4 \\ -2I_2 + 4I_3 &= 10. \end{aligned} \tag{33}$$

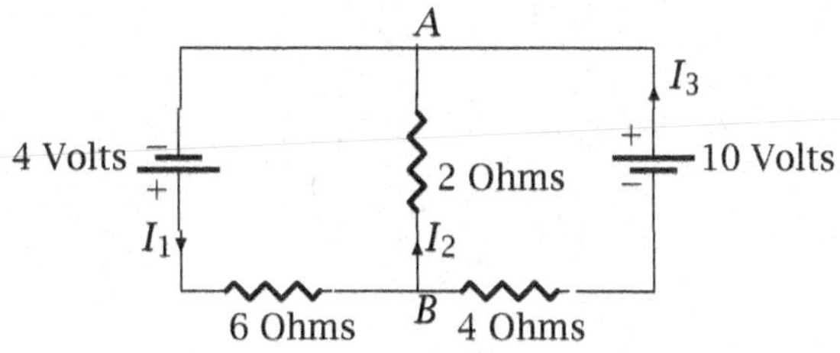

Figure 15

Gaussian elimination applied to the augmented matrix of coefficients proceeds

$$\left[\begin{array}{ccc|c} 1 & -1 & -1 & 0 \\ 6 & 2 & 0 & 4 \\ 0 & -2 & 4 & 10 \end{array}\right] \to \left[\begin{array}{ccc|c} 1 & -1 & -1 & 0 \\ 0 & 8 & 6 & 4 \\ 0 & -1 & 2 & 5 \end{array}\right]$$

$$\to \left[\begin{array}{ccc|c} 1 & -1 & -1 & 0 \\ 0 & 1 & -2 & -5 \\ 0 & 4 & 3 & 2 \end{array}\right] \to \left[\begin{array}{ccc|c} 1 & -1 & -1 & 0 \\ 0 & 1 & -2 & -5 \\ 0 & 0 & 11 & 22 \end{array}\right].$$

Back substitution now gives $11I_3 = 22$, so $I_3 = 2$, $I_2 - 2I_3 = -5$, so $I_2 = -5 + 2I_3 = -1$ and $I_1 = I_2 + I_3 = 1$. Don't worry about the negative I_2. This simply means we put the arrow on the middle wire in the wrong direction. The current is actually flowing from A to B.

12.3 Remark. The observant student may point out that there is a third circuit in Figure 15, the one that goes around the outer edge. What does Kirchoff's Circuit Law say about this? The sum of the voltage drops is $6I_1 + 4I_3$ and the total voltage supplied by the two batteries is 14, so we get $6I_1 + 4I_3 = 14$. Since this is satisfied by $I_1 = 1$, $I_3 = 2$, it is implied by the three equations we used in (33). This should always be the case.

12.4 Example. We present in Figure 16 what is known in electrical engineering as a "Wheatstone Bridge Circuit." If the ratio of the resistances to the left of the "bridge" BD is the same as the corresponding ratio on the right, there should be no current through the bridge. Thus, testing that $I_7 = 0$ provides a way to be sure that the ratios are the same. Lets see if this is the case.

There are four nodes, A, B, C, D, where Kirchoff's Node Law gives, respectively, these equations:

$$\begin{aligned} I_1 &= I_3 + I_4 \\ I_7 &= I_3 + I_5 \\ I_2 &= I_5 + I_6 \\ I_4 + I_6 + I_7 &= 0. \end{aligned}$$

There are three subcircuits. Kirchoff's Law applied to the one at the bottom says $4I_4 - 6I_6 = 20$. Applied to the two triangular subcircuits at the top, the law gives $8I_3 - 4I_4 = 0$ and

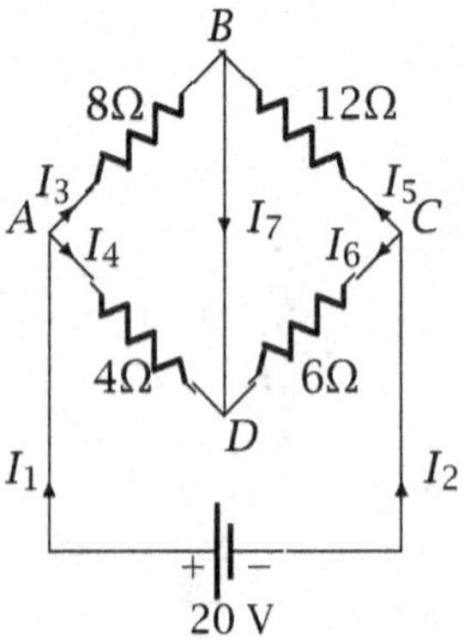

Figure 16: A Wheatstone Bridge Circuit

$6I_6 - 12I_5 = 0$. To find the individual currents, we solve the system

$$
\begin{array}{rcl}
I_1 - I_3 - I_4 & = & 0 \\
I_3 + I_5 - I_7 & = & 0 \\
I_2 - I_5 - I_6 & = & 0 \\
I_4 + I_6 + I_7 & = & 0 \\
4I_4 - 6I_6 & = & 20 \\
8I_3 - 4I_4 & = & 0 \\
-12I_5 + 6I_6 & = & 0.
\end{array}
$$

Gaussian elimination proceeds

$$
\left[\begin{array}{ccccccc|c}
1 & 0 & -1 & -1 & 0 & 0 & 0 & 0 \\
0 & 0 & 1 & 0 & 1 & 0 & -1 & 0 \\
0 & 1 & 0 & 0 & -1 & -1 & 0 & 0 \\
0 & 0 & 0 & 1 & 0 & 1 & 1 & 0 \\
0 & 0 & 0 & 4 & 0 & -6 & 0 & 20 \\
0 & 0 & 8 & -4 & 0 & 0 & 0 & 0 \\
0 & 0 & 0 & 0 & -12 & 6 & 0 & 0
\end{array}\right]
\to
\left[\begin{array}{ccccccc|c}
1 & 0 & -1 & -1 & 0 & 0 & 0 & 0 \\
0 & 1 & 0 & 0 & -1 & -1 & 0 & 0 \\
0 & 0 & 1 & 0 & 1 & 0 & -1 & 0 \\
0 & 0 & 0 & 1 & 0 & 1 & 1 & 0 \\
0 & 0 & 0 & 4 & 0 & -6 & 0 & 20 \\
0 & 0 & 8 & -4 & 0 & 0 & 0 & 0 \\
0 & 0 & 0 & 0 & -12 & 6 & 0 & 0
\end{array}\right]
$$

$$
\to
\left[\begin{array}{ccccccc|c}
1 & 0 & -1 & -1 & 0 & 0 & 0 & 0 \\
0 & 1 & 0 & 0 & -1 & -1 & 0 & 0 \\
0 & 0 & 1 & 0 & 1 & 0 & -1 & 0 \\
0 & 0 & 0 & 1 & 0 & 1 & 1 & 0 \\
0 & 0 & 0 & 4 & 0 & -6 & 0 & 20 \\
0 & 0 & 0 & -4 & -8 & 0 & 8 & 0 \\
0 & 0 & 0 & 0 & -12 & 6 & 0 & 0
\end{array}\right]
\to
\left[\begin{array}{ccccccc|c}
1 & 0 & -1 & -1 & 0 & 0 & 0 & 0 \\
0 & 1 & 0 & 0 & -1 & -1 & 0 & 0 \\
0 & 0 & 1 & 0 & 1 & 0 & -1 & 0 \\
0 & 0 & 0 & 1 & 0 & 1 & 1 & 0 \\
0 & 0 & 0 & 0 & 0 & -10 & -4 & 20 \\
0 & 0 & 0 & 0 & -8 & 4 & 12 & 0 \\
0 & 0 & 0 & 0 & -12 & 6 & 0 & 0
\end{array}\right]
$$

$$
\rightarrow \left[\begin{array}{ccccccc|c}
1 & 0 & -1 & -1 & 0 & 0 & 0 & 0 \\
0 & 1 & 0 & 0 & -1 & -1 & 0 & 0 \\
0 & 0 & 1 & 0 & 1 & 0 & -1 & 0 \\
0 & 0 & 0 & 1 & 0 & 1 & 1 & 0 \\
0 & 0 & 0 & 0 & -8 & 4 & 12 & 0 \\
0 & 0 & 0 & 0 & 0 & -10 & -4 & 20 \\
0 & 0 & 0 & 0 & -12 & 6 & 0 & 0
\end{array}\right]
\rightarrow \left[\begin{array}{ccccccc|c}
1 & 0 & -1 & -1 & 0 & 0 & 0 & 0 \\
0 & 1 & 0 & 0 & -1 & -1 & 0 & 0 \\
0 & 0 & 1 & 0 & 1 & 0 & -1 & 0 \\
0 & 0 & 0 & 1 & 0 & 1 & 1 & 0 \\
0 & 0 & 0 & 0 & 2 & -1 & -3 & 0 \\
0 & 0 & 0 & 0 & 0 & -10 & -4 & 20 \\
0 & 0 & 0 & 0 & -12 & 6 & 0 & 0
\end{array}\right]
$$

$$
\rightarrow \left[\begin{array}{ccccccc|c}
1 & 0 & -1 & -1 & 0 & 0 & 0 & 0 \\
0 & 1 & 0 & 0 & -1 & -1 & 0 & 0 \\
0 & 0 & 1 & 0 & 1 & 0 & -1 & 0 \\
0 & 0 & 0 & 1 & 0 & 1 & 1 & 0 \\
0 & 0 & 0 & 0 & 2 & -1 & -3 & 0 \\
0 & 0 & 0 & 0 & 0 & -10 & -4 & 20 \\
0 & 0 & 0 & 0 & 0 & 0 & -18 & 0
\end{array}\right].
$$

Back substitution now says that $I_7 = 0$ (oh boy!), $-10I_6 - 4I_7 = 20$, so $-10I_6 = 20$ and $I_6 = -2$ (the arrow on CD should be reversed), $2I_5 - I_6 - 3I_7 = 0$, so $2I_5 = I_6$ and $I_5 = -1$; $I_4 + I_6 + I_7 = 0$, so $I_4 = -I_6 = 2$; $I_3 + I_5 - I_7 = 0$, so $I_3 = -I_5 = 1$; $I_2 - I_5 - I_6 = 0$ so $I_2 = I_5 + I_6 = -3$ and $I_1 = I_3 + I_4 = 3$. Evidence that our solution is current can be obtained by checking the entire circuit, where Kirchoff's Circuit Law says that $8I_3 - 12I_5 = 20$ in agreement with our results. ⌣̈

%newpage

Some Exercises to Test Yourself

1. Find all indicated currents in each circuit. The standard symbol for ohms is the Greek symbol Ω, "Omega."

(a)

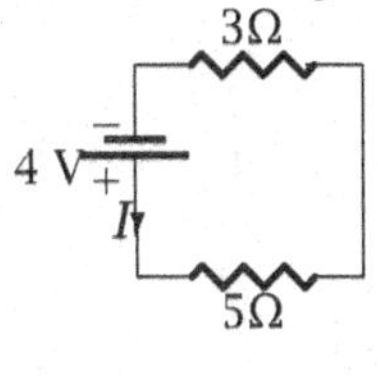

(b)

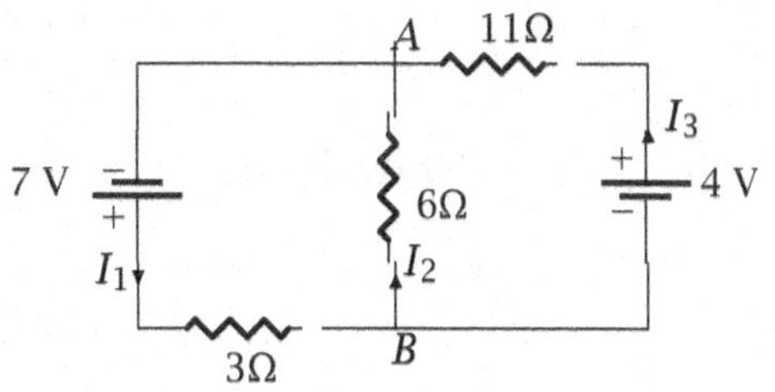

Using Matrices to Generate Codes

When information is transmitted electronically, it is usual for words to be converted to numbers, which are more easily manipulated. One can imagine many ways in which this conversion might take place, for example, by assigning to "A" the number 1, to "B" the number 2, ..., to Z the number 26, and to other characters such as punctuation marks and spaces, numbers 27 and higher. In practice, numbers are usually expressed in "base 2" and hence appear as sequences of 0s and 1s. Specifically, abc is the base 2 representation of the number $a(2^2) + b(2) + c(1)$, just as abc is the base 10 representation of the number $a(10^2) + b(10) + c(1)$. For example, in base ten, 327 means $3(10^2) + 2(10) + 7(1)$. In a similar way, in base two, 011 means $0(2^2) + 1(2) + 1(1) = 3$.

In base two, addition and multiplication are defined "modulo 2" ("mod 2," for short), that is, by these tables:

+	0	1
0	0	1
1	1	0

·	0	1
0	0	0
1	0	1

The only surprise here is that $1 + 1 = 0 \pmod 2$.

A *word* is a sequence of 0s and 1s, such as 001, and a *code* is a set of words. In this book, we assume that all the words of a code have the same length and that the sum of code words is also a code word. (In coding theory, such a code is said to be *linear.*) For example, a code might consist of all sequences of 0s and 1s of length three. There are eight such words, all written out below:

$$000, 001, 010, 011, 100, 101, 110, 111.$$

If we identify the word abc with the column vector $\begin{bmatrix} a \\ b \\ c \end{bmatrix}$, then the words of this code are just the matrix products $I\mathsf{v}$, where v is any of the eight column vectors $\begin{bmatrix} a \\ b \\ c \end{bmatrix}$, a, b, c each 0 or 1, and $I = \begin{bmatrix} 1 & 0 & 0 \\ 0 & 1 & 0 \\ 0 & 0 & 1 \end{bmatrix}$ is the 3×3 identity matrix. As used here, I is called a *generator matrix* for the code because the words of the code are generated by the columns, that is, they are linear combinations of the columns. Here's a more informative example.

12.5 Example. (The Hamming $(7, 4)$ Code) Let $G = \begin{bmatrix} 1 & 0 & 0 & 0 \\ 0 & 1 & 0 & 0 \\ 0 & 0 & 1 & 0 \\ 0 & 0 & 0 & 1 \\ 0 & 1 & 1 & 1 \\ 1 & 0 & 1 & 1 \\ 1 & 1 & 0 & 1 \end{bmatrix}$ and let C be the code whose words are linear combinations of the columns of G, that is, vectors of the form $G\mathsf{x}$, $\mathsf{x} = \begin{bmatrix} x_1 \\ x_2 \\ x_3 \\ x_4 \end{bmatrix}$. Since $G\mathsf{x} + G\mathsf{y} = G(\mathsf{x} + \mathsf{y})$, the sum of words is a word, so C is indeed a (linear) code. The sequence 1010101 is in C because $\begin{bmatrix} 1 \\ 0 \\ 1 \\ 0 \\ 1 \\ 0 \\ 1 \end{bmatrix} = G \begin{bmatrix} 1 \\ 0 \\ 1 \\ 0 \end{bmatrix}$ is the sum of the first and third columns of G. On the other hand, 0101011 is not in C because $G\begin{bmatrix} x_1 \\ x_2 \\ x_3 \\ x_4 \end{bmatrix} = \begin{bmatrix} x_1 \\ x_2 \\ x_3 \\ x_4 \\ x_2 + x_3 + x_4 \\ x_1 + x_3 + x_4 \\ x_1 + x_2 + x_4 \end{bmatrix}$.

If such a vector equals $\begin{bmatrix} 0 \\ 1 \\ 0 \\ 1 \\ 0 \\ 1 \\ 1 \end{bmatrix}$, then $x_1 = 0$, $x_2 = 1$, $x_3 = 0$, and $x_4 = 1$. Examining last components, we need $x_1 + x_2 + x_4 = 1$, which is not the case because $x_1 + x_2 + x_4 = 0$. (Remember that $1 + 1 = 0$.) ☺

In real communication, over wires or through space, reliability of transmission is a problem. Words get garbled, 0s become 1s and 1s are changed to 0s. How can one be sure that the message received was the message sent? Such concerns make the construction of codes that can detect and correct errors an important (and lucrative!) activity. A common approach in the theory of "error-correcting codes" is to append to each code word being sent another sequence of *bits* (that is, another sequence of one or more 0s and 1s), the purpose of which is to make errors obvious and, ideally, to make it possible to correct errors. Thus each transmitted word consists of a sequence of information bits, the message word itself, followed by a sequence of error correction bits.

Suppose the message words we wish to transmit each have length four. We might attach to each sequence of four 0s and 1s a fifth bit which is the sum of the first four (mod 2). Thus, if the message is 1010, we send 10100, the final 0 being the sum $1 + 0 + 1 + 0 \pmod 2$ of the bits in the message. If 11111 is received, an error is immediately detected because the sum of the first four digits is $1 + 1 + 1 + 1 = 0 \neq 1 \pmod 2$. This simple attaching of a "parity check" digit makes any error in a single digit obvious, but not correctable, because the recipient has no idea which of the first four digits is wrong.

Each word in the Hamming code illustrated in Example 12.5 actually consists of four information bits followed by **three** error correction bits. For example, the actual message in 1010101 is 1010, the first four bits. The final 101 permits not only to detect an error, but also to identify and hence correct it. We have essentially seen already how this works. A word of the Hamming code corresponds to a vector of the form

$$\begin{bmatrix} x_1 \\ x_2 \\ x_3 \\ x_4 \\ x_2 + x_3 + x_4 \\ x_1 + x_3 + x_4 \\ x_1 + x_2 + x_4 \end{bmatrix},$$

so, if the fifth bit of the transmitted sequence is not the sum of the second, third, and fourth, or if the sixth bit is not the sum of the first, third, and fourth, or if the last transmitted bit is not the sum of the first, second, and fourth, the received word is not correct. Algebraically, $\begin{bmatrix} x_1 \\ x_2 \\ x_3 \\ x_4 \\ x_5 \\ x_6 \\ x_7 \end{bmatrix}$ is in C if and only if

$$\begin{aligned} x_5 &= x_2 + x_3 + x_4 \\ x_6 &= x_1 + x_3 + x_4 \\ x_7 &= x_1 + x_2 + x_4, \end{aligned} \tag{34}$$

equations which (using the fact that $-1 = 1 \pmod 2$) can be written

$$\begin{aligned} x_2 + x_3 + x_4 + x_5 &= 0 \\ x_1 + x_3 + x_4 + x_6 &= 0 \\ x_1 + x_2 + x_4 + x_7 &= 0. \end{aligned} \tag{35}$$

Let $H = \begin{bmatrix} 0 & 1 & 1 & 1 & 1 & 0 & 0 \\ 1 & 0 & 1 & 1 & 0 & 1 & 0 \\ 1 & 1 & 0 & 1 & 0 & 0 & 1 \end{bmatrix}$ and notice that $H \begin{bmatrix} x_1 \\ x_2 \\ x_3 \\ x_4 \\ x_5 \\ x_6 \\ x_7 \end{bmatrix} = \begin{bmatrix} x_2 + x_3 + x_4 + x_5 \\ x_1 + x_3 + x_4 + x_6 \\ x_1 + x_2 + x_4 + x_7 \end{bmatrix}$. So a vector $\mathsf{x} = \begin{bmatrix} x_1 \\ x_2 \\ x_3 \\ x_4 \\ x_5 \\ x_6 \\ x_7 \end{bmatrix}$ corresponds to a code word if and only if $H\mathsf{x} = \mathsf{0}$. The equations in (35) are called *parity check equations* and the matrix H is called a *parity check matrix.*

12.6 Definition. If C is a code with $n \times k$ generator matrix G, a *parity check matrix* for C is an $(n - k) \times n$ matrix H with the property that x is in C if and only if $H\mathsf{x} = \mathsf{0}$.

For example, in the Hamming Code, the generator matrix G is 7×4 ($n = 7$, $k = 4$) and the parity check matrix H is 3×7 with the property that x is a code word if and only if $H\mathsf{x} = \mathsf{0}$.

Neither parity check equations nor a parity check matrix are unique, and some are more useful than others. One can show that the parity check equations in (35) are *equivalent* to

$$\begin{aligned} x_4 + x_5 + x_6 + x_7 &= 0 \\ x_2 + x_3 + x_6 + x_7 &= 0 \\ x_1 + x_3 + x_5 + x_7 &= 0 \end{aligned} \tag{36}$$

in the sense that a vector $\mathsf{x} = \begin{bmatrix} x_1 \\ x_2 \\ x_3 \\ x_4 \\ x_5 \\ x_6 \\ x_7 \end{bmatrix}$ satisfies equations (35) if and only if it satisfies equations

(36). The parity check matrix corresponding to the system (36) is

$$H = \begin{bmatrix} 0 & 0 & 0 & 1 & 1 & 1 & 1 \\ 0 & 1 & 1 & 0 & 0 & 1 & 1 \\ 1 & 0 & 1 & 0 & 1 & 0 & 1 \end{bmatrix}. \tag{37}$$

(Notice that the columns of H are, in order, just the numbers $1, 2, 3, 4, 5, 6, 7$ written in base 2.) Remember that if x is really a code word, then $H\mathsf{x} = 0$. Since $H\mathsf{x} = \begin{bmatrix} x_4 + x_5 + x_6 + x_7 \\ x_2 + x_3 + x_6 + x_7 \\ x_1 + x_3 + x_5 + x_7 \end{bmatrix}$, if x_1 is transmitted incorrectly (but all other bits are correct), then $H\mathsf{x} = \begin{bmatrix} 0 \\ 0 \\ 1 \end{bmatrix}$. Notice that the number 001 in base 2 is 1. If x_2 is in error (but all other bits are correct), then $H\mathsf{x} = \begin{bmatrix} 0 \\ 1 \\ 0 \end{bmatrix}$. The number 010 in base 2 is 2. If an error is made in x_3 (but all other bits are correct), then $H\mathsf{x} = \begin{bmatrix} 0 \\ 1 \\ 1 \end{bmatrix}$. The number 011 in base 2 is 3. Generally, if there is just one error in transmission, and that is in digit i, then $H\mathsf{x}$ is the number i in base 2. This particular parity check matrix not only tells us whether or not a transmitted message is correct, but it also allows us to make the correction. Clever? The $(7, 4)$ Hamming code is an example of a "single error correcting code."

Some Exercises to Test Yourself

3. Let C be the code with generator matrix $G = \begin{bmatrix} 1 & 0 & 0 \\ 0 & 1 & 0 \\ 0 & 0 & 1 \\ 1 & 0 & 1 \\ 1 & 1 & 1 \end{bmatrix}$. Determine whether or not 10011 is in C

4. Let G be the generator matrix of Exercise 3. Find a parity check matrix H, that is, a matrix H with the property that x is in the code generated by G if and only if $H\mathsf{x} = 0$.

Solutions to Test Yourself Exercises

1. This is a single circuit with no subcircuits. The total resistance is $R = 8\Omega$ and the total voltage is $E = 4$ V. So $I = \frac{E}{R} = \frac{4}{8} = \frac{1}{2}$ amps.

2. There are two nodes. At both A and B, Kirchoff's Node Law say $I_1 = I_2 + I_3$. There are two basic subcircuits (neither properly containing another circuit). Applied to the one on the left, Kirchoff's Circuit Law says $3I_1 + 6I_2 = 7$ and, to the one on the right,

$11I_3 - 6I_2 = 4$. So the currents are the solutions to the system

$$\begin{aligned} I_1 - I_2 - I_3 &= 0 \\ 3I_1 + 6I_2 &= 7 \\ -6I_2 + 11I_3 &= 4 \end{aligned}$$

Gaussian elimination applied to the augmented matrix of coefficients proceeds

$$\left[\begin{array}{ccc|c} 1 & -1 & -1 & 0 \\ 3 & 6 & 0 & 7 \\ 0 & -6 & 11 & 4 \end{array}\right] \to \left[\begin{array}{ccc|c} 1 & -1 & -1 & 0 \\ 0 & 9 & 3 & 7 \\ 0 & -6 & 11 & 4 \end{array}\right] \to \left[\begin{array}{ccc|c} 1 & -1 & -1 & 0 \\ 0 & 9 & 3 & 7 \\ 0 & 0 & 13 & \frac{26}{3} \end{array}\right]$$

so $I_3 = \frac{2}{3}$, $9I_2 = 7 - 3I_3 = 5$, so $I_2 = \frac{5}{9}$ and $I_1 = I_2 + I_3 = \frac{11}{9}$. These results are supported by checking Kirchoff's Circuit Law on the entire circuit where we should have $3I_1 + 11I_3 = 7 + 4 = 11$, and we do.

3. Note that $G\begin{bmatrix} x_1 \\ x_2 \\ x_3 \end{bmatrix} = \begin{bmatrix} x_1 \\ x_2 \\ x_3 \\ x_1 + x_3 \\ x_1 + x_2 + x_3 \end{bmatrix}$. So 10011 is in C because $x_1 = 1$ and $x_2 = x_3 = 0$, so $x_1 + x_3 = 1$ and $x_1 + x_2 + x_3 = 1$.

4. A word is in C if and only if its bits $x_1x_2x_3x_4x_5$ are of the form $G\begin{bmatrix} x_1 \\ x_2 \\ x_3 \\ x_4 \\ x_5 \end{bmatrix}$, that is, if and only if $x_4 = x_1 + x_3$ and $x_5 = x_1 + x_2 + x_3$. The parity check equations are

$$\begin{aligned} x_1 + x_3 + x_4 &= 0 \\ x_1 + x_2 + x_3 + x_5 &= 0, \end{aligned}$$

so $H = \begin{bmatrix} 1 & 0 & 1 & 1 & 0 \\ 1 & 1 & 1 & 0 & 1 \end{bmatrix}$.

Things I Must Remember

Over the years, students who have finished their linear algebra course have told me there were some important things they wished they had known when they started to use this book. There are indeed some recurring themes in the approach to linear algebra adopted here, and we suggest that all users of this text would find it helpful to consult this short note regularly. Points are raised in order of appearance in this book; page references point to the spot where the concept was first introduced.

1. Memorize definitions (the glossary in the back of these notes was put there to assist you with this). If you don't know what "similar" matrices are or what it means for a set of vectors to "span," or if you don't know that virtually any argument intended to establish the "linear independence" of vectors $\mathsf{u}, \mathsf{v}, \mathsf{w}, \ldots$ **must** begin "Suppose $a\mathsf{u} + b\mathsf{v} + c\mathsf{w} + \cdots = 0$," then you are in trouble!

2. Vectors in this book are always written as columns. Euclidean n-space, denoted R^n, is the set of all n-dimensional vectors, that is, the set of all vectors with n components. For example, $\begin{bmatrix} 1 \\ 7 \end{bmatrix}$ is a two-dimensional vector, that is, a vector in R^2 (also called *the plane*) and $\begin{bmatrix} -3 \\ 2 \\ 6 \end{bmatrix}$ is a three-dimensional vector, that is, a vector in R^3 (also called 3-*space*).

3. Since vectors are columns, vectors in R^n can also be regarded as matrices. For example, is $\begin{bmatrix} 1 \\ 2 \\ 3 \end{bmatrix}$ a vector or a 3×1 matrix? Often it doesn't matter; sometimes it's convenient to change points of view. The dot product of vectors of u and v in R^n is also the matrix product $\mathsf{u}^T\mathsf{v}$, thinking of u and v as $n \times 1$ matrices. For example, if $\mathsf{u} = \begin{bmatrix} 1 \\ 2 \\ 3 \end{bmatrix}$ and $\mathsf{v} = \begin{bmatrix} -5 \\ 4 \\ 8 \end{bmatrix}$, then

$$\mathsf{u} \cdot \mathsf{v} = 1(-5) + 2(4) + 3(8) = 27 = \begin{bmatrix} 1 & 2 & 3 \end{bmatrix} \begin{bmatrix} -5 \\ 4 \\ 8 \end{bmatrix} = \mathsf{u}^T\mathsf{v}.$$

   In particular, the square of the length of a vector u is $\|\mathsf{u}\|^2 = \mathsf{u} \cdot \mathsf{u} = \mathsf{u}^T\mathsf{u}$. Also, if u is a *unit vector* (that is, $\|\mathsf{u}\| = 1$), then the $n \times n$ matrix $Y = \mathsf{u}\mathsf{u}^T$ satisfies $Y^2 = Y$

because

$$Y^2 = (\mathsf{u}\mathsf{u}^T)(\mathsf{u}\mathsf{u}^T) = \mathsf{u}(\mathsf{u}^T\mathsf{u})\mathsf{u}^T = \mathsf{u}\mathsf{u}^T = Y$$

since $\mathsf{u}^T\mathsf{u} = \mathsf{u} \cdot \mathsf{u} = \|\mathsf{u}\|^2 = 1$.

4. When presented with the product AB of matrices, it is often useful to view B as a series of columns. If $B = \begin{bmatrix} \mathsf{b}_1 & \mathsf{b}_2 & \dots & \mathsf{b}_p \\ \downarrow & \downarrow & & \downarrow \end{bmatrix}$ has columns $\mathsf{b}_1, \mathsf{b}_2, \dots, \mathsf{b}_p$, the product AB is the matrix

$$AB = A\begin{bmatrix} \mathsf{b}_1 & \mathsf{b}_2 & \dots & \mathsf{b}_p \\ \downarrow & \downarrow & & \downarrow \\ & & & \end{bmatrix} = \begin{bmatrix} A\mathsf{b}_1 & A\mathsf{b}_2 & \dots & A\mathsf{b}_p \\ \downarrow & \downarrow & & \downarrow \\ & & & \end{bmatrix}$$

whose columns are $A\mathsf{b}_1, A\mathsf{b}_2, \dots, A\mathsf{b}_p$.

Suppose $A = \begin{bmatrix} 1 & -2 & 0 \\ -2 & 0 & 2 \\ 0 & 2 & -1 \end{bmatrix}$, $\mathsf{x} = \begin{bmatrix} 2 \\ 1 \\ 2 \end{bmatrix}$, $\mathsf{y} = \begin{bmatrix} 1 \\ 1 \\ 1 \end{bmatrix}$, and $\mathsf{z} = \begin{bmatrix} -2 \\ 2 \\ 1 \end{bmatrix}$. You should check that $A\mathsf{x} = \mathsf{0}$, $A\mathsf{y} = \begin{bmatrix} -1 \\ 0 \\ 1 \end{bmatrix}$, and $A\mathsf{z} = 3\mathsf{z}$. This means that if $B = \begin{bmatrix} \mathsf{x} & \mathsf{y} & \mathsf{z} \\ \downarrow & \downarrow & \downarrow \end{bmatrix} = \begin{bmatrix} 2 & 1 & -2 \\ 1 & 1 & 2 \\ 2 & 1 & 1 \end{bmatrix}$, then $AB = \begin{bmatrix} A\mathsf{x} & A\mathsf{y} & A\mathsf{z} \\ \downarrow & \downarrow & \downarrow \end{bmatrix} = \begin{bmatrix} 0 & -1 & -6 \\ 0 & 0 & 6 \\ 0 & 1 & 3 \end{bmatrix}$.

5. The product $A\mathsf{x}$ of an $m \times n$ matrix A and a vector x in R^n is a linear combination of the columns of A with coefficients the components of the vector. For example,

$$\begin{bmatrix} 1 & 4 & 7 \\ 2 & 5 & 8 \\ 3 & 6 & 9 \end{bmatrix}\begin{bmatrix} x_1 \\ x_2 \\ x_3 \end{bmatrix} = x_1\begin{bmatrix} 1 \\ 2 \\ 3 \end{bmatrix} + x_2\begin{bmatrix} 4 \\ 5 \\ 6 \end{bmatrix} + x_3\begin{bmatrix} 7 \\ 8 \\ 9 \end{bmatrix}.$$

So, if b is a vector in R^m, the linear system $A\mathsf{x} = \mathsf{b}$ has a solution if and only if b is a linear combination of the columns of A—that is, if and only if b is in the column space of A. The column space of a matrix is the set of all vectors $A\mathsf{x}$. *p. 66*

6. The inverse of the product of invertible matrices A and B is the product of the inverses **order reversed**: $(AB)^{-1} = B^{-1}A^{-1}$. Transpose works the same way: the transpose of the product of A and B is the product of the transposes **order reversed**: $(AB)^T = B^TA^T$.

7. Most students know a symmetric matrix when they see one—how about $\begin{bmatrix} 1 & 2 & -7 \\ 2 & -5 & 6 \\ -7 & 6 & 0 \end{bmatrix}$, for instance? On the other hand, knowing what a symmetric matrix "looks like" doesn't help much with exercises that involve the concept. Could you prove that AA^T is always symmetric, for any matrix A? Once again, we ask you to take seriously our request to **learn definitions**. A matrix X is *symmetric* if $X^T = X$. Thus, $X = AA^T$ is symmetric because $X^T = (AA^T)^T = (A^T)^TA^T = AA^T = X$.

8. A (square) matrix A is invertible if and only if $\det A \neq 0$. Thus, if $\det A \neq 0$, then $A\mathsf{x} = \mathsf{b}$ has a solution for any vector b, and a unique one, namely, $\mathsf{x} = A^{-1}\mathsf{b}$. In particular, the homogeneous system $A\mathsf{x} = \mathsf{0}$ has only the trivial solution, $\mathsf{x} = \mathsf{0}$. For example, the system $\begin{bmatrix} -1 & 2 \\ 3 & 4 \end{bmatrix} \begin{bmatrix} x \\ y \end{bmatrix} = \begin{bmatrix} 0 \\ 0 \end{bmatrix}$ has only the trivial solution $\begin{bmatrix} x \\ y \end{bmatrix} = \begin{bmatrix} 0 \\ 0 \end{bmatrix}$ because $\det \begin{bmatrix} -1 & 2 \\ 3 & 4 \end{bmatrix} = -10 \neq 0$.

Suggested Homework Exercises

1. If $B = (1,4)$ and $\overrightarrow{AB} = \begin{bmatrix} -1 \\ 2 \end{bmatrix}$, find A.

2. Find A, given $B = (1,4)$ and $\overrightarrow{AB} = \begin{bmatrix} -1 \\ 2 \end{bmatrix}$

3. Given $A = (1,-3)$ and $B = (-3,2)$, find the vector $\overrightarrow{AB}$ and illustrate with a picture.

4. Given $\mathsf{u} = \begin{bmatrix} -4 \\ 2 \end{bmatrix}$ and $\mathsf{v} = \begin{bmatrix} 1 \\ -4 \end{bmatrix}$, find $\mathsf{u} - \mathsf{v}$ and illustrate with a picture.

5. If possible, express $\mathsf{x} = \begin{bmatrix} 6 \\ -2 \\ 8 \end{bmatrix}$ as a scalar multiple of $\mathsf{u} = \begin{bmatrix} 4 \\ -\frac{4}{3} \\ \frac{16}{3} \end{bmatrix}$. Repeat using first $\mathsf{v} = \begin{bmatrix} \frac{1}{2} \\ -\frac{1}{6} \\ \frac{4}{3} \end{bmatrix}$ and $\mathsf{w} = \begin{bmatrix} -3 \\ 1 \\ -4 \end{bmatrix}$ and then $\mathsf{y} = \begin{bmatrix} 15 \\ -5 \\ 20 \end{bmatrix}$ instead of u. Justify your answers.

6. Let $\mathsf{u} = \begin{bmatrix} 1 \\ 0 \\ -1 \end{bmatrix}$ and $\mathsf{v} = \begin{bmatrix} 0 \\ 0 \\ 0 \end{bmatrix}$. Is either of these vectors a scalar multiple of the other? Explain.

7. Express each of the following as a single vector.

   (a) $3\begin{bmatrix} 1 \\ 0 \\ -2 \end{bmatrix} - 4\begin{bmatrix} 6 \\ 1 \\ 5 \end{bmatrix} + 2\begin{bmatrix} -1 \\ 1 \\ 2 \end{bmatrix}$

   (b) $a\begin{bmatrix} -1 \\ 1 \\ 7 \end{bmatrix} + 2\begin{bmatrix} a \\ a \\ 0 \end{bmatrix} - 4\begin{bmatrix} 0 \\ -1 \\ a \end{bmatrix}$

   (c) $a\begin{bmatrix} -1 \\ 5 \end{bmatrix} - 3\begin{bmatrix} -a \\ 2 \end{bmatrix}$

   (d) $2\begin{bmatrix} -1 \\ 0 \\ 1 \end{bmatrix} + 5\begin{bmatrix} -3 \\ 2 \\ -1 \end{bmatrix} + 3\begin{bmatrix} 3 \\ -1 \\ 3 \end{bmatrix}$

   (e) $3\begin{bmatrix} 2 \\ 1 \\ 3 \end{bmatrix} - 2\begin{bmatrix} 1 \\ 0 \\ -5 \end{bmatrix} - 4\begin{bmatrix} 0 \\ -1 \\ 2 \end{bmatrix}$

   (f) $a\begin{bmatrix} -1 \\ 0 \\ 1 \end{bmatrix} + b\begin{bmatrix} -3 \\ 2 \\ -1 \end{bmatrix} + c\begin{bmatrix} 3 \\ -1 \\ 3 \end{bmatrix}$

8. Vectors are defined to be *parallel* if one is a scalar multiple of the other. Is this the same as "the vectors are scalar multiples of each other?" Explain.

9. Which of the following vectors are parallel to other vectors in the list? In the case of parallel vectors state whether the vectors have the same or the opposite direction, if possible. Justify your answers.

$$\mathsf{u}_1 = \begin{bmatrix} -4 \\ 2 \\ 2 \end{bmatrix}, \ \mathsf{u}_2 = \begin{bmatrix} 3 \\ -6 \\ 12 \end{bmatrix}, \ \mathsf{u}_3 = \begin{bmatrix} -6 \\ 3 \\ 3 \end{bmatrix}, \ \mathsf{u}_4 = \begin{bmatrix} 0 \\ 0 \\ 0 \end{bmatrix}, \ \mathsf{u}_5 = \begin{bmatrix} -5 \\ 10 \\ -20 \end{bmatrix}.$$

10. Let $\mathsf{u} = \begin{bmatrix} 1 \\ 0 \\ -1 \end{bmatrix}$ and $\mathsf{v} = \begin{bmatrix} 0 \\ 0 \\ 0 \end{bmatrix}$. Determine whether or not each vector is a scalar multiple of the other. Are these vectors parallel? Explain.

11. Show that if some nontrivial linear combination of vectors u and v is 0, then u and v are parallel.

12. Express $\begin{bmatrix} 7 \\ 7 \end{bmatrix}$ as a linear combination of $\begin{bmatrix} -1 \\ 1 \end{bmatrix}$ and $\begin{bmatrix} 5 \\ 2 \end{bmatrix}$;

13. (a) Is it possible to express $\begin{bmatrix} 1 \\ 2 \end{bmatrix}$ as a linear combination of $\begin{bmatrix} -2 \\ -2 \end{bmatrix}$ and $\begin{bmatrix} 3 \\ 3 \end{bmatrix}$? Explain.

    (b) Does your answer to part (a) contradict 1.21? Explain.

14. Is it possible to express the zero vector as a linear combination of $\begin{bmatrix} 1 \\ 4 \end{bmatrix}$ and $\begin{bmatrix} 2 \\ -3 \end{bmatrix}$ in more than one way? Justify your answer.

15. In each of the following cases, either express p as a linear combination of u, v, w or explain why there is no such linear combination.

    (a) $\mathsf{p} = \begin{bmatrix} -4 \\ 7 \\ 5 \end{bmatrix}, \mathsf{u} = \mathsf{i} = \begin{bmatrix} 1 \\ 0 \\ 0 \end{bmatrix}, \mathsf{v} = \mathsf{j} = \begin{bmatrix} 0 \\ 1 \\ 0 \end{bmatrix}, \mathsf{w} = \mathsf{k} = \begin{bmatrix} 0 \\ 0 \\ 1 \end{bmatrix}.$

    (b) $\mathsf{p} = \begin{bmatrix} -1 \\ 2 \\ 4 \\ 0 \end{bmatrix}, \mathsf{u} = \begin{bmatrix} 3 \\ 7 \\ 0 \\ -4 \end{bmatrix}, \mathsf{v} = \begin{bmatrix} 0 \\ 2 \\ 0 \\ 9 \end{bmatrix}, \mathsf{w} = \begin{bmatrix} 3 \\ 1 \\ 0 \\ 5 \end{bmatrix}.$

    (c) $\mathsf{p} = \begin{bmatrix} 4 \\ 6 \\ 8 \\ 10 \\ 2 \end{bmatrix}, \mathsf{u} = \begin{bmatrix} 3 \\ 4 \\ 5 \\ 1 \\ 2 \end{bmatrix}, \mathsf{v} = \begin{bmatrix} 2 \\ 3 \\ 4 \\ 5 \\ 1 \end{bmatrix}, \mathsf{w} = \begin{bmatrix} -1 \\ -2 \\ -3 \\ -4 \\ -5 \end{bmatrix}.$

16. Find a and b given $a\begin{bmatrix} 1 \\ -1 \\ 1 \\ 2 \end{bmatrix} + b\begin{bmatrix} -1 \\ 1 \\ 2 \\ 3 \end{bmatrix} = \begin{bmatrix} -3 \\ a+1 \\ 12 \\ 4b-1 \end{bmatrix}.$

17. Suppose c is a nonzero scalar and x is a nonzero vector in R^n. Explain why $c\mathsf{x} \neq 0$. item Given that u, v, w are vectors with x = 2u + 3v − w, y = u − v and z = v + w, express u, v and w in terms of x, y and z.

18. Show that any linear combination of $\begin{bmatrix} 1 \\ \frac{3}{2} \\ 0 \end{bmatrix}$ and $\begin{bmatrix} 0 \\ 3 \\ 6 \end{bmatrix}$ is also a linear combination of $\begin{bmatrix} 2 \\ 3 \\ 0 \end{bmatrix}$ and $\begin{bmatrix} 0 \\ 1 \\ 2 \end{bmatrix}$.

19. Suppose u and v are vectors. Show that any linear combination of u and v is a linear combination of 2u and −3v.

20. Let u, v, and w be vectors. Show that any linear combination of u and v is also a linear combination of u, v and w.

21. Let $O(0,0)$, $A(2,2)$, $C(2,-1)$, and $H(8,2)$ be the indicated points in the xy-plane. Let $\mathsf{u} = \overrightarrow{OA}$ be the vector pictured by the arrow from O to A and $\mathsf{v} = \overrightarrow{OC}$ the vector pictured by the arrow from O to C.

    (a) Find u and v.

    (b) Express $\overrightarrow{OH}$, $\overrightarrow{HC}$ and $\overrightarrow{CA}$ as linear combinations of u and v.
    For example, $\overrightarrow{OE} = \begin{bmatrix} 6 \\ 3 \end{bmatrix} = 2\mathsf{u} + \mathsf{v}$.

22. Given three points $A(-1,0)$, $B(2,3)$, $C(4,-1)$ in the plane, find all points D such that the four points A, B, C, D are the vertices of a parallelogram.

23. Shown to the right are two nonparallel vectors u and v and four other vectors $\mathsf{w}_1, \mathsf{w}_2, \mathsf{w}_3, \mathsf{w}_4$. Reproduce u, v and w_2 in a picture by themselves and exhibit w_2 as the diagonal of a parallelogram with sides parallel to u and v. Guess values of a and b so that $\mathsf{w}_2 = a\mathsf{u} + b\mathsf{v}$.

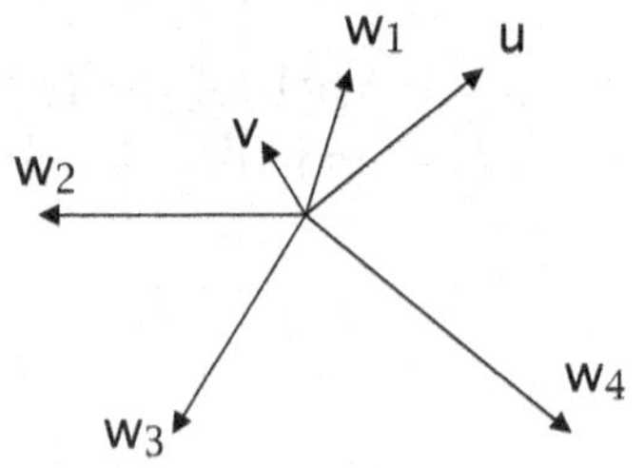

24. Given $P(2,-3,6)$ and $Q(2,2,-4)$, find a point R on the line segment from P to Q that is two fifths of the way from P to Q.

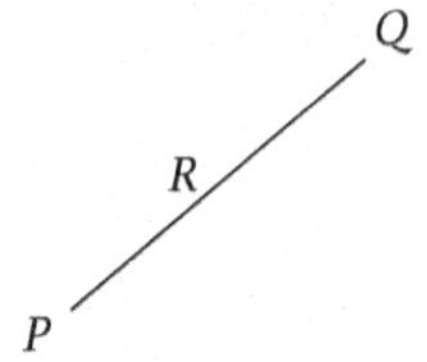

25. Suppose ABC is a triangle. Let D be the midpoint of AB and E be the midpoint of AC. Use vectors to show that DE is parallel to BC and one half its length.

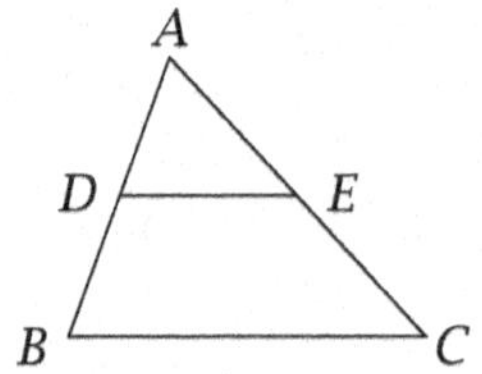

26. Use vectors to show that the mid-point of the line joining $A(x_1, y_1, z_1)$ to $B(x_2, y_2, z_2)$ is the point $C(\frac{x_1+x_2}{2}, \frac{y_1+y_2}{2}, \frac{z_1+z_2}{2})$.

27. Given the points $A(-1,0)$ and $B(2,3)$ in the plane, find all points C such that A, B, and C are the vertices of a right angle triangle with right angle at A and AC of length 2.

28. We have observed that every vector in R^n is a linear combination of the standard basis vectors $\mathsf{e}_1, \mathsf{e}_2, \ldots, \mathsf{e}_n$. Show that the coefficients used in such a linear combination are unique; that is, with if a vector x is written $\mathsf{x} = x_1\mathsf{e}_1 + x_2\mathsf{e}_2 + \cdots + x_n\mathsf{e}_n$ and also $\mathsf{x} = y_1\mathsf{e}_1 + y_2\mathsf{e}_2 + \cdots + y_n\mathsf{e}_n$, then $x_1 = y_1$, $x_2 = y_2$, and so on.

29. Determine whether each of the given sets of vectors is linearly independent or linearly dependent. In the case of linear dependence, write down a specific nontrivial linear combination of the vectors which equals the zero vector.

(a) $v_1 = \begin{bmatrix} -3 \\ 5 \end{bmatrix}$, $v_2 = \begin{bmatrix} 7 \\ 9 \end{bmatrix}$

(b) $v_1 = \begin{bmatrix} 1 \\ 0 \\ 0 \end{bmatrix}$, $v_2 = \begin{bmatrix} -2 \\ 1 \\ 2 \end{bmatrix}$, $v_3 = \begin{bmatrix} 3 \\ 2 \\ 4 \end{bmatrix}$

(c) $v_1 = \begin{bmatrix} 1 \\ 0 \\ 0 \end{bmatrix}$, $v_2 = \begin{bmatrix} -2 \\ 1 \\ 2 \end{bmatrix}$, $v_3 = \begin{bmatrix} 3 \\ 2 \\ 4 \end{bmatrix}$, $v_4 = \begin{bmatrix} -3 \\ 4 \\ 1 \end{bmatrix}$

(d) $v_1 = \begin{bmatrix} 1 \\ 0 \\ 0 \end{bmatrix}$, $v_2 = \begin{bmatrix} -2 \\ 1 \\ 2 \end{bmatrix}$, $v_3 = \begin{bmatrix} 3 \\ 2 \\ 5 \end{bmatrix}$

(e) $v_1 = \begin{bmatrix} -1 \\ 2 \end{bmatrix}$, $v_2 = \begin{bmatrix} -\frac{3}{2} \\ 3 \end{bmatrix}$

(f) $v_1 = \begin{bmatrix} -1 \\ 2 \\ 3 \end{bmatrix}$, $v_2 = \begin{bmatrix} 4 \\ 0 \\ 1 \end{bmatrix}$, $v_3 = \begin{bmatrix} 0 \\ 8 \\ 13 \end{bmatrix}$

(g) $v_1 = \begin{bmatrix} 1 \\ 0 \\ 1 \\ 0 \end{bmatrix}$, $v_2 = \begin{bmatrix} 2 \\ 1 \\ -1 \\ 3 \end{bmatrix}$, $v_3 = \begin{bmatrix} 1 \\ 3 \\ 2 \\ -1 \end{bmatrix}$

(h) $v_1 = \begin{bmatrix} 1 \\ 0 \\ 0 \\ 0 \end{bmatrix}$, $v_2 = \begin{bmatrix} 1 \\ 1 \\ 0 \\ 0 \end{bmatrix}$, $v_3 = \begin{bmatrix} 1 \\ 1 \\ 1 \\ 0 \end{bmatrix}$, and $v_4 = \begin{bmatrix} 1 \\ 1 \\ 1 \\ 1 \end{bmatrix}$

(i) $u = \begin{bmatrix} -2 \\ 3 \\ 1 \end{bmatrix}$, $v = \begin{bmatrix} -1 \\ 3 \\ -3 \end{bmatrix}$, $w = \begin{bmatrix} 1 \\ 0 \\ -4 \end{bmatrix}$

30. Suppose vectors u and v are linearly independent. Show that the vectors $w = 2u + 3v$ and $z = u + v$ are also linearly independent.

31. Suppose $\{v_1, v_2, \ldots, v_k\}$ is any set of vectors and we add the zero vector to this set. Is the set $\{0, v_1, v_2, \ldots, v_k\}$ linearly independent? linearly dependent? can't say?

32. Let $u = \begin{bmatrix} 1 \\ -4 \\ 2 \end{bmatrix}$. Find a unit vector in the direction opposite u and a vector of length 3 with the same direction as u.

33. Let $v = \begin{bmatrix} -1 \\ 2 \\ 2 \end{bmatrix}$. Find a vector of length 2 with the same direction as v and a vector of length 6 with direction opposite to v.

34. Let $u = \begin{bmatrix} 4 \\ 7 \\ -4 \end{bmatrix}$ and $v = \begin{bmatrix} 2 \\ -6 \\ -1 \end{bmatrix}$. Find the following:

(a) a unit vector in the direction of v;

(b) a vector of length 5 in the direction of v;

(c) a vector of length 3 in the direction opposite to u.

35. Find all vectors of length 3 that are parallel to $\mathsf{v} = \begin{bmatrix} 1 \\ -2 \\ 1 \end{bmatrix}$.

36. Find a vector of length 7 that has direction opposite to $\mathsf{u} = \begin{bmatrix} 0 \\ -3 \\ 1 \\ 1 \\ 2 \end{bmatrix}$.

37. Let A and P be the points $A(1, -1, 2)$, $P(4, 0, -1)$. Describe (geometrically) the set of all points (x, y, z) in R^3 for which $\left\| \overrightarrow{AP} - \overrightarrow{AQ} \right\| = 4$.

38. If u is a vector in R^n that is orthogonal to every vector in R^n, then u must be the zero vector. Why?

39. Let $\mathsf{u} = \begin{bmatrix} 1 \\ 2 \end{bmatrix}$ and $\mathsf{v} = \begin{bmatrix} 1 \\ -1 \end{bmatrix}$. Find all numbers k such that $k\mathsf{u} - 3\mathsf{v}$ is orthogonal to $\begin{bmatrix} -4 \\ 5 \end{bmatrix}$.

40. Suppose u and v are vectors that are not orthogonal, k is a scalar, and $\mathsf{u} + k\mathsf{v}$ is orthogonal to u. Find k.

41. Find the angle between each of the given pairs of vectors in radians (to two decimal places) and in degrees (to the nearest degree).

(a) $\mathsf{u} = \begin{bmatrix} 1 \\ 3 \\ 2 \end{bmatrix}$ and $\mathsf{v} = \begin{bmatrix} -4 \\ -1 \\ 1 \end{bmatrix}$

(b) $\mathsf{u} = \begin{bmatrix} 0 \\ 3 \\ 1 \\ -2 \end{bmatrix}$ and $\mathsf{v} = \begin{bmatrix} -3 \\ 6 \\ 1 \\ 1 \end{bmatrix}$

(c) $\mathsf{u} = \begin{bmatrix} 1 \\ 2 \\ 1 \\ 0 \\ 3 \end{bmatrix}$ and $\mathsf{v} = \begin{bmatrix} 1 \\ -1 \\ 1 \\ 1 \\ 1 \end{bmatrix}$

42. Let $\mathsf{u} = \begin{bmatrix} 1 \\ 1 \\ 0 \end{bmatrix}$ and $\mathsf{v} = \begin{bmatrix} x \\ 1 \\ 1 \end{bmatrix}$. Find all numbers x (if any) such that the angle between u and v is 60°.

43. Can $\mathsf{u} \cdot \mathsf{v} = -7$ if $\|\mathsf{u}\| = 3$ and $\|\mathsf{v}\| = 2$? Justify your answer.

44. Find $\mathsf{u} \cdot \mathsf{v}$ given $\|\mathsf{u}\| = 4$, $\|\mathsf{v}\| = 7$ and the angle between u and v is 2 radians.

45. Let $\mathsf{u} = \begin{bmatrix} -1 \\ 2 \\ 1 \end{bmatrix}$ and $\mathsf{v} = \begin{bmatrix} 2 \\ -3 \\ -5 \end{bmatrix}$.

(a) Find all numbers k, if any, so that $\|k\mathsf{u} - \mathsf{v}\| = 4$. Give k accurate to two decimal places.

(b) Find all numbers k, if any, so that $k\mathsf{u} + 2\mathsf{v}$ is orthogonal to u.

(c) Find (approximately) the angle between the vectors (Give your answer in radians to two decimal places and in degrees to the nearest degree.)

46. Given the points $A(0, -2)$ and $B(1, 1)$, find all points C so that A, B and C are the vertices of a right angled triangle with right angle at B and hypotenuse of length 10.

47. Let ABC be a triangle inscribed in a semicircle as shown. Use vectors to show that the angle at B is a right angle.

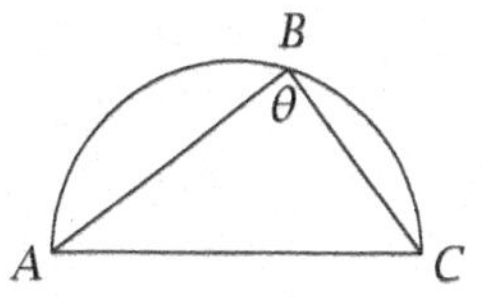

48. Let $A(1, 2)$, $B(-3, -1)$ and $C(4, -2)$ be three points in the Euclidean plane. Find a fourth point D such that the A, B, C and D are the vertices of a square **and justify your answer**.

49. Let u and v be vectors of lengths 3 and 5 respectively. Suppose also that $\mathsf{u} \cdot \mathsf{v} = 8$. Find $(-3\mathsf{u} + 4\mathsf{v}) \cdot (2\mathsf{u} + 5\mathsf{v})$.

50. Let u, v, and w be vectors of lengths 3 2, and 6, respectively. Suppose $\mathsf{u} \cdot \mathsf{v} = 3$, $\mathsf{u} \cdot \mathsf{w} = 5$ and $\mathsf{v} \cdot \mathsf{w} = 2$. Find

(a) $(8\mathsf{w} - 12\mathsf{v}) \cdot (37\mathsf{v} + 24\mathsf{w})$ and

(b) $\|\mathsf{u} - \mathsf{v} - \mathsf{w}\|^2$.

51. Let $\mathsf{u} = \begin{bmatrix} 3 \\ 4 \\ -2 \end{bmatrix}$, $\mathsf{w} = \begin{bmatrix} -2 \\ 5 \\ 7 \end{bmatrix}$ and $\mathsf{v} = \begin{bmatrix} 4 \\ -2 \\ 2 \end{bmatrix}$.

(a) Show that u is orthogonal to $\mathsf{v} - \mathsf{w}$.

(b) Show that u is orthogonal to $a\mathsf{v} + b\mathsf{w}$ for any scalars a and b.

52. Let u, v, and w be vectors of lengths 3, 2, and 6, respectively. Suppose $\mathsf{u} \cdot \mathsf{v} = 3$, $\mathsf{u} \cdot \mathsf{w} = 5$ and $\mathsf{v} \cdot \mathsf{w} = 2$. Find $(3\mathsf{w} - 2\mathsf{v}) \cdot (5\mathsf{v} - \mathsf{w})$.

53. Let $\mathsf{u} = \begin{bmatrix} 1 \\ 0 \\ -1 \end{bmatrix}$ and $\mathsf{v} = \begin{bmatrix} 2 \\ 1 \\ 0 \end{bmatrix}$. Find a real number k so that $\mathsf{u} + k\mathsf{v}$ is orthogonal to u or explain why no such k exists.

54. Let $\mathsf{u} = \begin{bmatrix} -1 \\ a \\ b \\ 2 \\ 0 \end{bmatrix}$, $\mathsf{v} = \begin{bmatrix} c \\ 1 \\ 1 \\ a \\ -4 \end{bmatrix}$ and $\mathsf{x} = \begin{bmatrix} 1 \\ 2 \\ 3 \\ 4 \\ 5 \end{bmatrix}$. Find a, b and c, if possible, in each of the following situations.

(a) u and x are orthogonal; (b) $\|\mathsf{v}\| = \sqrt{c^2 + 27}$.

55. Use the the important formula $u \cdot v = \|u\| \|v\| \cos\theta$ to derive the *Law of Cosines*, that goes as follows: If $\triangle ABC$ has sides of lengths a, b, c as shown, and $\theta = \angle BCA$, then $c^2 = a^2 + b^2 - 2ab\cos\theta$.

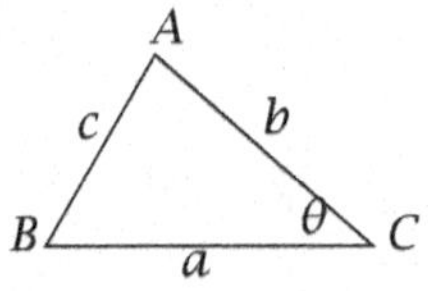

56. Prove that $\|u+v\|^2 + \|u-v\|^2 = 2\|u\|^2 + 2\|v\|^2$ and interpret this result geometrically when u and v lie in a plane.

57. In each case, verify the Cauchy-Schwarz and triangle inequalities.

(a) $u = \begin{bmatrix} -1 \\ 0 \\ 1 \\ 2 \end{bmatrix}$ and $v = \begin{bmatrix} 3 \\ -1 \\ 1 \\ 0 \end{bmatrix}$ (b) $u = \begin{bmatrix} 1 \\ 2 \\ 3 \\ 4 \\ 1 \end{bmatrix}$ and $v = \begin{bmatrix} 0 \\ 1 \\ 3 \\ -4 \\ 0 \end{bmatrix}$

58. (a) Use the Cauchy–Schwarz inequality to prove the generalized Theorem of Pythagoras, also known as the triangle inequality: For vectors u and v, $\|u+v\| \le \|u\| + \|v\|$.

(b) Verify the triangle inequality for $u = \begin{bmatrix} 1 \\ 1 \\ -2 \end{bmatrix}$ and $v = \begin{bmatrix} 1 \\ -3 \\ -2 \end{bmatrix}$.

59. Let a, b, c, d be real numbers. Use the Cauchy-Schwarz Inequality to prove that

$$(3ac + bd)^2 \le (3a^2 + b^2)(3c^2 + d^2).$$

60. Do there exist vectors u and v with $u \cdot v = -7$, $\|u\| = 3$ and $\|v\| = 2$? Explain.

61. Can $u \cdot v = 9$ if $\|u\| = 5$ and $\|v\| = 1$? Justify your answer.

62. Show that $\sqrt{\sin 2\theta} \le \frac{\sqrt{2}}{2}(\sin\theta + \cos\theta)$ for any θ, $0 \le \theta \le \frac{\pi}{2}$.

[*Hint.* Arithmetic-Geometric Mean inequality]

63. How many triangles have sides that have integer lengths and perimeter 15?

[*Hint.* First ask yourself what's the length of the longest side of such a triangle.]

64. Find a vector of length 5 which is perpendicular to both $\begin{bmatrix} 1 \\ 3 \\ -1 \end{bmatrix}$ and $\begin{bmatrix} 5 \\ 0 \\ 1 \end{bmatrix}$.

65. In each case, find the area of $\triangle ABC$.

(a) $A(-2,3)$, $B(4,4)$ and $C(5,-3)$.

(b) $A(1,2)$, $B(-3,0)$ and $C(2,-2)$.

66. Are there vectors u and v such that $\|u \times v\| = 15$, $\|u\| = 3$ and $\|v\| = 4$? Explain.

67. Let u and v be vectors. Verify that $\|u \times v\| = \|u\|\,\|v\| \sin\theta$ by the method suggested in the notes; that is, first show that $\|u \times v\|^2 + (u\cdot v)^2 = \|u\|^2\,\|v\|^2$.

68. (a) Find the equation of the plane that consists of all linear combinations of $u = \begin{bmatrix} 1 \\ 0 \\ 4 \end{bmatrix}$ and $v = \begin{bmatrix} -1 \\ 2 \\ 0 \end{bmatrix}$.

    (b) Find the equation of the plane through $(1, 2, 3)$ which is parallel to the plane in part (a).

69. (a) Find the equation of the plane consisting of all linear combinations of $\begin{bmatrix} 1 \\ -3 \\ 0 \end{bmatrix}$ and $\begin{bmatrix} 1 \\ 2 \\ 6 \end{bmatrix}$.

    (b) Find the equation of the plane parallel to the plane found in part (a) and which passes through the point $(-1, 1, 7)$.

70. Let $u = \begin{bmatrix} -3 \\ -1 \\ 1 \end{bmatrix}$ and $v = \begin{bmatrix} 2 \\ -6 \\ -2 \end{bmatrix}$ and let a and b denote arbitrary scalars. Find $au + bv$. Then find the equation of the plane that consists of all such linear combinations of u and v.

71. Find the equation of the plane each point of which is equidistant from the points $P(2, -1, 3)$ and $Q(1, 1, -1)$.

72. Let $u = \begin{bmatrix} -1 \\ 2 \\ 3 \end{bmatrix}$ and $v = \begin{bmatrix} 3 \\ 4 \\ 0 \end{bmatrix}$.

    (a) These vectors are not parallel. Why not?

    (b) Describe the vectors in the plane spanned by u and v.

    (c) Is $w = \begin{bmatrix} -5 \\ 0 \\ 6 \end{bmatrix}$ in this plane?

73. Is $w = \begin{bmatrix} 1 \\ 1 \\ 1 \end{bmatrix}$ in the plane spanned by $u = \begin{bmatrix} -1 \\ 0 \\ 3 \end{bmatrix}$ and $v = \begin{bmatrix} 3 \\ -1 \\ 0 \end{bmatrix}$? Explain.

74. Determine whether or not $u = \begin{bmatrix} 1 \\ 2 \\ 3 \end{bmatrix}$ is in the plane spanned by $v = \begin{bmatrix} 4 \\ 5 \\ 6 \end{bmatrix}$ and $w = \begin{bmatrix} 7 \\ 8 \\ 0 \end{bmatrix}$.

75. Find an equation of the plane spanned by $u = \begin{bmatrix} -1 \\ 2 \\ 3 \end{bmatrix}$ and $v = \begin{bmatrix} 2 \\ 0 \\ 5 \end{bmatrix}$.

76. Find the equation of the plane parallel to the plane with equation $18x+6y-5z=0$ and passing through the point $(-1,1,7)$.

77. Find the equation of the plane passing through $A(3,2,0)$, $B(-1,0,4)$ and $C(0,-3,2)$.

78. (a) Suppose u, v and w are three vectors and w = u + v. Show that every linear combination of u, v, w is actually just a linear combination of u and v.

    (b) Let $\mathsf{u} = \begin{bmatrix} 1 \\ 0 \\ -1 \end{bmatrix}$, $\mathsf{v} = \begin{bmatrix} 2 \\ 1 \\ 0 \end{bmatrix}$ and $\mathsf{w} = \begin{bmatrix} 3 \\ 1 \\ -1 \end{bmatrix}$. Show that the set of all linear combinations of these vectors is a plane and find the equation of this plane. [Hint: w = u + v.]

79. Find the equation of the plane which passes through $A(2,1,3)$, $B(3,-1,5)$ and $C(0,2,-4)$.

80. Find the projection of u on v and the projection of v on u, given $\mathsf{u} = \begin{bmatrix} 1 \\ 2 \\ 3 \end{bmatrix}$ and $\mathsf{v} = \begin{bmatrix} 4 \\ 5 \\ 6 \end{bmatrix}$.

81. Given $P(1,-1,1)$ and a plane π with equation $2x-y+z=2$.

    (a) find the distance from point P to the plane π;

    (b) find the point of π closest to P.

82. (a) Find the distance from $P(1,1,1)$ to the plane π with equation $x-3y+4z=10$.

    (b) Find the point of π which is closest to P.

83. Think about the procedure we have outlined for finding the distance from a point P to a plane π. Suppose two students choose different points Q_1, Q_2 in π. Explain why they will (should!) get the same answer for the distance.

84. Describe in geometrical terms the set of all points one unit away from the plane with equation $2x+y+2z=5$. Find an equation for this set of points.

85. In each case, find the distance from the point to the line.

    (a) from $P(-1,2,1)$ to the line ℓ with vector equation $\begin{bmatrix} x \\ y \\ z \end{bmatrix} = \begin{bmatrix} 1 \\ 2 \\ 3 \end{bmatrix} + t\begin{bmatrix} 1 \\ 1 \\ 1 \end{bmatrix}$

    (b) from the point $P(4,7,3)$ to the line with equation $\begin{bmatrix} x \\ y \\ z \end{bmatrix} = \begin{bmatrix} 3 \\ 1 \\ 1 \end{bmatrix} + t\begin{bmatrix} 3 \\ -2 \\ 3 \end{bmatrix}$

86. Let ℓ be the line with equation $\begin{bmatrix} x \\ y \\ z \end{bmatrix} = \begin{bmatrix} -1 \\ 1 \\ 1 \end{bmatrix} + t\begin{bmatrix} 5 \\ 7 \\ 4 \end{bmatrix}$ and let π be the plane with equation $2x-2y+z+5=0$.

(a) Explain why ℓ and π are parallel.

(b) Find the (shortest) distance from ℓ to π.

87. Do there exist vectors u and v such that $\|\mathsf{u} \times \mathsf{v}\| = 15$, $\|\mathsf{u}\| = 3$ and $\|\mathsf{v}\| = 4$? Explain.

88. Given $\mathsf{u} = \begin{bmatrix} 1 \\ 2 \\ 3 \end{bmatrix}$, $\mathsf{v} = \begin{bmatrix} -2 \\ 3 \\ 0 \end{bmatrix}$ and $\mathsf{w} = \begin{bmatrix} -1 \\ 1 \\ 1 \end{bmatrix}$, find $(\mathsf{u} \times \mathsf{v}) \times \mathsf{w}$ and $\mathsf{u} \times (\mathsf{v} \times \mathsf{w})$. Should the answers be the same?

89. Given $\mathsf{u} = \begin{bmatrix} 1 \\ 2 \\ 3 \end{bmatrix}$ and $\mathsf{v} = \begin{bmatrix} 0 \\ -1 \\ 1 \end{bmatrix}$

i. Verify that $\|\mathsf{u} \times \mathsf{v}\|^2 + (\mathsf{u} \cdot \mathsf{v})^2 = \|\mathsf{u}\|^2 \, \|\mathsf{v}\|^2$.

ii. Find the sine and the cosine of the angle between u and v.

90. Find two vectors which are perpendicular to both $\begin{bmatrix} 1 \\ 2 \\ 0 \end{bmatrix}$ and $\begin{bmatrix} 0 \\ 0 \\ 3 \end{bmatrix}$.

91. Find the equation of the plane which contains the points $A(1, 0, -1)$, $B(0, 2, 3)$ and $C(-2, 1, 1)$.

(a) Find the area of triangle ABC, with A, B, C as in part (a).

(b) Find a point which is one unit away from the plane in part (a).

92. (a) Find an equation of the plane passing through $A(1, 2, 3)$, $B(-1, 5, 2)$, $C(3, -1, 1)$.

(b) Find a point one unit away from the plane in part (a).

(c) What's the area of the triangle with vertices $A(1, 2, 3)$, $B(-1, 5, 2)$, $C(3, -1, 1)$?

93. In each case, find the equation of the plane containing P and the given line.

(a) $P(3, 1, -1)$; the line has equation $\begin{bmatrix} x \\ y \\ z \end{bmatrix} = \begin{bmatrix} 1 \\ 0 \\ 2 \end{bmatrix} + t \begin{bmatrix} -4 \\ 3 \\ 1 \end{bmatrix}$

(b) $P(5, 1, 2)$ and the line has equation $\begin{bmatrix} x \\ y \\ z \end{bmatrix} = \begin{bmatrix} 1 \\ 4 \\ -2 \end{bmatrix} + t \begin{bmatrix} 1 \\ 0 \\ 1 \end{bmatrix}$.

94. Find the equation of the plane π which contains $P(5, 4, 8)$ and the line with equation $\begin{bmatrix} x \\ y \\ z \end{bmatrix} = \begin{bmatrix} 9 \\ 3 \\ 5 \end{bmatrix} + t \begin{bmatrix} 4 \\ 5 \\ -3 \end{bmatrix}$.

95. Find an equation of the plane spanned by $\mathsf{u} = \begin{bmatrix} -3 \\ -1 \\ 1 \end{bmatrix}$ and $\mathsf{v} = \begin{bmatrix} 2 \\ -6 \\ -2 \end{bmatrix}$. (Note that this plane contains the origin.)

96. Find the equation of the line of intersection of the planes with equations $x + 2y + 6z = 5$ and $x - y - 3z = -1$.

97. Let ℓ_1 and ℓ_2 be the lines with equations

$$\ell_1: \begin{bmatrix} x \\ y \\ z \end{bmatrix} = \begin{bmatrix} 3 \\ 0 \\ -1 \end{bmatrix} + t \begin{bmatrix} 4 \\ 3 \\ 1 \end{bmatrix}, \qquad \ell_2: \begin{bmatrix} x \\ y \\ z \end{bmatrix} = \begin{bmatrix} 2 \\ -1 \\ -3 \end{bmatrix} + t \begin{bmatrix} 2 \\ 6 \\ 7 \end{bmatrix}.$$

Show that ℓ_1 and ℓ_2 are not parallel and that they do not intersect. (Such lines are called *skew*).

98. Let ℓ be the line with equation $\begin{bmatrix} x \\ y \\ z \end{bmatrix} = \begin{bmatrix} 3 \\ 5 \\ -4 \end{bmatrix} + t \begin{bmatrix} 2 \\ 1 \\ 3 \end{bmatrix}$ and let π be the plane with equation $-2x + y + z = 3$.

   (a) Explain why ℓ is parallel to π.

   (b) Find a point P on ℓ and a point Q on π.

   (c) Find the (shortest) distance from ℓ to π.

99. Are the lines with vector equations

$$\begin{bmatrix} x \\ y \\ z \end{bmatrix} = \begin{bmatrix} -1 \\ 4 \\ 4 \end{bmatrix} + t \begin{bmatrix} -1 \\ 5 \\ 2 \end{bmatrix} \quad \text{and} \quad \begin{bmatrix} x \\ y \\ z \end{bmatrix} = \begin{bmatrix} 1 \\ -6 \\ 0 \end{bmatrix} + t \begin{bmatrix} 2 \\ -10 \\ -4 \end{bmatrix}$$

are the same?

100. Let ℓ_1 be the line with vector equation $\begin{bmatrix} x \\ y \\ z \end{bmatrix} = \begin{bmatrix} 2 \\ -3 \\ -4 \end{bmatrix} + t \begin{bmatrix} 1 \\ 2 \\ 3 \end{bmatrix}$ and let ℓ_2 be the line through $P(2, -2, 5)$ in the direction $\mathsf{d} = \begin{bmatrix} 2 \\ 3 \\ -3 \end{bmatrix}$.

   (a) Show that ℓ_1 and ℓ_2 intersect by finding the point of intersection.

   (b) Give an easy reason why ℓ_1 and the plane with equation $3x - 4y + z = 18$ must intersect. (No calculation is required.)

   (c) Find the point of intersection of ℓ_1 and the plane in (b).

101. Do the lines with equations

$$\begin{bmatrix} x \\ y \\ z \end{bmatrix} = \begin{bmatrix} 1 \\ 0 \\ -2 \end{bmatrix} + t \begin{bmatrix} -3 \\ 1 \\ 1 \end{bmatrix} \quad \text{and} \quad \begin{bmatrix} x \\ y \\ z \end{bmatrix} = \begin{bmatrix} -4 \\ 1 \\ 1 \end{bmatrix} + t \begin{bmatrix} 11 \\ -3 \\ -5 \end{bmatrix}$$

intersect? If so, where?

102. Does the line with equation $\begin{bmatrix} x \\ y \\ z \end{bmatrix} = \begin{bmatrix} -1 \\ 1 \\ 2 \end{bmatrix} + t \begin{bmatrix} 2 \\ 1 \\ -4 \end{bmatrix}$ intersect the plane with equation $3x - 10y - z = 8$? If yes, find the point of intersection.

103. Find the projection of $\mathsf{w} = \begin{bmatrix} x \\ y \\ z \end{bmatrix}$ on the plane π with equation $3x - 4y + 2z = 0$.

104. In each case, find two orthogonal vectors in the plane π and the projection of w on π.

(a) π has equation $4x + y - 3z = 0$ and $\mathsf{w} = \begin{bmatrix} -1 \\ 1 \\ 2 \end{bmatrix}$

(b) π has equation $2x - y + 3z = 0$ and $\mathsf{w} = \begin{bmatrix} 1 \\ -1 \\ 2 \end{bmatrix}$.

105. Let ℓ be the line with equation $\begin{bmatrix} x \\ y \\ z \end{bmatrix} = \begin{bmatrix} -1 \\ 1 \\ 2 \end{bmatrix} + t \begin{bmatrix} -4 \\ 3 \\ 5 \end{bmatrix}$.

(a) Find the distance from $P(2, -5, 1)$ to ℓ.

(b) Find the point of ℓ closest to P.

106. (a) Find two orthogonal vectors in the plane π with equation $2x - y + z = 0$.

(b) Use your answer to part (a) to find the projection of $\mathsf{w} = \begin{bmatrix} 1 \\ 1 \\ 1 \end{bmatrix}$ on π.

107. (a) The lines with vector equations $\begin{bmatrix} x \\ y \\ z \end{bmatrix} = \begin{bmatrix} 0 \\ 1 \\ 4 \end{bmatrix} + t \begin{bmatrix} -2 \\ 1 \\ 2 \end{bmatrix}$ and $\begin{bmatrix} x \\ y \\ z \end{bmatrix} = \begin{bmatrix} 4 \\ -2 \\ 2 \end{bmatrix} + t \begin{bmatrix} 6 \\ -3 \\ -6 \end{bmatrix}$ are parallel. Why?

(b) Find the distance between the lines in (a).

108. Find x, y, a and b if such numbers exist, given $\begin{bmatrix} a-b & x+a \\ y & 2 \\ -x & x+y \end{bmatrix} = \begin{bmatrix} 4 & b \\ x-2 & x-y \\ a-b & b-a \end{bmatrix}$.

109. Write down the 2×3 matrix A for which

(a) a_{ij} is the larger of i and j

(b) $a_{ij} = ij$.

110. Write down the 3×2 matrix $A = [a_{ij}]$ with $a_{ij} = 2ij - \cos \frac{\pi j}{3}$.

111. Let $\mathsf{v} = \begin{bmatrix} 1 \\ 2 \\ -3 \end{bmatrix}$ and $\mathsf{w} = \begin{bmatrix} 0 \\ 4 \\ 5 \end{bmatrix}$. Let $A = \begin{bmatrix} \mathsf{v} & \mathsf{w} \\ \downarrow & \downarrow \end{bmatrix}$ be the 3×2 matrix whose columns are v and w and let $B = \begin{bmatrix} \mathsf{v}^T & \rightarrow \\ \mathsf{w}^T & \rightarrow \end{bmatrix}$ be the 2×3 matrix whose rows are v^T and w^T. Find a_{11}, a_{13}, a_{21}, b_{32}, b_{12} and b_{22} if possible.

112. If A is any $n \times n$ matrix and x is a vector in R^n, what is the size of $\mathsf{x}^T A \mathsf{x}$ and why?

113. Let $A = \begin{bmatrix} 1 & 3 \\ 2 & 1 \end{bmatrix}$ and $B = \begin{bmatrix} 1 & 3 \\ 0 & 1 \end{bmatrix}$. Compute AB, $(AB)^T$, $A^T B^T$ and $B^T A^T$. These calculations illustrate what important property of the transpose of a matrix?

114. Let A be any $m \times n$ matrix. Let x be a vector in R^n and let y be a vector in R^m. Prove that $\|A\mathsf{x}\|^2 = \mathsf{x}^T A^T A\mathsf{x}$.

115. Express the system $\begin{aligned} x - y + z - 2w &= 0 \\ 3x + 5y - 2z + 2w &= 1 \\ 3y + 4z - 7w &= -1 \end{aligned}$ in the form $A\mathsf{x} = \mathsf{b}$.

116. Write $-2\begin{bmatrix} 1 \\ 1 \\ -2 \end{bmatrix} + 0\begin{bmatrix} 2 \\ 0 \\ 1 \end{bmatrix} + \begin{bmatrix} -3 \\ 1 \\ 1 \end{bmatrix} - \begin{bmatrix} 4 \\ 2 \\ 7 \end{bmatrix} + 3\begin{bmatrix} 6 \\ 5 \\ 4 \end{bmatrix}$ in the form $A\mathsf{x}$ for a suitable matrix A and vector x.

117. Write the vector $5\begin{bmatrix} 1 \\ 3 \\ -5 \\ 7 \\ 9 \end{bmatrix} + \begin{bmatrix} -2 \\ 1 \\ 1 \\ 0 \\ 7 \end{bmatrix} - 3\begin{bmatrix} 2 \\ 4 \\ 6 \\ 0 \\ -1 \end{bmatrix}$ in the form $A\mathsf{x}$ for a suitable matrix A and vector x. What is the important principle being reinforced here?

118. Express $\begin{bmatrix} -1 \\ 14 \\ 2 \end{bmatrix}$ as a linear combination of the columns of $A = \begin{bmatrix} 1 & 2 & 0 \\ 0 & 3 & 1 \\ 0 & 0 & 1 \end{bmatrix}$.

119. Find AB and BA (if defined) given $A = \begin{bmatrix} 1 & 2 & -1 \\ -1 & 1 & 1 \\ 4 & 3 & -1 \end{bmatrix}$ and $B = \begin{bmatrix} 2 & 1 \\ -1 & 4 \\ 2 & 3 \end{bmatrix}$

120. If A and B are $m \times n$ matrices and $A\mathsf{x} = B\mathsf{x}$ for each vector x in R^n, show that $A = B$.

121. Let $A = \begin{bmatrix} 1 & 2 & 3 \\ 4 & 5 & 6 \\ 7 & 8 & 9 \end{bmatrix}$ and $D = \begin{bmatrix} 2 & 0 & 0 \\ 0 & 3 & 0 \\ 0 & 0 & -1 \end{bmatrix}$. Find AD and DA without explicitly multiplying matrices. Explain your reasoning.

122. Suppose A and B are matrices with linearly independent columns. Prove that AB also has linearly independent columns (assuming this matrix product is defined).

123. Let $A = \begin{bmatrix} 1 & 2 \\ 3 & 4 \end{bmatrix}$ and $B = \begin{bmatrix} 0 & -1 \\ 5 & -2 \end{bmatrix}$. Compute $(A+B)^2$ and $A^2 + 2AB + B^2$. Are these equal? What is the correct expansion of $(A+B)^2$?

124. Suppose $A = \begin{bmatrix} a & b \\ c & d \end{bmatrix}$ is a 2×2 matrix that *commutes* with $B = \begin{bmatrix} 0 & -1 \\ 1 & 0 \end{bmatrix}$; that is, $AB = BA$. This surely implies some conditions on a, b, c, d. Discover what A must look like.

125. Suppose A and B are matrices with B invertible and $A = B^{-1}AB$.

(a) Show that A and B commute.

(b) Like many theorems in mathematics, part (a) is an *implication*; that is, a statement of the form "If $\mathcal{X}$, then $\mathcal{Y}$." What are $\mathcal{X}$ and $\mathcal{Y}$? The *converse* of "If $\mathcal{X}$, then $\mathcal{Y}$" is the implication "If $\mathcal{Y}$, then $\mathcal{X}$." What is the converse of the implication in part (a)? Is the converse true? Explain.

126. (a) Suppose A and B are matrices such that $AB = 0$. Does this imply $A = 0$ or $B = 0$? If you say "yes," give a proof. If you say "no," give an example of two nonzero matrices A and B for which $AB = 0$.

(b) If A is a 2×2 matrix, $B = \begin{bmatrix} 1 & 2 \\ 0 & -1 \end{bmatrix}$ and $AB = 0$, show that $A = 0$. Does this result contradict your answer to part (a)?

(c) If X and Y are any 2×2 matrices and B is the matrix of part (b), and if $XB = YB$, show that $X = Y$.

127. Suppose A is an invertible matrix and, for some matrix B (of compatible size), $AB = 0$, the zero matrix. Explain why $B = 0$.

128. If a matrix A is invertible, show that A^2 is also invertible.

129. If A and B are matrices with both AB and B invertible, prove that A is invertible.

130. (a) Given two $n \times n$ matrices X and Y, how do you determine whether or not $Y = X^{-1}$?

Let A be an $n \times n$ matrix and let I denote the $n \times n$ identity matrix.

(b) If $A^3 = 0$, verify that $(I - A)^{-1} = I + A + A^2$.

(c) Use part (b) to find the inverse of $\begin{bmatrix} 1 & 2 & -1 \\ 0 & 1 & 3 \\ 0 & 0 & 1 \end{bmatrix}$.

131. In each case, determine whether the given matrices are inverses.

(a) $A = \begin{bmatrix} 2 & 0 & -\frac{1}{2} \\ -1 & 0 & \frac{1}{2} \end{bmatrix}$, $B = \begin{bmatrix} 1 & 1 \\ -1 & -2 \\ 2 & 4 \end{bmatrix}$

(b) $A = \begin{bmatrix} 1 & 0 & 0 & 0 \\ 0 & 2 & 0 & 0 \\ 0 & 0 & 3 & 0 \\ 0 & 0 & 0 & 4 \end{bmatrix}$, $B = \frac{1}{24}\begin{bmatrix} 24 & 0 & 0 & 0 \\ 0 & 12 & 0 & 0 \\ 0 & 0 & 8 & 0 \\ 0 & 0 & 0 & 6 \end{bmatrix}$

(c) $A = \begin{bmatrix} 1 & 2 & 3 \\ 4 & 5 & 6 \end{bmatrix}$, quad $B = \begin{bmatrix} -1 & 1 \\ 0 & -1 \\ \frac{2}{3} & \frac{1}{3} \end{bmatrix}$.

132. Find a formula for $((AB)^T)^{-1}$ in terms of $(A^T)^{-1}$ and $(B^T)^{-1}$.

133. Suppose A and B are $n \times n$ matrices and $I - AB$ is invertible. Show that $I - BA$ is invertible by showing that $(I - BA)^{-1} = I + B(I - AB)^{-1}A$.

134. Suppose A is an $n \times n$ matrix such that $A + A^2 = I$. Show that A is invertible.

135. Suppose we want to find a quadratic polynomial $p(x) = a + bx + cx^2$ that passes through the points $(3, 0)$, $(-1, 4)$, $(0, 6)$.

   (a) Write the system of equations that must be solved in order to determine a, b, and c.

   (b) Find the desired polynomial.

136. The parabola with equation $y = ax^2 + bx + c$ passes through the points $(1, 4)$ and $(2, 8)$. Write down a matrix equation whose solution is the vector $\begin{bmatrix} a \\ b \\ c \end{bmatrix}$. (You are not being asked to find a, b, c.)

137. It is conjectured that the points $(\frac{\pi}{3}, 2)$ and $(-\frac{\pi}{4}, 1)$ lie on a curve with equation of the form $y = a\sin x + b\cos x$. Assuming this is the case, write down a matrix equation whose solution is $\begin{bmatrix} a \\ b \end{bmatrix}$.

138. Solve $A\mathsf{x} = \mathsf{b}$ with $A = \begin{bmatrix} 2 & 2 & 2 & -8 \\ 4 & 6 & 6 & 0 \\ 6 & 6 & 10 & -4 \end{bmatrix}$, $\mathsf{x} = \begin{bmatrix} x_1 \\ x_2 \\ x_3 \\ x_4 \end{bmatrix}$ and $\mathsf{b} = \begin{bmatrix} 2 \\ 4 \\ 2 \end{bmatrix}$. Express your solution as a vector, or as a linear combination of vectors as appropriate.

139. Solve each of the following systems of linear equations or explain why the system has no solution. Write any solution as a vector or as a linear combination of vectors.

   (a) $$\begin{aligned} x + y + 7z &= 2 \\ 2x - 4y + 14z &= -1 \\ 5x + 11y - 7z &= 8 \\ 2x + 5y - 4z &= -3 \end{aligned}$$

   (b) $$\begin{aligned} 2x_1 - 7x_2 + x_3 + x_4 &= 0 \\ x_1 - 2x_2 + x_3 &= 0 \\ 3x_1 + 6x_2 + 7x_3 - 4x_4 &= 0 \end{aligned}$$

   (c) $x - y + 2z = 4$

   (d) $$\begin{aligned} x - y + z - w &= 0 \\ 2x - 2z + 3w &= 11 \\ 5x - 2y + z - w &= 6 \\ -x + y + w &= 0 \end{aligned}$$

   (e) $$\begin{aligned} 2x - y + z &= 2 \\ 3x + y - 6z &= -9 \\ -x + 2y - 5z &= -4 \end{aligned}$$

   (f) $$\begin{aligned} -x_1 + 2x_2 + 3x_3 + 5x_4 - x_5 &= 0 \\ -2x_1 + 5x_2 + 10x_3 + 13x_4 - 4x_5 &= -5 \\ -3x_1 + 7x_2 + 13x_3 + 19x_4 - 11x_5 &= 1 \\ -x_1 + 4x_2 + 11x_3 + 11x_4 - 5x_5 &= -10 \end{aligned}$$

(g)
$$\begin{aligned} x \qquad\quad + 3z &= 6 \\ 3x + 4y - z &= 4 \\ 2x + 5y - 4z &= -3 \end{aligned}$$

(h)
$$\begin{aligned} -x_1 + x_2 + x_3 + 2x_4 &= 4 \\ 2x_3 - x_4 &= -7 \\ 3x_1 - 3x_2 - 7x_3 - 4x_4 &= 2 \end{aligned}$$

(i)
$$\begin{aligned} -x_1 + x_3 + 3x_4 &= 4 \\ x_1 + x_2 - 2x_3 - 3x_4 &= 5 \\ -2x_1 - 3x_2 + 5x_3 + 3x_4 &= -15 \\ x_2 - x_3 - 6x_4 &= 18 \end{aligned}$$

(j)
$$\begin{aligned} 2x - y + 2z &= -4 \\ 3x + 2y &= 1 \\ x + 3y - 6z &= 5 \end{aligned}$$

(k)
$$\begin{aligned} 2x_1 - 3x_2 + 4x_3 - x_4 &= 5 \\ -x_1 + x_2 + x_4 &= -1 \\ 2x_2 - x_3 - 3x_4 &= 1 \\ 3x_1 + x_3 + 4x_4 &= 7 \end{aligned}$$

140. Determine whether $\begin{bmatrix} 3 \\ -2 \\ 4 \end{bmatrix}$ is a linear combination of $\begin{bmatrix} -1 \\ 2 \\ 3 \end{bmatrix}$, $\begin{bmatrix} 0 \\ 1 \\ 1 \end{bmatrix}$, and $\begin{bmatrix} 5 \\ -3 \\ 1 \end{bmatrix}$.

141. Write $\begin{bmatrix} 3 \\ 5 \\ 10 \end{bmatrix}$ as a linear combination of the columns of $A = \begin{bmatrix} 3 & 0 & 6 \\ 3 & 1 & 9 \\ -2 & 4 & 10 \end{bmatrix}$.

142. In each case, determine whether or not the given vector is a linear combination of the columns of the given matrix.

(a) $\begin{bmatrix} 2 \\ -11 \\ -3 \end{bmatrix}$; $\begin{bmatrix} 0 & -1 \\ -1 & 4 \\ 5 & 9 \end{bmatrix}$

(b) $\begin{bmatrix} 8 \\ -11 \\ -3 \end{bmatrix}$; $\begin{bmatrix} 2 & -1 \\ -1 & 4 \\ 5 & 9 \end{bmatrix}$.

143. Let $A = \begin{bmatrix} 2 & 5 \\ 1 & 3 \end{bmatrix}$ and $\mathsf{b} = \begin{bmatrix} b_1 \\ b_2 \end{bmatrix}$.

(a) Solve the system $A\mathsf{x} = \mathsf{b}$ for $\mathsf{x} = \begin{bmatrix} x_1 \\ x_2 \end{bmatrix}$.

(b) Write b as a linear combination of the columns of A.

144. Determine whether or not each of the given sets of vectors is linearly independent or linearly dependent. In the case of linear dependence, write down a specific nontrivial linear combination of the vectors which equals the zero vector.

(a) $\mathsf{v}_1 = \begin{bmatrix} 1 \\ 1 \\ 2 \\ 1 \end{bmatrix}, \mathsf{v}_2 = \begin{bmatrix} 3 \\ 3 \\ 3 \\ 1 \end{bmatrix}, \mathsf{v}_3 = \begin{bmatrix} 2 \\ 5 \\ 10 \\ 4 \end{bmatrix}, \mathsf{v}_4 = \begin{bmatrix} 1 \\ 2 \\ 3 \\ 1 \end{bmatrix}$

(b) $\mathsf{v}_1 = \begin{bmatrix} -1 \\ 5 \\ -1 \\ 3 \\ 1 \end{bmatrix}, \mathsf{v}_2 = \begin{bmatrix} -2 \\ 9 \\ 1 \\ 6 \\ 1 \end{bmatrix}, \mathsf{v}_3 = \begin{bmatrix} 3 \\ -11 \\ -1 \\ -9 \\ -3 \end{bmatrix}$

(c) $\mathsf{v}_1 = \begin{bmatrix} -2 \\ 1 \\ 6 \\ 1 \\ 1 \end{bmatrix}, \mathsf{v}_2 = \begin{bmatrix} 0 \\ -1 \\ 2 \\ 9 \\ 7 \end{bmatrix}, \mathsf{v}_3 = \begin{bmatrix} 1 \\ 3 \\ 4 \\ 0 \\ -1 \end{bmatrix}, \mathsf{v}_4 = \begin{bmatrix} 1 \\ 1 \\ 1 \\ 2 \\ 1 \end{bmatrix}$

145. Find values of a, b, and c such that the graph of the function $f(x) = a2^x + b2^{2x} + c2^{3x}$ passes through the points $(-1, 2)$, $(0, 0)$, and $(1, -2)$.

146. Find a condition on a, b and c in order that the system $\begin{aligned} x + y - z &= a \\ 2x - 3y + 5z &= b \\ 5x \qquad + 2z &= c \end{aligned}$ have a solution. In the case of a solution, is this unique? Explain.

147. Find conditions on a, b, and c (if any) such that the system

$$\begin{aligned} x \qquad + z &= -1 \\ 2x - y \qquad &= 2 \\ y + 2z &= -4 \\ ax + by + cz &= 3 \end{aligned}$$

- has a unique solution;
- has no solution;
- has infinitely many solutions.

148. Find a condition on a, b and c which implies that the system

$$\begin{aligned} 7x_1 + x_2 - 4x_3 + 5x_4 &= a \\ x_1 + x_2 + 2x_3 - x_4 &= b \\ 3x_1 + x_2 \qquad + x_4 &= c \end{aligned}$$

does **not** have a solution.

149. Consider the system $\begin{aligned} 2x_1 + 3x_2 + 4x_3 &= 1 \\ 4x_1 + 7x_2 + 5x_3 &= 6 \\ - x_2 + 3x_3 &= -4. \end{aligned}$

(a) Find one particular solution x_p of this system.

(b) Express $\begin{bmatrix} 1 \\ 6 \\ -4 \end{bmatrix}$ as a linear combination of the columns of $\begin{bmatrix} 2 & 3 & 4 \\ 4 & 7 & 5 \\ 0 & -1 & 3 \end{bmatrix}$.

(c) Find the solution x_h of the corresponding homogeneous system.

(d) Show that the solution of the given system is $\mathsf{x}_p + \mathsf{x}_h$.

150. Express the solution to
$$\begin{aligned} 3x_1 - x_2 + x_3 + 2x_4 &= 4 \\ -4x_1 + x_2 + 2x_3 + 7x_4 &= 3 \\ x_1 - 2x_2 + 3x_3 + 7x_4 &= 1 \\ 3x_2 + x_3 + 6x_4 &= 18 \end{aligned}$$
in the form $\mathsf{x} = \mathsf{x}_p + \mathsf{x}_h$ where x_p is a particular solution and x_h is a solution of the corresponding homogeneous system.

151. (a) Suppose A and B are invertible matrices and X is a matrix such that $AXB = A + B$. What is X?

(b) Suppose A, B, C, and X are matrices, X and C invertible, such that $X^{-1}A = C - X^{-1}B$. What is X?

(c) Suppose A, B, and X are invertible matrices such that $BAX = XABX$. What is X?

152. Let $A = \begin{bmatrix} 1 & 1 \\ 2 & 4 \end{bmatrix}$ and $C = \begin{bmatrix} 5 & 3 \\ 2 & 2 \end{bmatrix}$.

Given that B is a 2×2 matrix and that $ABC^{-1} = I$, the identity matrix, find B.

153. Find a matrix X satisfying $\begin{bmatrix} 4 & 7 & 8 \\ 2 & -3 & 1 \\ 0 & 5 & -3 \end{bmatrix}^{-1} X = \begin{bmatrix} 1 & 2 & 3 \\ -4 & 0 & 6 \\ -2 & 1 & 1 \end{bmatrix} \begin{bmatrix} 2 & 0 & 0 \\ 1 & 4 & -1 \\ 1 & 3 & 0 \end{bmatrix}^{-1}$.

154. Given that A is a 2×2 matrix and $\begin{bmatrix} 1 & 2 \\ 3 & 0 \end{bmatrix}^{-1} A \begin{bmatrix} 5 & 1 \\ -1 & 1 \end{bmatrix}^{-1} = \begin{bmatrix} -3 & 4 \\ 0 & 2 \end{bmatrix}$, find A.

155. Consider the system of equations
$$\begin{aligned} x_2 - x_3 &= 8 \\ x_1 + 2x_2 + x_3 &= 5 \\ x_1 \qquad + x_3 &= -7. \end{aligned}$$

(a) Write this in the form $A\mathsf{x} = \mathsf{b}$.

(b) Solve the system by finding A^{-1}.

156. Consider the system
$$\begin{aligned} x_1 - 2x_2 + 2x_3 &= 3 \\ 2x_1 + x_2 + x_3 &= 0 \\ x_1 \qquad + x_3 &= -2. \end{aligned}$$

i. Write this in the form $A\mathsf{x} = \mathsf{b}$.

ii. Solve the system by finding A^{-1}.

iii. Express $\begin{bmatrix} 3 \\ 0 \\ -2 \end{bmatrix}$ as a linear combination of the vectors $\begin{bmatrix} 1 \\ 2 \\ 1 \end{bmatrix}, \begin{bmatrix} -2 \\ 1 \\ 0 \end{bmatrix}, \begin{bmatrix} 2 \\ 1 \\ 1 \end{bmatrix}$.

157. Determine whether or not each of the following matrices has an inverse. Find the inverse whenever this exists.

(a) $\begin{bmatrix} 1 & 1 & 2 \\ 3 & 2 & 3 \\ 4 & 2 & 1 \end{bmatrix}$ (b) $\begin{bmatrix} 0 & -1 & 2 \\ 2 & 1 & 4 \\ 1 & -1 & 5 \end{bmatrix}$ (c) $\begin{bmatrix} -2 & 3 & 4 \\ 1 & 0 & 1 \\ -1 & 5 & 8 \end{bmatrix}$

(d) $\begin{bmatrix} 1 & 1 & 1 \\ 0 & 2 & 3 \\ 5 & 5 & 1 \end{bmatrix}$ (e) $\begin{bmatrix} 0 & 1 & 0 & 0 & 0 \\ 0 & 0 & 0 & 1 & 0 \\ 1 & 0 & 0 & 0 & 0 \\ 0 & 0 & 0 & 0 & 1 \\ 0 & 0 & 1 & 0 & 0 \end{bmatrix}$

158. Let $A = \begin{bmatrix} 1 & 2 & 3 \\ 4 & 5 & 6 \\ 6 & 8 & 9 \end{bmatrix}$. Write down an elementary matrix E so that $EA = \begin{bmatrix} 1 & 2 & 3 \\ 0 & -3 & -6 \\ 6 & 8 & 9 \end{bmatrix}$.

159. Let $A = \begin{bmatrix} 1 & 0 & 0 \\ 0 & 1 & 0 \\ 0 & 5 & 1 \end{bmatrix}$. Find A^{-1} without calculation and explain your reasoning.

160. (a) Find an elementary matrix E so that $E \begin{bmatrix} 1 & 2 & 3 \\ 4 & 5 & 6 \\ 6 & 8 & 9 \end{bmatrix} = \begin{bmatrix} 1 & 2 & 3 \\ 4 & 5 & 6 \\ 0 & \frac{1}{2} & 0 \end{bmatrix}$.

(b) Write down E^{-1} without calculation but explaining your reasoning.

161. Write down the inverse of each of the following matrices without calculation. Explain your reasoning.

(a) $A = \begin{bmatrix} 1 & 0 & 0 \\ 0 & 1 & -2 \\ 0 & 0 & 1 \end{bmatrix}$ (b) $A = \begin{bmatrix} 0 & 0 & 0 & 1 \\ 0 & 1 & 0 & 0 \\ 0 & 0 & 1 & 0 \\ 1 & 0 & 0 & 0 \end{bmatrix}$

162. Write $L = \begin{bmatrix} 1 & 0 & 0 \\ a & 1 & 0 \\ b & c & 1 \end{bmatrix}$ as the product of elementary matrices (without calculation) and use this factorization to find L^{-1}.

163. Write $U = \begin{bmatrix} 1 & -1 & 9 \\ 0 & 1 & 4 \\ 0 & 0 & 1 \end{bmatrix}$ as the product of elementary matrices (without calculation) and use this factorization to find U^{-1}. [Hint: You might start with $\begin{bmatrix} 1 & -1 & 0 \\ 0 & 1 & 0 \\ 0 & 0 & 1 \end{bmatrix}$, which is elementary, and then use your knowledge of how elementary matrices work to transform this to U.]

164. Reduce $\begin{bmatrix} 2 & 1 & 4 & 2 \\ 3 & 0 & 2 & 1 \\ 5 & 2 & 3 & 5 \end{bmatrix}$ to row echelon form using only the third elementary row operation. Identify the pivots and the pivot columns.

165. Is the product of elementary matrices invertible? Explain.

166. Suppose A is an invertible matrix and its first two rows are interchanged to give a matrix B. Is B invertible? Explain.

167. Express each invertible matrix as the product of elementary matrices.

(a) $\begin{bmatrix} 0 & -1 \\ 2 & 1 \end{bmatrix}$ (b) $\begin{bmatrix} -1 & 2 \\ 3 & 1 \end{bmatrix}$ (c) $\begin{bmatrix} 1 & -1 & 1 \\ 0 & 2 & -1 \\ 2 & 1 & 0 \end{bmatrix}$

(d) $\begin{bmatrix} 0 & -2 & 1 \\ 0 & 1 & 0 \\ 1 & -5 & 2 \end{bmatrix}$ (e) $\begin{bmatrix} -1 & 1 & 3 \\ -2 & 2 & 7 \\ 4 & -3 & -12 \end{bmatrix}$

168. (a) You are given that $U = \begin{bmatrix} 3 & 2 & 1 \\ 0 & 1 & 2 \\ 0 & 0 & 0 \end{bmatrix}$ is a row echelon form of $A = \begin{bmatrix} 3 & 2 & 1 \\ 6 & 5 & 4 \\ 9 & 8 & 7 \end{bmatrix}$. Find elementary matrices whose product M satisfies $U = MA$. [Hint: Use Gaussian elimination to move A to U.]

(b) Use the idea of part (a) to show that an invertible matrix is the product of elementary matrices.

169. Find an LU factorization of each of the following matrices or explain why no such factorization is possible.

(a) $A = \begin{bmatrix} 2 & -6 & -2 & 4 \\ -1 & 3 & 3 & 2 \\ -1 & 3 & 7 & 10 \end{bmatrix}$ (b) $A = \begin{bmatrix} -1 & 2 & 2 & 3 & 4 \\ 5 & -10 & -6 & -18 & -12 \\ 2 & -2 & -3 & -5 & -7 \end{bmatrix}$

(c) $A = \begin{bmatrix} 1 & 3 & 8 \\ 2 & 5 & 21 \\ 1 & 7 & -5 \end{bmatrix}$ (d) $\begin{bmatrix} -3 & 8 & 10 & 1 \\ 3 & -2 & -1 & 0 \\ -3 & 4 & 4 & 0 \\ 6 & -4 & -2 & 1 \\ 9 & 0 & 1 & 0 \end{bmatrix}$

(e) $\begin{bmatrix} -2 & 4 & 6 & -8 \\ 1 & 2 & -8 & 2 \\ -4 & 0 & -5 & 3 \\ 2 & 1 & -4 & 1 \end{bmatrix}$ (f) $\begin{bmatrix} -2 & 4 & 6 & -8 & 1 \\ 6 & -8 & -23 & 22 & -3 \\ -2 & -12 & -1 & 15 & 4 \\ 4 & 12 & 26 & -26 & -3 \end{bmatrix}$

170. Let $A = \begin{bmatrix} 2 & 4 & 6 & 18 \\ 4 & 5 & 6 & 24 \\ 3 & 1 & -2 & 4 \end{bmatrix}$.

(a) Find an LU factorization of A by row reduction to an upper triangular matrix keeping track of the elementary matrices. Express L as the product of elementary matrices.

(b) Find an LU factorization A by using only the third elementary row operation and simply writing L down.

(c) Find an LDU factorization of A.

171. Find an LU and an LDU factorization of each matrix.

(a) $A = \begin{bmatrix} 1 & -1 & -2 \\ 2 & -3 & -5 \\ -1 & 3 & 5 \end{bmatrix}$ (b) $A = \begin{bmatrix} -2 & 4 & 6 \\ 1 & 2 & -8 \\ 4 & 0 & -5 \end{bmatrix}$.

172. Show that $\begin{bmatrix} 0 & 1 \\ 1 & 0 \end{bmatrix}$ does not have an LU decomposition.

173. Explain how an LU factorization of a matrix A can be used to obtain an LU factorization of A^T.

174. Suppose that A and P are $n \times n$ matrices and A is symmetric. Prove that P^TAP is symmetric.

175. Find a PLU factorization of each matrix.

(a) $A = \begin{bmatrix} 1 & -2 & 1 \\ 0 & 1 & 2 \\ 0 & 4 & 8 \\ 3 & -6 & 3 \\ 1 & 2 & 5 \end{bmatrix}$ (b) $A = \begin{bmatrix} 2 & -1 & 2 \\ 6 & -3 & 1 \\ 4 & -2 & 4 \\ -2 & 1 & 8 \end{bmatrix}$.

176. Find a U^TDU factorization of the symmetric matrix $A = \begin{bmatrix} 5 & 2 & 2 \\ 2 & 2 & -4 \\ 2 & -4 & 2 \end{bmatrix}$.

177. Let $A = \begin{bmatrix} 1 & 6 & 7 \\ 5 & 25 & 36 \\ -2 & -27 & -4 \end{bmatrix} = \begin{bmatrix} 1 & 0 & 0 \\ 5 & 1 & 0 \\ -2 & 3 & 1 \end{bmatrix} \begin{bmatrix} 1 & 6 & 7 \\ 0 & -5 & 1 \\ 0 & 0 & 7 \end{bmatrix} = LU$.

(a) Use the given factorization of A to solve the linear system $A\mathsf{x} = \mathsf{b}$, where $\mathsf{b} = \begin{bmatrix} 1 \\ 2 \\ 3 \end{bmatrix}$.

(b) Express b as a linear combination of the columns of A.

178. Use the factorization $\begin{bmatrix} 2 & 2 & 2 \\ 4 & 6 & 6 \\ 4 & 8 & 10 \end{bmatrix} = \begin{bmatrix} 1 & 0 & 0 \\ 2 & 1 & 0 \\ 2 & 2 & 1 \end{bmatrix} \begin{bmatrix} 2 & 2 & 2 \\ 0 & 2 & 2 \\ 0 & 0 & 2 \end{bmatrix}$ to solve $A\mathsf{x} = \mathsf{b}$ for $\mathsf{x} = \begin{bmatrix} x_1 \\ x_2 \\ x_3 \end{bmatrix}$, where $\mathsf{b} = \begin{bmatrix} 2 \\ 4 \\ 2 \end{bmatrix}$ and $A = \begin{bmatrix} 2 & 2 & 2 \\ 4 & 6 & 6 \\ 4 & 8 & 10 \end{bmatrix}$.

179. (a) Use the factorization

$$A = \begin{bmatrix} 1 & 4 & 3 & 4 \\ 3 & 10 & 10 & 17 \\ -4 & -20 & -14 & -1 \\ 7 & 38 & 28 & -6 \end{bmatrix} = \begin{bmatrix} 1 & 0 & 0 & 0 \\ 3 & 1 & 0 & 0 \\ -4 & 2 & 1 & 0 \\ 7 & -5 & -3 & 1 \end{bmatrix} \begin{bmatrix} 1 & 4 & 3 & 4 \\ 0 & -2 & 1 & 5 \\ 0 & 0 & -4 & 5 \\ 0 & 0 & 0 & 6 \end{bmatrix} = LU$$

to solve the linear system $A\mathsf{x} = \mathsf{b}$, where $\mathsf{b} = \begin{bmatrix} 3 \\ 4 \\ 2 \\ -2 \end{bmatrix}$.

(b) Express b as a linear combination of the columns of A.

180. In each case,

i. find the matrix M of minors, the matrix C of cofactors, and compute the products AC^T and C^TA,

ii. find $\det A$,

iii. if A is invertible, find A^{-1}.

(a) $A = \begin{bmatrix} 5 & 3 \\ 15 & 9 \end{bmatrix}$ (b) $A = \begin{bmatrix} 3 & 5 & 2 \\ 4 & 8 & 9 \\ -1 & 2 & 5 \end{bmatrix}$ (c) $A = \begin{bmatrix} -1 & 2 & 4 \\ 0 & 3 & 5 \\ 2 & -2 & 3 \end{bmatrix}$ (d) $A = \begin{bmatrix} 2 & 3 & 4 \\ 0 & -1 & 3 \\ 4 & 7 & 5 \end{bmatrix}$

181. Let $A = \begin{bmatrix} -4 & 2 & 1 \\ 0 & -3 & 5 \\ 6 & 4 & -1 \end{bmatrix}$.

(a) Find the determinant of A with a Laplace expansion down the first column.

(b) Find the determinant of A with a Laplace expansion along the third row.

182. Let $A = \begin{bmatrix} 1 & -1 & 2 \\ 3 & 1 & 1 \\ 2 & -1 & 3 \end{bmatrix}$.

(a) Find the determinant of A with a Laplace expansion down the second column.

(b) Find the determinant of A with a Laplace expansion along the second row.

183. Suppose A is a 2×2 matrix with $\det A = 5$ and cofactor matrix $C = \begin{bmatrix} 3 & 1 \\ -2 & 1 \end{bmatrix}$. What is A?

184. In each case, find the determinant by reducing to triangular form.

(a) $\begin{bmatrix} -2 & 1 & 1 & 2 \\ 1 & -2 & 1 & 2 \\ 4 & -3 & 5 & 1 \\ 0 & -2 & 2 & 3 \end{bmatrix}$ (b) $\begin{bmatrix} 1 & -1 & 2 & 0 & -2 \\ 0 & 1 & 0 & 4 & 1 \\ 1 & 1 & 3 & 0 & 0 \\ 0 & 0 & 0 & 3 & -1 \\ 0 & 0 & 0 & 1 & 1 \end{bmatrix}$

(c) $\begin{bmatrix} 1 & 0 & 0 & 0 & 5 \\ 0 & 0 & 1 & 0 & 1 \\ 2 & 5 & 0 & 0 & 0 \\ 0 & 0 & 1 & -4 & 0 \\ 0 & 1 & 0 & 1 & 0 \end{bmatrix}$ (d) $\begin{bmatrix} -1 & -1 & 1 & 0 \\ 2 & 1 & 1 & 3 \\ 0 & 1 & 1 & 2 \\ 1 & 3 & -1 & 2 \end{bmatrix}$

(e) $\begin{bmatrix} 10 & -8 & 13 & 16 \\ 2 & -1 & 1 & 4 \\ 6 & 3 & -13 & 24 \\ -4 & 5 & -4 & -5 \end{bmatrix}$

185. For what value(s) of x is $\begin{bmatrix} -1 & x & x \\ 2 & 0 & x \\ 4 & 4 & -2 \end{bmatrix}$ singular?

186. If P is a permutation matrix and

$$PA = \begin{bmatrix} 1 & 6 & 7 \\ 5 & 25 & 36 \\ -2 & -27 & -4 \end{bmatrix} = \begin{bmatrix} 1 & 0 & 0 \\ 5 & 1 & 0 \\ -2 & 3 & 1 \end{bmatrix} \begin{bmatrix} 1 & 6 & 7 \\ 0 & -5 & 1 \\ 0 & 0 & 7 \end{bmatrix} = LU,$$

what are the possible values for $\det A$? Explain.

187. Let P be an $n \times n$ permutation matrix. Let x, y be n-dimensional vectors. Show that $(P\mathsf{x}) \cdot (P\mathsf{y}) = \mathsf{x} \cdot \mathsf{y}$. [Hint: The dot product $\mathsf{u} \cdot \mathsf{v}$ of vectors u and v is a certain matrix product.]

188. The determinant is a *multiplicative* function, meaning that $\det AB = (\det A)(\det B)$. The proof is well within your abilities to understand, but it's a bit long. See if you can put together your own argument, step by step.

    (a) Show that $\det(AB) = (\det A)(\det B)$ if A is not invertible.

    It remains to obtain the same result when A is invertible.

    (b) Show that $\det(EB) = (\det E)(\det B)$ for any elementary matrix E.

    (c) If $E_1, E_2, \ldots, E_k$ are all elementary matrices, then $\det(E_1 E_2 \cdots E_k B) = (\det E_1)(\det E_2) \cdots (\det E_k)(\det B)$.

    (d) If A is an invertible matrix, show that $\det AB = (\det A)(\det B)$.

189. If A and B are 3×3 matrices, $\det A = 2$ and $\det B = -5$, find $\det A^T B^{-1} A^3(-B)$.

190. If A and B are 3×3 matrices, $\det A = 2$ and $\det B = -5$, find $\det A^T B^{-1} A^3(-B)$.

191. Let A and B be 5×5 matrices with $\det(-3A) = 4$ and $\det B^{-1} = 2$. Find $\det A$, $\det B$ and $\det AB$.

192. Suppose A and B are 4×4 matrices with $\det A = 2$ and $\det B = -5$. Let $X = A^T B^{-1} A^3(-B)$. Find $\det X$ and $\det(-X)$.

193. Suppose A is a 4×4 matrix and P is an invertible 4×4 matrix such that $PAP^{-1} = D = \begin{bmatrix} 1 & 0 & 0 & 0 \\ 0 & -1 & 0 & 0 \\ 0 & 0 & -1 & 0 \\ 0 & 0 & 0 & -1 \end{bmatrix}$. Find (a) $\det A$, (b) A^2 and (c) A^{101}.

194. If A is a 20×20 matrix with every entry an odd integer, explain what $\det A$ must be divisible (evenly) by 2^{19}.

195. Let $B = \begin{bmatrix} 1 & 2 & 1 \\ 3 & -2 & 0 \\ -1 & 4 & 1 \end{bmatrix}$.

    (a) Find $\det B$, $\det \frac{1}{3}B$ and $\det B^{-1}$.

    (b) Suppose A is a matrix whose inverse is B. Find the cofactor matrix of A.

196. Let $A = \begin{bmatrix} a & b & c \\ p & q & r \\ u & v & w \end{bmatrix}$ and suppose $\det A = 5$. Find $\begin{vmatrix} 2p & -a+u & 3u \\ 2q & -b+v & 3v \\ 2r & -c+w & 3w \end{vmatrix}$.

197. Suppose $\begin{vmatrix} a & b & c \\ d & e & f \\ g & h & i \end{vmatrix} = -3$. Find $\begin{vmatrix} a & -g & 2d \\ b & -h & 2e \\ c & -i & 2f \end{vmatrix}$.

198. Given that $\begin{vmatrix} a & b & c \\ u & v & w \\ x & y & z \end{vmatrix} = 5$, find $\begin{vmatrix} x & y & z \\ a+u & b+v & c+w \\ a-u & b-v & c-w \end{vmatrix}$.

199. Let $A = \begin{bmatrix} 5 & -7 & 7 \\ 4 & -3 & 4 \\ 4 & -1 & 2 \end{bmatrix}$; $\mathsf{v}_1 = \begin{bmatrix} 0 \\ 0 \\ 0 \end{bmatrix}$, $\mathsf{v}_2 = \begin{bmatrix} 1 \\ 0 \\ 0 \end{bmatrix}$, $\mathsf{v}_3 = \begin{bmatrix} 0 \\ 2 \\ 2 \end{bmatrix}$, $\mathsf{v}_4 = \begin{bmatrix} 1 \\ 0 \\ -1 \end{bmatrix}$, $\mathsf{v}_5 = \begin{bmatrix} 3 \\ 3 \\ 3 \end{bmatrix}$ and $\mathsf{v}_6 = \begin{bmatrix} 1 \\ 0 \\ 1 \end{bmatrix}$. Determine whether or not the given vectors v_i are eigenvectors of A. Justify your answers.

200. Find the characteristic polynomial, the (real) eigenvalues and the corresponding eigenspaces of each of the following matrices A.

(a) $\begin{bmatrix} -1 & 3 \\ 2 & 0 \end{bmatrix}$ (b) $\begin{bmatrix} 1 & 2 \\ 3 & 2 \end{bmatrix}$ (c) $\begin{bmatrix} 1 & -2 & 3 \\ 2 & 6 & -6 \\ 1 & 2 & -1 \end{bmatrix}$

(d) $\begin{bmatrix} 3 & 4 & -4 & -4 \\ 4 & 3 & -4 & -4 \\ 0 & 4 & -1 & -4 \\ 4 & 0 & -4 & -1 \end{bmatrix}$

201. Let $A = \begin{bmatrix} 5 & -7 & 7 \\ 4 & -3 & 4 \\ 4 & -1 & 2 \end{bmatrix}$.

(a) Show that $\lambda = -2$ is an eigenvalue of A and find the corresponding eigenspace.

(b) Given that the eigenvalues of A are $1, -2, 5$, what is the characteristic polynomial of A?

(c) Given that $\begin{bmatrix} 1 \\ 1 \\ 1 \end{bmatrix}$ is an eigenvector of A corresponding to $\lambda = 5$ and $\begin{bmatrix} 0 \\ 1 \\ 1 \end{bmatrix}$ is an eigenvector of A corresponding to $\lambda = 1$, find a matrix P such that $P^{-1}AP = \begin{bmatrix} 1 & 0 & 0 \\ 0 & -2 & 0 \\ 0 & 0 & 5 \end{bmatrix}$.

(d) Let $Q = \begin{bmatrix} -1 & 1 & 0 \\ 0 & 1 & 1 \\ 1 & 1 & 1 \end{bmatrix}$. Find $Q^{-1}AQ$ without calculation and explain.

(e) Find A^{100}.

202. If x is an eigenvector of an invertible matrix A corresponding to λ, show that x is also an eigenvector of A^{-1}. What is the corresponding eigenvalue?

203. If 0 is an eigenvalue of a matrix A, then A cannot be invertible. Why not?

204. Let λ be an eigenvalue of an $n \times n$ matrix A and let u, v be eigenvectors corresponding to λ. If a and b are scalars, show that $A(a\mathsf{u} + b\mathsf{v}) = \lambda(a\mathsf{u} + b\mathsf{v})$. Thus $a\mathsf{u} + b\mathsf{v} \neq \mathsf{0}$ is either $\mathsf{0}$ or another eigenvector of A corresponding to λ.

205. Let $A = \begin{bmatrix} 5 & -7 & 7 \\ 4 & -3 & 4 \\ 4 & -1 & 2 \end{bmatrix}$.

(a) Show that $\lambda = -2$ is an eigenvalue of A and find the corresponding eigenspace.

(b) Given that the eigenvalues of A are $1, -2, 5$, what is the characteristic polynomial of A?

(c) Given that $\begin{bmatrix} 1 \\ 1 \\ 1 \end{bmatrix}$ is an eigenvector of A corresponding to $\lambda = 5$ and $\begin{bmatrix} 0 \\ 1 \\ 1 \end{bmatrix}$ is an eigenvector of A corresponding to $\lambda = 1$, find a matrix P such that $P^{-1}AP = \begin{bmatrix} 1 & 0 & 0 \\ 0 & -2 & 0 \\ 0 & 0 & 5 \end{bmatrix}$.

(d) Let $Q = \begin{bmatrix} -1 & 1 & 0 \\ 0 & 1 & 1 \\ 1 & 1 & 1 \end{bmatrix}$. Find $Q^{-1}AQ$ without calculation.

206. (a) Show that if two matrices are similar, then they have the same characteristic polynomial.

(b) Are the matrices $A = \begin{bmatrix} 2 & 1 \\ 1 & -1 \end{bmatrix}$ and $B = \begin{bmatrix} 3 & 0 \\ 1 & -1 \end{bmatrix}$ similar? Explain.

(c) State the converse of (a). Show that this converse is false by considering the matrices $A = \begin{bmatrix} 1 & 0 \\ 0 & 1 \end{bmatrix}$ and $B = \begin{bmatrix} 1 & 1 \\ 0 & 1 \end{bmatrix}$.

207. Explain why $A = \begin{bmatrix} 5 & -2 & 6 \\ 0 & -1 & 9 \\ 0 & 0 & 3 \end{bmatrix}$ is diagonalizable. Describe all the matrices to which A is similar.

208. Determine whether or not each of the matrices A given below is diagonalizable. If it is, find an invertible matrix P and a diagonal matrix D such that $P^{-1}AP = D$. If it isn't, explain why not.

(a) $A = \begin{bmatrix} 2 & 1 \\ -1 & 0 \end{bmatrix}$ (b) $A = \begin{bmatrix} 4 & -3 & 3 \\ 0 & 1 & 0 \\ -6 & 6 & -5 \end{bmatrix}$

209. Is $A = \begin{bmatrix} 3 & 0 & 6 \\ 0 & -3 & 0 \\ 5 & 0 & 2 \end{bmatrix}$ diagonalizable? If no, explain why not. If yes, find an invertible matrix P and a diagonal matrix D such that $P^{-1}AP = D$.

210. Answer Exercise 209 for the matrix $A = \begin{bmatrix} 2 & 1 & 1 \\ 0 & 1 & 0 \\ 1 & -1 & 2 \end{bmatrix}$.

211. Let $A = \begin{bmatrix} 5 & -7 & 7 \\ 4 & -3 & 4 \\ 4 & -1 & 2 \end{bmatrix}$.

(a) Show that $\lambda = -2$ is an eigenvalue of A and find the corresponding eigenspace.

(b) Given that the eigenvalues of A are $1, -2, 5$, what is the characteristic polynomial of A?

(c) Given that $\begin{bmatrix} 1 \\ 1 \\ 1 \end{bmatrix}$ is an eigenvector of A corresponding to $\lambda = 5$ and $\begin{bmatrix} 0 \\ 1 \\ 1 \end{bmatrix}$ is an eigenvector of A corresponding to $\lambda = 1$, find a matrix P such that $P^{-1}AP = \begin{bmatrix} 1 & 0 & 0 \\ 0 & -2 & 0 \\ 0 & 0 & 5 \end{bmatrix}$.

(d) Let $Q = \begin{bmatrix} -1 & 1 & 0 \\ 0 & 1 & 1 \\ 1 & 1 & 1 \end{bmatrix}$. Find $Q^{-1}AQ$ without calculation.

212. One of the most celebrated theorems in linear algebra says that a real symmetric matrix can be *orthogonally diagonalized*; that is, there is a matrix P with orthogonal columns such that $P^{-1}AP$ is diagonal. Illustrate this theorem with respect to each of the following matrices.

(a) $A = \begin{bmatrix} 3 & -5 \\ -5 & 3 \end{bmatrix}$ (b) $A = \begin{bmatrix} 1 & -2 & -1 \\ -2 & 4 & 2 \\ -1 & 2 & 1 \end{bmatrix}$

213. Orthogonally diagonalize $A = \begin{bmatrix} 1 & -3 & 2 \\ -3 & 1 & -2 \\ 2 & -2 & 2 \end{bmatrix}$; that is, find an orthogonal matrix Q (orthogonal columns of length 1) and a real diagonal matrix D such that $Q^TAQ = D$.

214. Find all indicated currents in each circuit. The standard symbol for ohms is the Greek symbol Ω, "Omega."

(a)

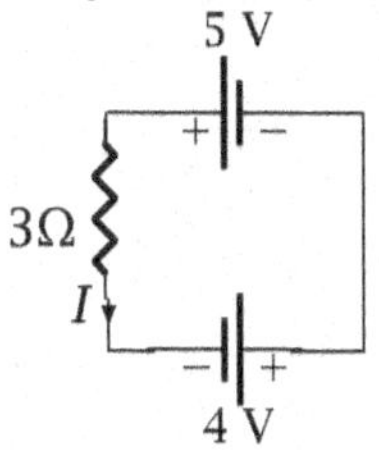

(b)

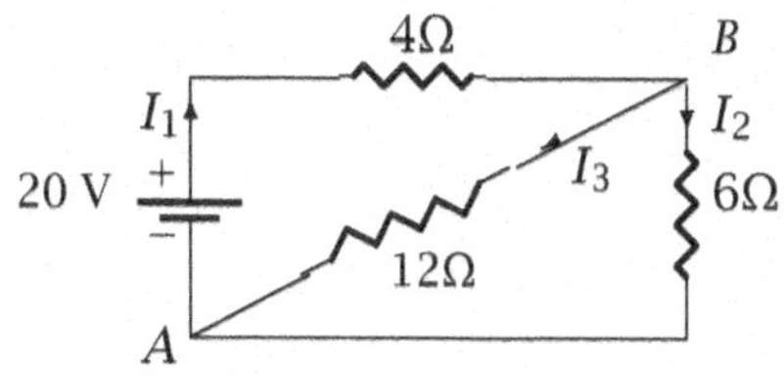

(c)

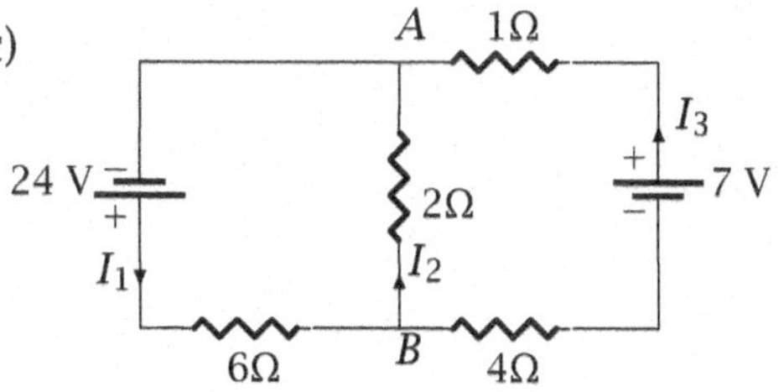

(d)

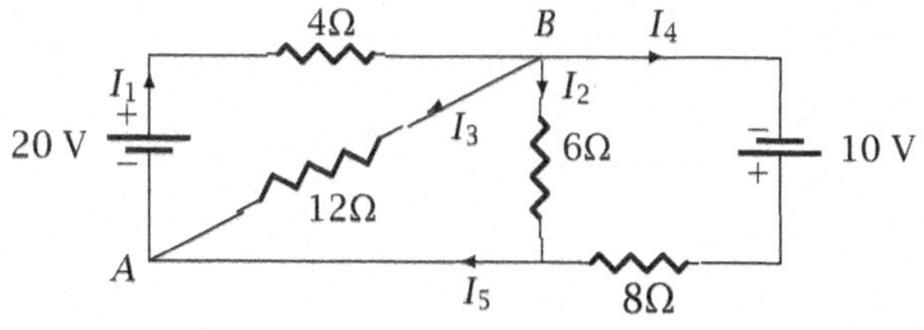

215. Let C be the code with generator matrix $G = \begin{bmatrix} 1 & 0 & 0 \\ 0 & 1 & 0 \\ 0 & 0 & 1 \\ 0 & 1 & 1 \\ 1 & 1 & 0 \end{bmatrix}$.

Determine which of the following words are in C.

(a) 00011 (b) 10101 (c) 01010 (d) 11010 (e) 01101

216. Let G be the generator matrix of Exercise 215. Find a parity check matrix H, that is, a matrix H with the property that x is in the code generated by G if and only if $H\mathsf{x} = 0$. Assuming that the information in the code which G generates is carried in the first three bits, try to find an H that permits the correction of a single error in an information bit.

Answers to True/False Questions

Week 1—True/False

1. False. If true, then $a + 1 = 1$, so $a = 0$, but then $a - 1 \neq 3$.

2. False. $\overrightarrow{AB} = \begin{bmatrix} -4 \\ 3 \end{bmatrix}$ (the coordinates of B less those of A).

3. False. A vector is a column $\begin{bmatrix} a \\ b \end{bmatrix}$; an arrow is a picture.

4. False. $\mathsf{u} = -\mathsf{v}$

5. True.

6. True. $\begin{bmatrix} 0 \\ 0 \\ 0 \end{bmatrix} = 0 \begin{bmatrix} 1 \\ 2 \\ 3 \end{bmatrix}$.

7. True.

8. True. One is a positive scalar multiple of the other.

9. False. This is commutativity.

10. False. $\mathsf{u} - \mathsf{v}$ runs from the head of v to the head of u.

11. False. A linear combination of $\begin{bmatrix} 1 \\ 0 \\ 1 \end{bmatrix}$ and $\begin{bmatrix} 2 \\ 0 \\ -1 \end{bmatrix}$ is a vector of the form $a \begin{bmatrix} 1 \\ 0 \\ 1 \end{bmatrix} + b \begin{bmatrix} 2 \\ 0 \\ -1 \end{bmatrix} = \begin{bmatrix} a + 2b \\ 0 \\ a - b \end{bmatrix}$. Since $\begin{bmatrix} 1 \\ 2 \\ 3 \end{bmatrix}$ does not have second component 0, it is not such a linear combination.

12. True. $4\mathsf{u} - 9\mathsf{v} = 2(2\mathsf{u}) + (-\frac{9}{7})(7\mathsf{v})$.

13. True. Given $\mathsf{u} = \mathsf{v} + \mathsf{w}$, we have $\mathsf{w} = \mathsf{u} - \mathsf{v}$.

14. True. u and v are not parallel.

15. False. $|c|$ times the length of v.

16. False. True just in two dimensions. See Example 1.18.

17. True. These vectors, also labelled $\mathsf{e}_1, \mathsf{e}_2, \mathsf{e}_3$, are the standard basis vectors for R^3.

Week 2—True/False

1. False. $\sqrt{4} = 2$.

2. False. For example, if $k = -2$, then $k^2 = 4$ and $\sqrt{4} = 2$, not k. The general rule is $\sqrt{k^2} = |k|$.

3. False: The solution is $\pm\sqrt{4} = \pm 2$.

4. True.

5. True. The length is $\sqrt{(-1)^2 + 2^2 + 2^2} = 3$.

6. False. A unit vector has length 1; this vector has length $\sqrt{3}$.

7. True: The length of this vector is $\sqrt{(\frac{1}{\sqrt{6}})^2 + (\frac{-2}{\sqrt{6}})^2 + (\frac{1}{\sqrt{6}})^2} = \sqrt{\frac{1}{6} + \frac{4}{6} + \frac{1}{6}} = \sqrt{1} = 1$.

8. True. This is **2.6**.

9. False. This is true if and only if the vectors are unit vectors. In general, $\cos\theta = \dfrac{\mathsf{u} \cdot \mathsf{v}}{\|\mathsf{u}\|\,\|\mathsf{v}\|}$.

10. True. $\mathsf{u} \cdot \mathsf{u} = \|\mathsf{u}\|^2$ is the sum of the squares of the components of u. If this is 0, each number being squared is 0, so each component of u is 0. This means $\mathsf{u} = \mathbf{0}$.

11. False. If u and v are unit vectors, $\mathsf{u} \cdot \mathsf{v}$ is the cosine of the angle between them, and this cannot be more than 1.

12. True. The dot product of these vectors is 0.

13. False. Try $\mathsf{u} = \mathsf{i}$ and $\mathsf{v} = \mathsf{j}$.

14. False. If $c = d = 0$, the alleged inequality would say $|ab| \le 0$ for any a, b, and this is silly.

15. False. Let $a = 1$ and $b = 4$.

16. True. The Cauchy-Schwarz inequality says $|\mathsf{u} \cdot \mathsf{v}| \le \|\mathsf{u}\|\,\|\mathsf{v}\|$. Since $\mathsf{u} \cdot \mathsf{v} \le |\mathsf{u} \cdot \mathsf{v}|$, the given statement also is true.

Week 3—True/False

1. True. The coordinates of the point satisfy the equation.

2. True. The coordinates of both $(0,0,0)$ and $(1,-2,2)$ satisfy the equation of the plane, so the tail and head of the arrow are in the plane.

3. False. The arrow from the origin to $(-1,2,0)$ is not in π since the plane does not pass through $(0,0,0)$.

4. True. The arrow from $(0,0,0)$ to $(-1,2,0)$ is in π.

5. True. The normal vectors $\begin{bmatrix}-2\\3\\-1\end{bmatrix}$ and $\begin{bmatrix}4\\-6\\2\end{bmatrix}$ are parallel.
6. True. The point $(-2,3,1)$ is on the plane.
7. False. $\begin{vmatrix}2 & -3\\5 & 4\end{vmatrix} = 8-(-15) = 23.$
8. True. $a-b=-(b-a)$ for any numbers a and b.
9. False. See **3.18**.
10. False. $\mathsf{u}\times\mathsf{v}$ is a vector and the absolute value of a vector is meaningless.
11. False. Removing the absolute value signs on the left, we obtain the correct formula $\mathsf{u}\cdot\mathsf{v} = \|\mathsf{u}\|\,\|\mathsf{v}\|\cos\theta$.
12. True. Since $\|\mathsf{u}\times\mathsf{v}\| = \|\mathsf{u}\|\,\|\mathsf{v}\|\sin\theta$, the given statement says $\sin\theta = 1$, so $\theta = \frac{\pi}{2}$ and the vectors are orthogonal.
13. False. The norm of a vector cannot be negative.
14. True. $\mathsf{i}\times(\mathsf{i}\times\mathsf{j}) = \mathsf{i}\times\mathsf{k} = -\mathsf{j}$.
15. True: $t=0$.
16. False: $-1,0,1$ are the components of a direction vector.
17. True: $t=1$.
18. False. The line is perpendicular to the plane since its direction is a normal to the plane.
19. False. The line has direction normal to the plane, so the line is perpendicular to the plane and certainly hits it!

Week 4—True/False

1. False. The projection of a vector u on a nonzero vector v is a multiple of v. The u at the end should be v.
2. True: If $\frac{\mathsf{u}\cdot\mathsf{v}}{\mathsf{v}\cdot\mathsf{v}}\mathsf{u} = 0$, the numerator of the fraction must be 0; that is, $\mathsf{u}\cdot\mathsf{v} = 0$, so the vectors are orthogonal.
3. False. If u and v are orthogonal, the projection of u on v is the zero vector.
4. False. The vectors e and f must be **orthogonal**. See Theorem 4.8.
5. True.
6. True: There are lots of linear combinations of u,v,w that are 0 without all coefficients being 0; for instance, $75\mathsf{u} + 0\mathsf{v} + 0\mathsf{w} = 0$.

7. True: This is the definition of "linear independence."

8. False: For any vectors u, v, w—linearly dependent or not—$0u + 0v + 0w = 0$.

9. False. The statement $0u + 0v = 0$ is true for any vectors u and v and has nothing to do with linear independence.

10. False. Some nontrivial linear combination of the vectors is 0.

11. True: $v = -3u$ and this gives a nontrivial linear combination of u and v that is 0: $3u + v = 0$.

Week 5—True/False

1. True. 3 rows; 2 columns.

2. True.

3. False. The $(2, 1)$ entry is 2, the entry in row two, column one.

4. False: Each entry of B is twice the corresponding entry of A.

5. True.

6. False. Take A to be 2×3 and B to be 3×2. Then AB is 2×2 and BA is 3×3, both of which are square.

7. False.

8. True.

9. True. Since $b = e_3$, the third standard basis vector, Ab is the third column of A.

10. True. The hypothesis implies, in particular, that $Ae_i = 0$, where $e_1, e_2, \dots$ denote the standard basis vectors. But Ae_i is column i of A, so each column of A is 0. Thus $A = 0$.

11. True. AB is 2×5.

12. True. To compute the product of the $m \times n$ matrix A and the $m \times n$ matrix A, $m = n$.

13. False. This says matrix multiplication is **associative**.

14. False. For example, let $A = \begin{bmatrix} 1 & 0 \end{bmatrix}$ and $B = \begin{bmatrix} 0 \\ 1 \end{bmatrix}$.

15. False. For example, let $A = \begin{bmatrix} 1 & 0 \end{bmatrix}$, $B = \begin{bmatrix} 0 \\ 1 \end{bmatrix}$, and $X = \begin{bmatrix} 1 & 1 \\ 1 & 1 \end{bmatrix}$.

16. True. This is the very important fact **5.23**.

17. True: This is again **5.23**.

Week 6—True/False

1. False. It's two elementary operations.
2. True.
3. True. The second system is the same as the first, except that the third equation has been replaced by its sum with −2 times the first.
4. True.
5. False. A system of linear equations has no solution, one solution, or infinitely many solutions.
6. True.
7. True.
8. False. Consider $\begin{bmatrix} 0 & 1 \\ 0 & 1 \end{bmatrix}$ as in the notes.
9. True.
10. True. Replacing row three by row three less one seventh of row two gives the row echelon matrix $\begin{bmatrix} 4 & 0 & 2 \\ 0 & 7 & 1 \\ 0 & 0 & \frac{6}{7} \end{bmatrix}$. Since Gaussian elimination did not require any multiplication of a row by a scalar, the pivots of the given matrix are, by definition, the pivots of the row echelon matrix.
11. False. This matrix is upper triangular but not row echelon. Row echelon form is $\begin{bmatrix} 0 & 0 & 2 & 1 & 5 \\ 0 & 0 & 0 & 0 & 1 \\ 0 & 0 & 0 & 0 & 0 \end{bmatrix}$, so the pivots are 2 and 1.
12. True.
13. False. The last row corresponds to the equation $0 = 1$, so there are no solutions.
14. False. Free variables correspond to the columns **without** pivots.

Week 7—True/False

1. True.
2. True. Since Ax is a linear combination of the columns of A, this is the definition of linear dependence.
3. False.
4. True.
5. False, unless $AB = BA$ (which is seldom the case).

6. False. $X = BA^{-1}$. (Since matrix multiplication is not commutative, there is no reason why $BA^{-1} = A^{-1}B$.)

7. True.

8. True.

9. True.

10. False.

11. False. The $(1,5)$ entry should be 0.

Week 8—True/False

1. False. If an elementary row operation applied to the identity matrix puts a 2 in the $(2,2)$ position, the operation is "multiply row two by 2". This operation leaves a 0, not a 1, in the $(2,3)$ position.

2. False. An elementary matrix corresponds to just one elementary row operation.

3. True. Since E and F are elementary, they are invertible, and the product of invertible matrices is invertible.

4. True. E is elementary, corresponding to the elementary row operation which subtracts twice row three from row two. The inverse of E adds twice row three to row two.

5. True. The elementary matrix E is obtained from the 3×3 identity matrix I by an elementary row operation that not does not change rows two or three of I.

6. True.

7. False. For example, $\begin{bmatrix} 0 & 1 \\ 1 & 0 \end{bmatrix}$ is not lower triangular.

8. True. Elementary matrices corresponding to elementary row operations of types 2 and 3 are lower triangular and the product of lower triangular matrices is lower triangular.

9. True. This is Theorem 8.12.

10. False. First solve $L\mathsf{y} = \mathsf{b}$ for y and then $U\mathsf{x} = \mathsf{y}$ for x.

11. True. Since E is a permutation matrix, $E^{-1} = E^T = E$.

12. True.

13. False. In reducing A to row echelon form, an interchange of rows is required.

Week 9—True/False

1. True. Strike out the first row and second column of the matrix. The determinant of what is left is 2, which is the $(1,2)$ minor. The $(1,2)$ cofactor is $(-1)^{1+2}2 = -2$.

2. False. $C = \begin{bmatrix} 2 & 3 \\ -4 & 1 \end{bmatrix}$.

3. False. Adjoint is the transpose of the matrix of cofactors.

4. True. $A(\operatorname{adj} A) = (\det A)I$.

5. True.

6. True.

7. True.

Week 10—True/False

1. False. $\det(-2A) = (-2)^4 \det A = 48$.

2. True. $\det A = (-2)(3)(-4) \neq 0$.

3. True. By linearity in row two.

4. True. The determinant of a permutation matrix is ± 1.

5. False.

6. True. $\det A^{-1}P = \det A^{-1} \det P = \dfrac{1}{\det A} \det P$.

7. False. The matrix $A = \begin{bmatrix} 1 & 2 \\ 1 & 2 \end{bmatrix}$ has determinant 0.

8. True. $\det AB = (\det A)(\det B) = (\det B)(\det A) = \det BA$.

9. False. The determinant is divided by 2 also.

10. True. The operation $R2 \to R2 + 2(R1)$ does not change the determinant.

11. True. Factor -1 from row two and 2 from row three.

12. True. The determinant is linear in row two (and all other rows too).

13. True. The determinant is the sum $\begin{vmatrix} a & b & c \\ d & e & f \\ x & y & z \end{vmatrix} + \begin{vmatrix} a & b & c \\ x & y & z \\ x & y & z \end{vmatrix}$ but the second determinant is 0 (it has two equal rows).

14. False. ± 1.

15. False. If and only if it's **determinant** is not 0.

16. True. "Not singular" means "invertible" and if A is invertible, so is A^{-1}, with $(A^{-1})^{-1} = A$.

Week 11—True/False

1. True.
2. False.
3. False. The eigenspace contains 0, which is not an admissible eigenvector.
4. False. $(A - \lambda \mathbf{I})\mathrm{x} = 0$.
5. True.
6. True.
7. True.
8. False. The eigen**values**.
9. False.
10. True. The characteristic polynomial is the determinant of $\begin{bmatrix} 1-\lambda & 2 & 3 \\ 0 & 4-\lambda & 5 \\ 0 & 0 & 6-\lambda \end{bmatrix}$.

    This matrix is triangular, so its determinant is the product of its diagonal entries.
11. True. The eigenvalues of the diagonal matrix $A^{10} = \begin{bmatrix} 1 & 0 \\ 0 & 2^{10} \end{bmatrix}$ are its diagonal entries.
12. False. Similar matrices have the same eigenvalues, but the given matrices have eigenvalues $2, 3, 5$, and $1, 6, 5$, respectively.
13. False. Similar matrices have the same trace, but these matrices have traces 1 and 4.
14. False.
15. False. The only matrix similar to $I = \begin{bmatrix} 1 & 0 \\ 0 & 1 \end{bmatrix}$ is I itself so $\begin{bmatrix} 1 & 0 \\ 0 & 2 \end{bmatrix}$, for example, isn't similar to I.
16. False: Similar matrices have the same determinant, but these matrices have determinants 16 and 11.
17. True. Any diagonal matrix is diagonalizable: Take $P = I$.
18. False. The identity matrix $\begin{bmatrix} 1 & 0 \\ 0 & 1 \end{bmatrix}$ is diagonalizable with 1 as its only eigenvalue.
19. True. This is part of Theorem 11.19.
20. False. The 3×3 matrix in Example 11.16 is diagonalizable, yet it has just two different eigenvalues.
21. False. The symmetric matrix $A = \begin{bmatrix} 0 & 1 \\ 1 & 0 \end{bmatrix}$ has eigenvalues ± 1.

Solutions to Test Yourself Exercises

Week 1—Test Yourself

1. They are the **components** of the vector.

2. $\overrightarrow{AB} = \begin{bmatrix} 3 \\ 3 \end{bmatrix}$.

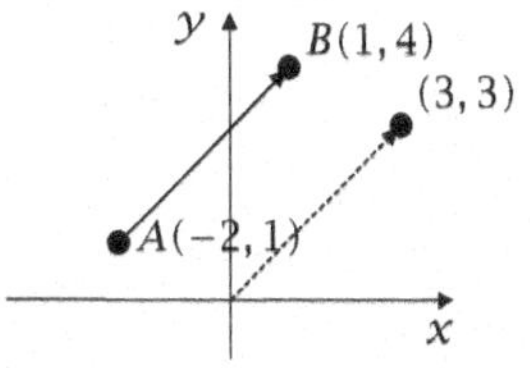

3. Let $B = (x, y)$. Then $\overrightarrow{AB} = \begin{bmatrix} x-1 \\ y-4 \end{bmatrix} = \begin{bmatrix} -1 \\ 2 \end{bmatrix}$, so $x - 1 = -1$, $y - 4 = 2$, and $B = (0, 6)$.

4. $\mathsf{x} = -\frac{1}{4}\mathsf{u} = -7\mathsf{w}$; x is not a scalar multiple of v.

5. One is a scalar multiple of the other.

6. $\mathsf{v} = c\mathsf{u}$ (or $\mathsf{u} = c\mathsf{v}$) for some **negative** number c.

7. The sum of vectors is a vector.

8. $\begin{bmatrix} 4 \\ 7 \\ 11 \end{bmatrix}$

9. We wish to find a and b such that $\begin{bmatrix} 7 \\ 7 \end{bmatrix} = a\begin{bmatrix} 2 \\ 6 \end{bmatrix} + b\begin{bmatrix} 14 \\ -8 \end{bmatrix}$.

   Thus we should have $\begin{aligned} 2a + 14b &= 7 \\ 6a - 8b &= 7 \end{aligned}$.

   The first equation gives $a = \frac{7}{2} - 7b$. Substituting in the second gives $21 - 42b - 8b = 7$, so $50b = 14$, $b = \frac{7}{25}$.

10. The picture at the right shows the position of the points. The answer is not unique. There are three possibilities for $D(x, y)$.
    If $\overrightarrow{CA} = \overrightarrow{DB}$, then $\begin{bmatrix} 2 \\ 4 \end{bmatrix} = \begin{bmatrix} 4-x \\ -3-y \end{bmatrix}$ and D is $(2, -7)$.
    If $\overrightarrow{DA} = \overrightarrow{CB}$, then $\begin{bmatrix} 1-x \\ 2-y \end{bmatrix} = \begin{bmatrix} 5 \\ -1 \end{bmatrix}$ and D is $(-4, 3)$.
    If $\overrightarrow{AD} = \overrightarrow{CB}$, then $\begin{bmatrix} x-1 \\ y-2 \end{bmatrix} = \begin{bmatrix} 5 \\ -1 \end{bmatrix}$ and D is $(6, 1)$.

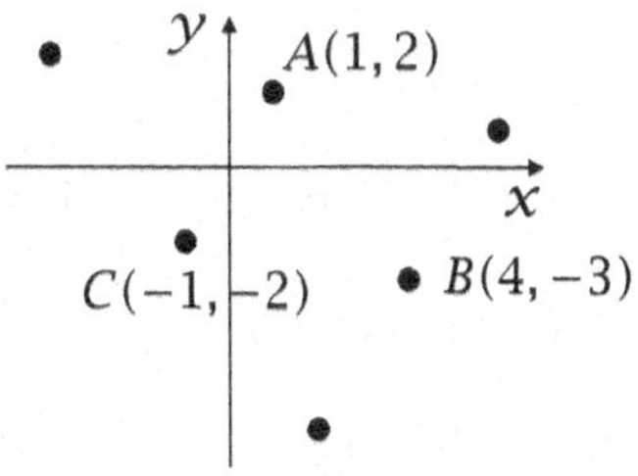

11. A linear combination of $\begin{bmatrix}1\\ \frac{3}{2}\\ 0\end{bmatrix}$ and $\begin{bmatrix}0\\3\\6\end{bmatrix}$ is a vector of the form $a\begin{bmatrix}1\\ \frac{3}{2}\\ 0\end{bmatrix} + b\begin{bmatrix}0\\3\\6\end{bmatrix}$. This is $\frac{1}{2}\begin{bmatrix}2\\3\\0\end{bmatrix} + 3b\begin{bmatrix}0\\1\\2\end{bmatrix}$, which is a linear combination of $\begin{bmatrix}2\\3\\0\end{bmatrix}$ and $\begin{bmatrix}0\\1\\2\end{bmatrix}$.

12. $\mathsf{i} = \begin{bmatrix}1\\0\\0\end{bmatrix}$, $\mathsf{j} = \begin{bmatrix}0\\1\\0\end{bmatrix}$ and $\mathsf{k} = \begin{bmatrix}0\\0\\1\end{bmatrix}$.

13. $\begin{bmatrix}1\\3\\2\\4\end{bmatrix} = \mathsf{e}_1 + 3\mathsf{e}_2 + 2\mathsf{e}_3 + 4\mathsf{e}_4.$

Week 2—Test Yourself

1. We must find scalars c_1 and c_2 so that $\begin{bmatrix}1\\0\end{bmatrix} = c_1\begin{bmatrix}1\\1\end{bmatrix} + c_2\begin{bmatrix}-2\\1\end{bmatrix}$. We get the equations $\begin{aligned} c_1 - 2c_2 &= 1 \\ c_1 - c_2 &= 0 \end{aligned}$. So $c_2 = -c_1$ and $c_1 + 2c_1 = 1$, so $c_1 = \frac{1}{3}$, $c_2 = -\frac{1}{3}$.

2. Since $\begin{bmatrix}-18\\5\\9\end{bmatrix} = 3\begin{bmatrix}-1\\0\\3\end{bmatrix} - 5\begin{bmatrix}3\\-1\\0\end{bmatrix}$, $\mathsf{w} = 3\mathsf{u} - 5\mathsf{v}$, the answer is yes.

3. $\mathsf{u} \cdot \mathsf{v} = \mathsf{v} \cdot \mathsf{u}$ for any vectors u and v.

4. $c\mathsf{u} = \begin{bmatrix}2\\4\\6\end{bmatrix}$, so $(c\mathsf{u}) \cdot \mathsf{v} = 2(-2) + 4(1) + 6(1) = 6$. On the other hand, $\mathsf{u} \cdot \mathsf{v} = 1(-2) + 2(1) + 3(1) = 3$, so $c(\mathsf{u} \cdot \mathsf{v}) = 2(3) = 6$ as well. This is an example of "scalar associativity."

5. (a) $\|\mathsf{u}\| = \sqrt{0^2 + 3^2 + 4^2} = 5$ and $\|\mathsf{v}\| = \sqrt{1^2 + 2^2 + 2^2} = 3$.

   (b) Since $\|\mathsf{u}\| = 5$, $\frac{1}{5}\mathsf{u} = \frac{1}{5}\begin{bmatrix}0\\3\\4\end{bmatrix} = \begin{bmatrix}0\\ \frac{3}{5}\\ \frac{4}{5}\end{bmatrix}$ is a unit vector in the direction of u. Since $\|\mathsf{v}\| = 3$, $-\frac{1}{3}\mathsf{v} = \begin{bmatrix}-\frac{1}{3}\\ -\frac{2}{3}\\ -\frac{2}{3}\end{bmatrix}$ is a unit vector in the direction opposite v.

6. A better approach would be to say that $\mathsf{u} = \frac{4}{7}\mathsf{v}$, with $\mathsf{v} = \begin{bmatrix}-2\\1\\5\end{bmatrix}$, so $\|\mathsf{u}\| = \frac{4}{7}\|\mathsf{v}\| = \frac{4}{7}\sqrt{30}$.

7. (a) $\mathsf{u} \cdot \mathsf{v} = -3 + 4 = 1$. Also $\|\mathsf{u}\| = \sqrt{3^2 + 4^2} = 5$, $\|\mathsf{v}\| = \sqrt{1^2 + 1^2} = \sqrt{2}$, so $\cos\theta = \dfrac{\mathsf{u} \cdot \mathsf{v}}{\|\mathsf{u}\|\,\|\mathsf{v}\|} = \dfrac{1}{5\sqrt{2}} \approx 0.1414$. $\theta \approx \cos^{-1}(.1414) \approx 1.43$ rads $\approx 82°$.

(b) $u \cdot v = 2 - 1 + 2 = 3$. Also $\|u\| = \sqrt{6} = \|v\|$, so $\cos\theta = \dfrac{u \cdot v}{\|u\|\,\|v\|} = \dfrac{3}{6} = \dfrac{1}{2}$. $\theta = \cos^{-1}(\frac{1}{2}) = \frac{\pi}{3}$.

8. $\|u\| = \sqrt{3^2 + 4^2 + 0^2} = 5$; $u + v = \begin{bmatrix} 5 \\ 5 \\ 2 \end{bmatrix}$, so $\|u + v\| = \sqrt{5^2 + 5^2 + 2^2} = \sqrt{54}$;

$\left\| \frac{w}{\|w\|} \right\| = 1$ (this is true for any $w \neq 0$).

9. We are given $u \cdot u = 9$, $v \cdot v = 4$ and $w \cdot w = 36$. Thus $(u - 2v) \cdot (3w + 2u) = 3u \cdot w - 6v \cdot w + 2u \cdot u - 4u \cdot v = 3(5) - 6(2) + 2(9) - 4(3) = 9$.

10. (a) Since $u \cdot v = -2 + 6 + 3 \neq 0$, u and v are not orthogonal.

(b) Since $u \cdot v = -3 + 0 - 2 + 5 = 0$, u and v are orthogonal.

11. We have $\cos\theta = \dfrac{u \cdot v}{\|u\|\,\|v\|} = \dfrac{-8 - 2 - 2}{\sqrt{24}\sqrt{6}} = \dfrac{-12}{\sqrt{144}} = -1$, so $\theta = \pi$, a fact consistent with the observation that $u = -2v$.

12. (a) Since $\|cu\| = |c|\,\|u\| = |c|\sqrt{6}$, we want $|c| = \dfrac{8}{\sqrt{6}}$, so $c = \pm\frac{4}{3}\sqrt{6}$.

(b) Since $u + cv = \begin{bmatrix} 1 - 3c \\ 1 \\ 0 \\ -2 + c \end{bmatrix}$ and we want $u \cdot (u + cv) = 0$, we must have $(1 - 3c) + 1 + 0 - 2(-2 + c) = 0$, so $6 - 5c = 0$ and $c = \frac{6}{5}$.

13. We have $\cos 60° = \frac{1}{2}$, $\|u\| = \sqrt{2}$, $\|v\| = \sqrt{x^2 + 2}$ and $u \cdot v = x + 1$.

We seek those x such that $u \cdot v = \|u\|\,\|v\| \cos 60°$; that is, so that $x + 1 = \frac{\sqrt{2x^2+4}}{2}$. Squaring both sides gives $x^2 + 2x + 1 = \frac{2x^2+4}{4}$, so $4x^2 + 8x + 4 = 2x^2 + 4$ and $2x^2 + 8x = 0$; i.e., $x^2 + 4x = 0 = x(x + 4)$. So $x = 0$ and $x = -4$ are possibilities. When $x = -4$, $u \cdot v = -3$, whereas $\|u\|\,\|v\| \cos 60° = \sqrt{2}\sqrt{18}(\frac{1}{2}) = +3$, so $x = -4$ is not a solution. When $x = 0$, $u \cdot v = 1$ and $\|u\|\,\|v\| \cos 60° = \sqrt{2}\sqrt{2}(\frac{1}{2}) = 1$, so $x = 0$ is a solution, and the only one.

14. (a) We have $u \cdot v = -3$, $\|u\| = \sqrt{5}$, $\|v\| = \sqrt{10}$. The Cauchy-Schwarz inequality is the statement $3 \leq \sqrt{5}\sqrt{10}$. Note that $\sqrt{5}\sqrt{10} \approx 7.1$. Since $u + v = \begin{bmatrix} -2 \\ 1 \\ -2 \end{bmatrix}$, $\|u + v\| = \sqrt{9} = 3$. The triangle inequality says $3 \leq \sqrt{5} + \sqrt{10}$. Note that $\sqrt{5} + \sqrt{10} \approx 5.4$.

(b) We have $u \cdot v = -2$, $\|u\| = \sqrt{6}$, $\|v\| = \sqrt{11}$. The Cauchy-Schwarz inequality is the statement $2 \leq \sqrt{6}\sqrt{11}$. Note that $\sqrt{6}\sqrt{11} \approx 8.1$. Since $u + v = \begin{bmatrix} 2 \\ -1 \\ 2 \\ 2 \end{bmatrix}$, $\|u + v\| = \sqrt{13}$. The triangle inequality says $\sqrt{13} \leq \sqrt{6} + \sqrt{11}$. Note that $\sqrt{13} \approx 3.6$, while $\sqrt{6} + \sqrt{11} \approx 5.8$.

15. Let $\mathsf{u} = \begin{bmatrix} a \\ b \end{bmatrix}$ and $\mathsf{v} = \begin{bmatrix} \cos\theta \\ \sin\theta \end{bmatrix}$. Then $\mathsf{u} \cdot \mathsf{v} = a\cos\theta + b\sin\theta$, $\|\mathsf{u}\|^2 = a^2 + b^2$ and $\|\mathsf{v}\|^2 = \cos^2\theta + \sin^2\theta = 1$. Squaring the Cauchy-Schwarz inequality gives $(\mathsf{u} \cdot \mathsf{v})^2 \le \|\mathsf{u}\|^2 \|\mathsf{v}\|^2$, which is the given inequality.

16. No. 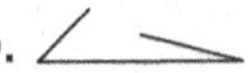

The "Triangle Inequality" is so-named because the sum of the lengths of any two sides of a triangle can never be less than the length of the third: $\|a\| + \|b\| \ge \|a + b\|$.

Week 3—Test Yourself

1. (a) Yes: The plane contains the arrow from $(0,0,0)$ to $(2,-1,3)$.

   (b) No: The plane does not pass through the origin.

   (c) Yes: The plane passes through the origin and contains the point $(2,-1,3)$.

2. The coefficients of x, y, z are the components of a normal, so one normal is $\begin{bmatrix} -1 \\ 2 \\ 1 \end{bmatrix}$.

   Any triple (x, y, z) satisfying the equation gives a point on the plane, for example, $(-4,0,0)$.

3. The vector $\begin{bmatrix} x \\ y \\ z \end{bmatrix}$ is in the plane if and only if $x = \frac{3}{2}y - 2z$, that is, if and only if $\begin{bmatrix} x \\ y \\ z \end{bmatrix} = \begin{bmatrix} \frac{3}{2}y - 2z \\ y \\ z \end{bmatrix} = y\begin{bmatrix} \frac{3}{2} \\ 1 \\ 0 \end{bmatrix} + z\begin{bmatrix} -2 \\ 0 \\ 1 \end{bmatrix}$. Thus the plane is the set of all linear combinations of $\begin{bmatrix} \frac{3}{2} \\ 1 \\ 0 \end{bmatrix}$ and $\begin{bmatrix} -2 \\ 0 \\ 1 \end{bmatrix}$. (Multiplying the first vector by 2, the plane is also the set of all linear combinations of $\begin{bmatrix} 3 \\ 2 \\ 0 \end{bmatrix}$ and $\begin{bmatrix} -2 \\ 0 \\ 1 \end{bmatrix}$.)

4. $$\mathsf{u} \times \mathsf{v} = \begin{vmatrix} \mathsf{i} & \mathsf{j} & \mathsf{k} \\ 4 & -2 & -1 \\ 5 & 6 & -3 \end{vmatrix} = \begin{vmatrix} -2 & -1 \\ 6 & -3 \end{vmatrix} \mathsf{i} - \begin{vmatrix} 4 & -1 \\ 5 & -3 \end{vmatrix} \mathsf{j} + \begin{vmatrix} 4 & -2 \\ 5 & 6 \end{vmatrix} \mathsf{k}$$
   $$= 12\mathsf{i} - (-7)\mathsf{j} + 34\mathsf{k} = \begin{bmatrix} 12 \\ 7 \\ 34 \end{bmatrix}.$$
   $$\mathsf{v} \times \mathsf{u} = \begin{vmatrix} \mathsf{i} & \mathsf{j} & \mathsf{k} \\ 5 & 6 & -3 \\ 4 & -2 & -1 \end{vmatrix} = \begin{vmatrix} 6 & -3 \\ -2 & -1 \end{vmatrix} \mathsf{i} - \begin{vmatrix} 5 & -3 \\ 4 & -1 \end{vmatrix} \mathsf{j} + \begin{vmatrix} 5 & 6 \\ 4 & -2 \end{vmatrix} \mathsf{k}$$
   $$= -12\mathsf{i} - 7\mathsf{j} - 34\mathsf{k} = \begin{bmatrix} -12 \\ -7 \\ -34 \end{bmatrix} = -(\mathsf{u} \times \mathsf{v}).$$

5. $\mathsf{u} \times \mathsf{v} = \begin{vmatrix} \mathsf{i} & \mathsf{j} & \mathsf{k} \\ -2 & -4 & 1 \\ 1 & 5 & -2 \end{vmatrix} = \begin{vmatrix} -4 & 1 \\ 5 & -2 \end{vmatrix} \mathsf{i} - \begin{vmatrix} -2 & 1 \\ 1 & -2 \end{vmatrix} \mathsf{j} + \begin{vmatrix} -2 & -4 \\ 1 & 5 \end{vmatrix} \mathsf{k}$

$= 3\mathsf{i} - 3\mathsf{j} - 6\mathsf{k} = \begin{bmatrix} 3 \\ -3 \\ -6 \end{bmatrix} = 3\begin{bmatrix} 1 \\ -1 \\ -2 \end{bmatrix}$. Thus $\|\mathsf{u} \times \mathsf{v}\| = 3\sqrt{1+1+4} = 3\sqrt{6}$. Also $\|\mathsf{u}\| = \sqrt{4+16+1} = \sqrt{21}$, $\|\mathsf{v}\| = \sqrt{1+25+4} = \sqrt{30}$ and $\mathsf{u} \cdot \mathsf{v} = -2-20-2 = -24$. The cosine of the angle θ between u and v is $\cos\theta = \frac{\mathsf{u} \cdot \mathsf{v}}{\|\mathsf{u}\|\,\|\mathsf{v}\|} = -\frac{24}{\sqrt{630}} = -\frac{8}{\sqrt{70}}$. So $\sin^2\theta = 1 - \cos^2\theta = \frac{6}{70} = \frac{3}{35}$. Assuming $0 \le \theta \le \pi$, $\sin\theta \ge 0$, so $\sin\theta = \sqrt{\frac{3}{35}}$. Thus

$\|\mathsf{u}\|\,\|\mathsf{v}\| \sin\theta$

$$= \sqrt{21}\sqrt{30}\sqrt{\frac{3}{35}} = \sqrt{\frac{3(7)(3)(2)(5)(3)}{5(7)}} = \sqrt{3(3)(2)(3)} = 3\sqrt{6} = \|\mathsf{u} \times \mathsf{v}\|.$$

6. The answer is an equation of the form $18x + 6y - 5z = d$. Substituting $x = -1$, $y = 1$, $z = 7$, we get $d = -18 + 6 - 35 = -47$, so the plane has equation $18x + 6y - 5z = -47$.

7. $\overrightarrow{AB} = \begin{bmatrix} 1 \\ -1 \\ 0 \end{bmatrix}$ and $\overrightarrow{AC} = \begin{bmatrix} 8 \\ -5 \\ -1 \end{bmatrix}$.

A normal vector is $\overrightarrow{AB} \times \overrightarrow{AC} = \begin{vmatrix} \mathsf{i} & \mathsf{j} & \mathsf{k} \\ 1 & -1 & 0 \\ 8 & -5 & -1 \end{vmatrix} = 1\mathsf{i} - (-1)\mathsf{j} + 3\mathsf{k} = \begin{bmatrix} 1 \\ 1 \\ 3 \end{bmatrix}$.

The plane has equation of the form $x + y + 3z = d$. Since the coordinates of A satisfy the equation, we have $-1 + 2 + 3 = d$, so $d = 4$ and the equation is $x + y + 3z = 4$.

8. $c_1\mathsf{u} + c_2\mathsf{v} = \begin{bmatrix} c_1 + c_2 \\ -3c_1 + 2c_2 \\ 6c_2 \end{bmatrix}$. A normal is $\mathsf{u} \times \mathsf{v}$

$$= \begin{vmatrix} \mathsf{i} & \mathsf{j} & \mathsf{k} \\ 1 & -3 & 0 \\ 1 & 2 & 6 \end{vmatrix} = \begin{vmatrix} -3 & 0 \\ 1 & 6 \end{vmatrix} \mathsf{i} - \begin{vmatrix} 1 & 0 \\ 1 & 6 \end{vmatrix} + \begin{vmatrix} 1 & -3 \\ 1 & 2 \end{vmatrix}$$

$$= -18\mathsf{i} - 6\mathsf{j} + 5\mathsf{k} = \begin{bmatrix} -18 \\ -6 \\ 5 \end{bmatrix}.$$

Since the plane in question passes through $(0,0,0)$, an equation is $-18x - 6y + 5z = 0$. Another equation is $18x + 6y - 5z = 0$.

9. $\begin{vmatrix} \mathsf{i} & \mathsf{j} & \mathsf{k} \\ 4 & 3 & -1 \\ 2 & 3 & 0 \end{vmatrix} = 3\mathsf{i} - 2\mathsf{j} + 6\mathsf{k} = \begin{bmatrix} 3 \\ -2 \\ 6 \end{bmatrix}$ is a vector n perpendicular to both the given

vectors. Since $\|n\| = \sqrt{3^2 + (-2)^2 + 6^2} = \sqrt{49} = 7$, one possible answer is $\frac{1}{7}\begin{bmatrix} 3 \\ -2 \\ 6 \end{bmatrix}$.

The other possibility is $-\frac{1}{7}\begin{bmatrix} 3 \\ -2 \\ 6 \end{bmatrix}$.

10. $j \times i = k$, $j \times k = i$ and $j \times j = 0$.

11. $\begin{bmatrix} x \\ y \\ z \end{bmatrix} = \begin{bmatrix} 1 \\ 2 \\ 3 \end{bmatrix} + t \begin{bmatrix} -2 \\ 1 \\ 7 \end{bmatrix}$.

12. The point $(1, 2, 3)$ does not lie on the line because the equations

$$\begin{aligned} 1 &= 3 + 2t \\ 2 &= -4 + t \\ 3 &= 1 + 5t \end{aligned}$$

have no solution.

The point $(7, -2, 11)$ does lie on the line because the equations

$$\begin{aligned} 7 &= 3 + 2t \\ -2 &= -4 + t \\ 11 &= 1 + 5t \end{aligned}$$

have the solution $t = 2$.

13. (a) The line has direction $d = \begin{bmatrix} -1 \\ 1 \\ 2 \end{bmatrix}$, the plane has normal $n = \begin{bmatrix} 2 \\ 1 \\ -1 \end{bmatrix}$ and these directions are not perpendicular because $d \cdot n = -3 \neq 0$.

(b) The point of intersection has coordinates $(1-t, 1+t, 2+2t)$ and these satisfy the equation of the plane. So $2(1-t)+(1+t)-(2+2t) = 3$, that is, $-3t+1 = 7$, $t = -2$ and the point is $(3, -1, -2)$.

14. If you got the correct answer to Exercise 13(a), you know that $(3, -1, -2)$ is on ℓ, and this is "obviously" on the other line here: take $t = 0$.

Alternatively, the equations

$$\begin{bmatrix} x \\ y \\ z \end{bmatrix} = \begin{bmatrix} 1 \\ 1 \\ 2 \end{bmatrix} + \begin{bmatrix} -1 \\ 1 \\ 2 \end{bmatrix} = \begin{bmatrix} 3 \\ -1 \\ -2 \end{bmatrix} + s \begin{bmatrix} 1 \\ 0 \\ 1 \end{bmatrix}$$

have solution $t = -2$, $s = 0$.

15. The direction of the line is $d = \begin{bmatrix} 2 \\ 1 \\ -4 \end{bmatrix}$. A normal to the plane is $n = \begin{bmatrix} 3 \\ -10 \\ -1 \end{bmatrix}$. Since $d \cdot n = 0$, d and n are orthogonal, the line is parallel to the plane. There is no point of intersection.

Week 4—Test Yourself

1. The projection of u on v is $\operatorname{proj}_{\mathsf{v}} \mathsf{u} = \frac{\mathsf{u}\cdot\mathsf{v}}{\mathsf{v}\cdot\mathsf{v}}\mathsf{v} = \frac{1}{26}\begin{bmatrix}3\\1\\-4\end{bmatrix}$.

   The projection of v on u is $\operatorname{proj}_{\mathsf{u}} \mathsf{v} = \frac{\mathsf{v}\cdot\mathsf{u}}{\mathsf{u}\cdot\mathsf{u}}\mathsf{u} = \frac{1}{74}\begin{bmatrix}7\\0\\5\end{bmatrix}$.

2. See the figure on 47.

3. With reference to the diagram in the notes accompany Example 4.6, we let Q be any point on the line, say $Q(-1,2,0)$. Then $\overrightarrow{QP} = \begin{bmatrix}5\\-6\\17\end{bmatrix}$. The line has direction $\mathsf{d} = \begin{bmatrix}4\\-1\\8\end{bmatrix}$ and the distance we want is the length of $\overrightarrow{QP} - \mathsf{p}$, where $\mathsf{p} = \operatorname{proj}_{\mathsf{d}} \overrightarrow{QP}$ is the projection of $\overrightarrow{QP}$ on d. We have

$$\mathsf{p} = \operatorname{proj}_{\mathsf{d}} \overrightarrow{QP} = \frac{\mathsf{w}\cdot\mathsf{d}}{\mathsf{d}\cdot\mathsf{d}}\mathsf{d} = \frac{162}{81}\begin{bmatrix}4\\-1\\8\end{bmatrix} = \begin{bmatrix}8\\-2\\16\end{bmatrix}, \text{ so } \overrightarrow{QP} - \mathsf{p} = \begin{bmatrix}-3\\-4\\1\end{bmatrix}$$

   and the required distance is $\left\|\begin{bmatrix}-3\\-4\\1\end{bmatrix}\right\| = \sqrt{26}$.

   If $A(x,y,z)$ is the point of ℓ closest to P, then $\overrightarrow{AP} = \overrightarrow{QP} - \mathsf{p}$, so $\begin{bmatrix}4-x\\-4-y\\17-z\end{bmatrix} = \begin{bmatrix}-3\\-4\\1\end{bmatrix}$.

   This gives $A(7,0,16)$.

4. Let Q be any point in the plane, say $Q(4,0,0)$. The required distance is the length of the projection p of $\overrightarrow{PQ}$ on n, a normal to the plane. We have $\overrightarrow{PQ} = \begin{bmatrix}7\\-1\\-12\end{bmatrix}$ and a normal $\mathsf{n} = \begin{bmatrix}2\\-1\\-4\end{bmatrix}$. Then $\operatorname{proj}_{\mathsf{n}} \overrightarrow{PQ} = \frac{\overrightarrow{PQ}\cdot\mathsf{n}}{\mathsf{n}\cdot\mathsf{n}}\mathsf{n} = \frac{63}{21}\mathsf{n}$. The required distance is $\left\|\frac{63}{21}\mathsf{n}\right\| = 3\,\|\mathsf{n}\| = 3\sqrt{21}$. The point of the plane closest to P is the point A with $\overrightarrow{PA} = \operatorname{proj}_{\mathsf{n}} \overrightarrow{PQ}$. Letting A have coordinates (x,y,z), we have $\begin{bmatrix}x+3\\y-1\\z-12\end{bmatrix} = 3\begin{bmatrix}2\\-1\\-4\end{bmatrix}$, so $x+3=6$, $y-1=-3$, $z-12=-12$, so A is $(3,-2,0)$.

5. Two orthogonal vectors in π are $\mathsf{e} = \begin{bmatrix}1\\1\\1\end{bmatrix}$ and $\mathsf{f} = \begin{bmatrix}1\\-1\\0\end{bmatrix}$, so

$$\operatorname{proj}_{\pi} \mathsf{w} = \frac{\mathsf{w}\cdot\mathsf{e}}{\mathsf{e}\cdot\mathsf{e}}\mathsf{e} + \frac{\mathsf{w}\cdot\mathsf{f}}{\mathsf{f}\cdot\mathsf{f}}\mathsf{f} = \frac{1}{3}\begin{bmatrix}1\\1\\1\end{bmatrix} - \frac{5}{2}\begin{bmatrix}1\\-1\\0\end{bmatrix} = \begin{bmatrix}-\frac{13}{6}\\\frac{17}{6}\\\frac{1}{3}\end{bmatrix} = \frac{1}{6}\begin{bmatrix}-13\\17\\2\end{bmatrix}.$$

6. With an eye to the figure on p. 46, we begin with a point on the plane, say $Q(0,0,1)$. A normal to the plane is $\mathsf{n} = \begin{bmatrix} 1 \\ -1 \\ 2 \end{bmatrix}$ and $\overrightarrow{PQ} = \begin{bmatrix} -2 \\ 0 \\ 2 \end{bmatrix}$. The projection of w on n is $\operatorname{proj}_{\mathsf{n}} \overrightarrow{PQ} = \frac{\overrightarrow{PQ}\cdot \mathsf{n}}{\mathsf{n}\cdot\mathsf{n}}\,\mathsf{n} = \frac{2}{6}\,\mathsf{n} = \frac{1}{3}\mathsf{n}$ and the distance from P to π is $\left\|\frac{1}{3}\,\mathsf{n}\right\| = \frac{1}{3}\,\|\mathsf{n}\| = \frac{1}{3}\sqrt{6}$.

7. (a) First we find two vectors in the plane; for example, $\mathsf{u} = \begin{bmatrix} 1 \\ -1 \\ 0 \end{bmatrix}$ and $\mathsf{v} = \begin{bmatrix} 1 \\ 0 \\ -1 \end{bmatrix}$. We take $\mathsf{e} = \mathsf{v}$ for one of the desired two vectors and use for the other

$$\mathsf{f} = \mathsf{u} - \operatorname{proj}_{\mathsf{v}} \mathsf{u} = \mathsf{u} - \frac{\mathsf{u}\cdot\mathsf{v}}{\mathsf{v}\cdot\mathsf{v}}\,\mathsf{v} = \begin{bmatrix} 1 \\ -1 \\ 0 \end{bmatrix} - \frac{1}{2}\begin{bmatrix} 1 \\ 0 \\ -1 \end{bmatrix} = \begin{bmatrix} \frac{1}{2} \\ -1 \\ \frac{1}{2} \end{bmatrix}.$$

Of course, e and $2\mathsf{f} = \begin{bmatrix} 1 \\ -2 \\ 1 \end{bmatrix}$ will also do (and is easier with which to work).

(b) With $\mathsf{e} = \begin{bmatrix} 1 \\ 0 \\ -1 \end{bmatrix}$ and $\mathsf{f} = \begin{bmatrix} 1 \\ -2 \\ 1 \end{bmatrix}$ from part (a), the projection of w on π is

$$\frac{\mathsf{w}\cdot\mathsf{e}}{\mathsf{e}\cdot\mathsf{e}}\,\mathsf{e} + \frac{\mathsf{w}\cdot\mathsf{f}}{\mathsf{f}\cdot\mathsf{f}}\,\mathsf{f} = \frac{-4}{2}\,\mathsf{e} + \frac{-4}{6}\,\mathsf{f} = -2\begin{bmatrix} 1 \\ 0 \\ -1 \end{bmatrix} - \frac{2}{3}\begin{bmatrix} 1 \\ -2 \\ 1 \end{bmatrix} = \begin{bmatrix} -\frac{8}{3} \\ \frac{4}{3} \\ \frac{4}{3} \end{bmatrix}.$$

8. (a) Since $\mathsf{v}_4 = -2\mathsf{v}_1$, the vectors are linearly dependent: $2\mathsf{v}_1 + 0\mathsf{v}_2 + 0\mathsf{v}_3 + \mathsf{v}_4 = \mathsf{0}$.

(b) If $c_1\mathsf{v}_1 + c_2\mathsf{v}_2 = \mathsf{0}$, then $\begin{bmatrix} -c_1 \\ c_1 + c_2 \\ 2c_1 + 4c_2 \end{bmatrix} = \begin{bmatrix} 0 \\ 0 \\ 0 \end{bmatrix}$. Equating corresponding components, we get $c_1 = 0$, $c_2 = 0$, so the vectors are linearly independent.

9. We are given that there are scalars a and b so that $\mathsf{u} = a\mathsf{v} + b\mathsf{w}$. Thus $\mathsf{u} - a\mathsf{v} - b\mathsf{w} = \mathsf{0}$ and not all coefficients are 0—the coefficient of u is 1.

Week 5—Test Yourself

1. $A = \begin{bmatrix} 1 & 1 & 1 & 1 \\ 2 & 4 & 8 & 16 \end{bmatrix}$.

2. $\mathsf{0} = \begin{bmatrix} 0 & 0 \\ 0 & 0 \\ 0 & 0 \end{bmatrix}$.

3. Matrices are equal if and only if corresponding corresponding entries are equal.

We must have
$$\begin{aligned} 5x - 4y &= 8 \\ -x &= -4 \\ 3x + 3y &= -33 \\ -3x - 3y &= 33 \\ \tfrac{1}{2}x + y &= -9 \\ x + y &= -18 \\ -x + y &= -3 \\ x - 2y &= 10. \end{aligned}$$

We find that $x = -4$, $y = -7$.

4. $\begin{bmatrix} 0 & -2 & -4 \\ 3 & 1 & -1 \end{bmatrix}$.

5. $A + B = \begin{bmatrix} 4 & -2 & 5 \\ -4 & 0 & 6 \end{bmatrix}$; $A - B = \begin{bmatrix} 0 & 4 & -3 \\ 2 & -2 & 2 \end{bmatrix}$; $3A^T + B$ is not defined because $3A^T$ is 3×2 and B is 2×3; $2A - B = \begin{bmatrix} 2 & 5 & -2 \\ 1 & -3 & 6 \end{bmatrix}$.

6. AB is not defined; $BA = \begin{bmatrix} -27 & 26 & 16 \\ -34 & 52 & -2 \end{bmatrix}$.

7. Since x is $n \times 1$, x^T is $1 \times n$, so $\mathsf{x}^T A\mathsf{x}$ is the product of a $1 \times n$, an $n \times n$ and an $n \times 1$ matrix. This is 1×1.

8. $\mathsf{b} = -4\begin{bmatrix} 1 \\ 4 \\ 7 \end{bmatrix} + \begin{bmatrix} 2 \\ 5 \\ 8 \end{bmatrix} + 7\begin{bmatrix} 3 \\ 6 \\ 9 \end{bmatrix}$.

9. $AB = \begin{bmatrix} 1 & 6 \\ 2 & 7 \end{bmatrix}$, $B^T A^T = \begin{bmatrix} 1 & 0 \\ 3 & 1 \end{bmatrix}\begin{bmatrix} 1 & 2 \\ 3 & 1 \end{bmatrix} = \begin{bmatrix} 1 & 2 \\ 6 & 7 \end{bmatrix} = (AB)^T$, $A^T B^T = \begin{bmatrix} 7 & 2 \\ 6 & 1 \end{bmatrix} \neq (AB)^T$.

10. $XY = \frac{1}{3}\begin{bmatrix} -3 & 5 & -2 \\ 4 & -21 & 16 \\ -1 & 14 & -8 \end{bmatrix}$ and $Y = \begin{bmatrix} 0 & 2 & 3 \\ 4 & 5 & 6 \\ 7 & 8 & 9 \end{bmatrix} = \frac{1}{3}\begin{bmatrix} 6 & 3 & 3 \\ 28 & 31 & 30 \\ 0 & 4 & 9 \end{bmatrix} = \begin{bmatrix} 2 & 1 & 1 \\ \frac{28}{3} & \frac{31}{3} & 10 \\ 0 & \frac{4}{3} & 3 \end{bmatrix}$. (Either form of the answer is correct and fine.)

Remark. It is easier to do the calculation this way than to first write $X = \begin{bmatrix} -1 & \frac{5}{3} & -\frac{2}{3} \\ \frac{4}{3} & -7 & \frac{16}{3} \\ -\frac{1}{3} & \frac{14}{3} & -\frac{8}{3} \end{bmatrix}$.

11. $AB = \begin{bmatrix} 2 & 4 \\ 7 & 7 \end{bmatrix}$, $BA = \begin{bmatrix} 10 & -4 \\ -1 & -1 \end{bmatrix}$, so $AB \neq BA$.

$A + B = \begin{bmatrix} 3 & 4 \\ -1 & 0 \end{bmatrix}$, so $(A + B)^2 = \begin{bmatrix} 5 & 12 \\ -3 & -4 \end{bmatrix}$ whereas $A^2 + 2AB + B^2 = \begin{bmatrix} 1 & 0 \\ 0 & 1 \end{bmatrix} + \begin{bmatrix} 4 & 8 \\ 14 & 14 \end{bmatrix} + \begin{bmatrix} -8 & 12 \\ -9 & -11 \end{bmatrix} = \begin{bmatrix} -3 & 20 \\ 5 & 4 \end{bmatrix}$.

12. It's the 3×3 identity matrix.

13. This is the 2×2 identity matrix: $AX = X$ and $YA = Y$ whenever these equations made sense; that is, $AX = X$ for any $2 \times n$ matrix X and $YA = Y$ for any $m \times 2$ matrix Y.

14. $A\mathbf{x} = -2\begin{bmatrix}1\\4\end{bmatrix} + 3\begin{bmatrix}2\\5\end{bmatrix} + 5\begin{bmatrix}3\\6\end{bmatrix}$.

15. $4\begin{bmatrix}1\\2\\3\end{bmatrix} - 3\begin{bmatrix}5\\6\\7\end{bmatrix} = \begin{bmatrix}1&5\\2&6\\3&7\end{bmatrix}\begin{bmatrix}4\\-3\end{bmatrix}$.

16. Using the idea of Example 5.25, we find that $\begin{bmatrix}1&2\\3&4\\5&6\end{bmatrix}\begin{bmatrix}10&0&0&0\\0&20&0&0\end{bmatrix}2 = \begin{bmatrix}10&40&0&0\\30&80&0&0\\50&120&0&0\end{bmatrix}$.

17. Suppose E is $m \times n$. If order for the product $E^2 = EE$ to be defined the second matrix E must be $n \times k$ and then this product if $m \times k$. Since E is $m \times n$, $k = n$ and E is $n \times n$.

18. Let A and U be $m \times n$. If order for the product LU to be defined, L has to be $k \times m$, and then LU is $k \times n$. But $LU = A$ is $m \times n$, so $k = m$ which says that L is square.

Week 6—Test Yourself

1. The system is $A\mathbf{x} = \mathbf{b}$ with $A = \begin{bmatrix}-2&1&5\\-8&7&19\end{bmatrix}$, $\mathbf{x} = \begin{bmatrix}x_1\\x_2\\x_3\end{bmatrix}$, and $\mathbf{b} = \begin{bmatrix}-10\\-42\end{bmatrix}$.

2. The systems have the same solutions.

3. See **6.1**.

4. 1 and 4.

5. (a), (c), and (d) are not in row echelon form. In the matrix of (a), the leading nonzero entries do not step to the right. In the matrix of (c), the zero rows are not all at the bottom. In the matrix of (d), the second row of is not all 0s, so its leading nonzero entry should be 1.

6. $\begin{bmatrix}1&-1&-2\\2&-3&-5\\-1&4&5\end{bmatrix} \to \begin{bmatrix}1&-1&-2\\0&-1&-1\\0&3&3\end{bmatrix} \to \begin{bmatrix}1&-1&-2\\0&-1&-1\\0&0&0\end{bmatrix}$.

   The pivots are 1 and -1. The pivot columns are columns one and two.

7. (a) $\left[\begin{array}{cccc|c}① & -2 & 3 & -1 & 5\end{array}\right]$. The free variables are x_2, x_3, and x_4, corresponding to the columns that are not pivot columns.

   (b) Let $x_2 = t$, $x_3 = s$, and $x_4 = r$. Then $x_1 - 2x_2 + 3x_3 - x_4 = 5$, so $x_1 = 5 + 2x_2 - 3x_3 + x_4 = 5 + 2t - 3s + r$.

   The solution is $\begin{bmatrix}x_1\\x_2\\x_3\\x_4\end{bmatrix} = \begin{bmatrix}5+2t-3s+r\\t\\s\\r\end{bmatrix} = \begin{bmatrix}5\\0\\0\\0\end{bmatrix} + t\begin{bmatrix}2\\1\\0\\0\end{bmatrix} + s\begin{bmatrix}-3\\0\\1\\0\end{bmatrix} + r\begin{bmatrix}1\\0\\0\\1\end{bmatrix}$.

8. $x_3 = t$ is free; $x_4 = \frac{1}{3}$, $x_2 + x_3 + 2x_4 = 3$, so $x_2 = \frac{7}{3} - t$; $x_1 + 3x_4 = 2$, so $x_1 = 1$.

Thus $\mathsf{x} = \begin{bmatrix} 1 \\ \frac{7}{3} - t \\ t \\ \frac{1}{3} \end{bmatrix} = \begin{bmatrix} 1 \\ \frac{7}{3} \\ 0 \\ \frac{1}{3} \end{bmatrix} + t\begin{bmatrix} 0 \\ -1 \\ 1 \\ 0 \end{bmatrix}$. There are infinitely many solutions.

9. (a) $\left[\begin{array}{cc|c} 2 & 5 & b_1 \\ 1 & 3 & b_2 \end{array}\right] \to \left[\begin{array}{cc|c} 1 & 3 & b_2 \\ 2 & 5 & b_1 \end{array}\right] \to \left[\begin{array}{cc|c} 1 & 3 & b_2 \\ 0 & -1 & b_1 - 2b_2 \end{array}\right]$

$\to \left[\begin{array}{cc|c} 1 & 3 & b_2 \\ 0 & 1 & 2b_2 - b_1 \end{array}\right]$. Thus $x_2 = 2b_2 - b_1$ and $x_1 + 3x_2 = b_2$, giving $x_1 = b_2 - 3x_2 = b_2 - 3(2b_2 - b_1) = 3b_1 - 5b_2$.

(b) The results of part (a) show that $A\begin{bmatrix} 3b_1 - 5b_2 \\ 2b_2 - b_1 \end{bmatrix} = \mathsf{b}$. So $\begin{bmatrix} b_1 \\ b_2 \end{bmatrix} = (3b_1 - 5b_2)\begin{bmatrix} 2 \\ 1 \end{bmatrix} + (2b_2 - b_1)\begin{bmatrix} 5 \\ 3 \end{bmatrix}$.

10. We attempt to solve $A\mathsf{x} = \mathsf{b}$, with $\mathsf{x} = \begin{bmatrix} x_1 \\ x_2 \\ x_3 \end{bmatrix}$ and $\mathsf{b} = \begin{bmatrix} 1 \\ 6 \\ -4 \end{bmatrix}$. Gaussian elimination proceeds

$$[A|\mathsf{b}] = \left[\begin{array}{ccc|c} 2 & 3 & 4 & 1 \\ 4 & 7 & 5 & 6 \\ 6 & -1 & 9 & -4 \end{array}\right] \to \left[\begin{array}{ccc|c} 2 & 3 & 4 & 1 \\ 0 & 1 & -3 & 4 \\ 0 & -10 & -3 & -7 \end{array}\right]$$

$$\to \left[\begin{array}{ccc|c} 2 & 3 & 4 & 1 \\ 0 & 1 & -3 & 4 \\ 0 & 0 & -33 & 33 \end{array}\right] \to \left[\begin{array}{ccc|c} 2 & 3 & 4 & 1 \\ 0 & 1 & -3 & 4 \\ 0 & 0 & 1 & -1 \end{array}\right],$$

so $x_3 = -1$, $x_2 = 4 + 3x_3 = 1$, $2x_1 = 1 - 3x_2 - 4x_3 = 2$, and $x_1 = 1$.

Thus $A\begin{bmatrix} 1 \\ 1 \\ -1 \end{bmatrix} = \begin{bmatrix} 1 \\ 6 \\ -4 \end{bmatrix}$. This says that $\begin{bmatrix} 1 \\ 6 \\ -4 \end{bmatrix} = 1\begin{bmatrix} 2 \\ 4 \\ 6 \end{bmatrix} + 1\begin{bmatrix} 3 \\ 7 \\ -1 \end{bmatrix} - \begin{bmatrix} 4 \\ 5 \\ 9 \end{bmatrix}$.

11. (a) The system is $A\mathsf{x} = \mathsf{b}$ with $A = \begin{bmatrix} 2 & 2 & 2 \\ 4 & 6 & 6 \\ 6 & 6 & 10 \end{bmatrix}$, $\mathsf{x} = \begin{bmatrix} x_1 \\ x_2 \\ x_3 \end{bmatrix}$ and $\mathsf{b} = \begin{bmatrix} 2 \\ 4 \\ 2 \end{bmatrix}$.

$$[A|\mathsf{b}] = \left[\begin{array}{ccc|c} 2 & 2 & 2 & 2 \\ 4 & 6 & 6 & 4 \\ 6 & 6 & 10 & 2 \end{array}\right] \to \left[\begin{array}{ccc|c} 2 & 2 & 2 & 2 \\ 0 & 2 & 2 & 0 \\ 0 & 0 & 4 & -4 \end{array}\right].$$

Thus $x_3 = -1$, $2x_2 + 2x_3 = 0$, so $x_2 = -x_3 = 1$, and $x_1 + x_2 + x_3 = 1$, so $x_1 = 1 - x_2 - x_3 = 1$. The solution is $\mathsf{x} = \begin{bmatrix} 1 \\ 1 \\ -1 \end{bmatrix}$.

(b) The system is $A\mathsf{x} = \mathsf{b}$ with $A = \begin{bmatrix} -2 & 1 & 5 \\ -8 & 7 & 19 \end{bmatrix}$, $\mathsf{x} = \begin{bmatrix} x_1 \\ x_2 \\ x_3 \end{bmatrix}$, and $\mathsf{b} = \begin{bmatrix} -10 \\ -42 \end{bmatrix}$.

$$[A|\mathsf{b}] = \left[\begin{array}{ccc|c} -2 & 1 & 5 & -10 \\ -8 & 7 & 19 & -42 \end{array}\right] \to \left[\begin{array}{ccc|c} -2 & 1 & 5 & -10 \\ 0 & 3 & -1 & -2 \end{array}\right]$$

so $x_3 = t$ is free,

- $3x_2 - x_3 = -2$, so $3x_2 = x_3 - 2 = t - 2$ and $x_2 = \frac{1}{3}t - \frac{2}{3}$,
- $-2x_1 + x_2 + 5x_3 = -10$, so $-2x_1 = -x_2 - 5x_3 - 10 = -\frac{16}{3}t - \frac{28}{3}$ and $x_1 = \frac{8}{3}t + \frac{14}{3}$.

The solution is $\mathsf{x} = \begin{bmatrix} \frac{8}{3}t + \frac{14}{3} \\ \frac{1}{3}t - \frac{2}{3} \\ t \end{bmatrix} = \begin{bmatrix} \frac{14}{3} \\ -\frac{2}{3} \\ 0 \end{bmatrix} + t\begin{bmatrix} \frac{8}{3} \\ \frac{1}{3} \\ 1 \end{bmatrix}$.

(c) $[A|\mathsf{b}] = \left[\begin{array}{rrr|r} 2 & -1 & 1 & 2 \\ 3 & 1 & -6 & -9 \\ -1 & 2 & -5 & -4 \end{array}\right] \to \left[\begin{array}{rrr|r} -1 & 2 & -5 & -4 \\ 2 & -1 & 1 & 2 \\ 3 & 1 & -6 & -9 \end{array}\right]$

$\to \left[\begin{array}{rrr|r} -1 & 2 & -5 & -4 \\ 0 & 3 & -9 & -6 \\ 0 & 7 & -21 & -21 \end{array}\right] \to \left[\begin{array}{rrr|r} -1 & 2 & -5 & -4 \\ 0 & 1 & -3 & -2 \\ 0 & 1 & -3 & -3 \end{array}\right] \to \left[\begin{array}{rrr|r} -1 & 2 & -5 & -4 \\ 0 & 1 & -3 & -2 \\ 0 & 0 & 0 & -1 \end{array}\right]$.

The last equation reads $0 = -1$. There is no solution.

12. Gaussian elimination begins $\left[\begin{array}{rr|r} 5 & 2 & a \\ -15 & -6 & b \end{array}\right] \to \left[\begin{array}{rr|c} 5 & 2 & a \\ 0 & 0 & b+3a \end{array}\right]$.

(a) If $b + 3a \neq 0$, the second equation is $0 = b + 3a \neq 0$, so there is no solution.

(b) Under no circumstances does the system have a unique solution since if $b + 3a = 0$, y is free and there are infinitely many solutions.

(c) If $b + 3a = 0$, there are infinitely many solutions.

Row echelon form is $\left[\begin{array}{rr|r} 1 & \frac{2}{5} & \frac{a}{5} \\ 0 & 0 & 0 \end{array}\right]$, so $y = t$ is free and $x = \frac{a}{5} - \frac{2}{5}y = \frac{a}{5} - \frac{2}{5}t$.

Week 7—Test Yourself

1. These systems have solutions because each is homogeneous: the zero vector is at least one solution in each case. We solve by applying Gaussian elimination and back substitution to A. It is not necessary to augment with a final column of 0s since 0s are unchanged by the elementary row operations.

(a) $A = \begin{bmatrix} 1 & -2 & 1 \\ 3 & -7 & 2 \end{bmatrix} \to \begin{bmatrix} 1 & -2 & 1 \\ 0 & -1 & -1 \end{bmatrix} \to \begin{bmatrix} 1 & -2 & 1 \\ 0 & 1 & 1 \end{bmatrix}$.

The variable $z = t$ is free. Back substitution gives $y + z = 0$, so $y = -z = -t$ and $x - 2y + z = 0$, so $x = 2y - z = -3t$.

The solution is $\begin{bmatrix} x \\ y \\ z \end{bmatrix} = \begin{bmatrix} -3t \\ -t \\ t \end{bmatrix} = t\begin{bmatrix} -3 \\ -1 \\ 1 \end{bmatrix}$.

(b) $A = \begin{bmatrix} -1 & 1 & 2 & 1 & 3 \\ 1 & -1 & 1 & 3 & -4 \\ -8 & 8 & 7 & -4 & 36 \end{bmatrix} \to \begin{bmatrix} -1 & 1 & 2 & 1 & 3 \\ 0 & 0 & 3 & 4 & -1 \\ 0 & 0 & -9 & -12 & 12 \end{bmatrix} \to \begin{bmatrix} -1 & 1 & 2 & 1 & 3 \\ 0 & 0 & 3 & 4 & -1 \\ 0 & 0 & 0 & 0 & 9 \end{bmatrix}$

so

$x_2 = t$ and $x_4 = s$ are free,
$9x_5 = 0$, so $x_5 = 0$,
$3x_3 + 4x_4 - x_5 = 0$, so $3x_3 = -4x_4 + x_5 = -4s$ and $x_3 = -\frac{4}{3}s$,
$-x_1 + x_2 + 2x_3 + x_4 + 3x_5 = 0$, so $x_1 = x_2 + 2x_3 + x_4 + 3x_5 = t - \frac{8}{3}s + s = t - \frac{5}{3}s$

$$\text{and } \mathsf{x} = \begin{bmatrix} t - \frac{5}{3}s \\ t \\ -\frac{4}{3}s \\ s \\ 0 \end{bmatrix} = t\begin{bmatrix} 1 \\ 1 \\ 0 \\ 0 \\ 0 \end{bmatrix} + s\begin{bmatrix} -\frac{5}{3} \\ 0 \\ -\frac{4}{3} \\ 1 \\ 0 \end{bmatrix}.$$

2. (a) This is $A\mathsf{x} = \mathsf{b}$ with $A = \begin{bmatrix} 1 & -2 & 3 & 1 & 3 & 4 \\ -3 & 6 & -8 & 2 & -11 & -15 \\ 1 & -2 & 2 & -4 & 6 & 9 \\ -2 & 4 & -6 & -2 & -6 & -7 \end{bmatrix}$, $\mathsf{x} = \begin{bmatrix} x_1 \\ x_2 \\ x_3 \\ x_4 \\ x_5 \\ x_6 \end{bmatrix}$, $\mathsf{b} = \begin{bmatrix} -1 \\ 2 \\ 3 \\ 1 \end{bmatrix}$.

(b) $$[A|\mathsf{b}] = \left[\begin{array}{cccccc|c} 1 & -2 & 3 & 1 & 3 & 4 & -1 \\ -3 & 6 & -8 & 2 & -11 & -15 & 2 \\ 1 & -2 & 2 & -4 & 6 & 9 & 3 \\ -2 & 4 & -6 & -2 & -6 & -7 & 1 \end{array}\right]$$

$$\to \left[\begin{array}{cccccc|c} 1 & -2 & 3 & 1 & 3 & 4 & -1 \\ 0 & 0 & 1 & 5 & -2 & -3 & -1 \\ 0 & 0 & -1 & -5 & 3 & 5 & 4 \\ 0 & 0 & 0 & 0 & 0 & 1 & -1 \end{array}\right] \to \left[\begin{array}{cccccc|c} 1 & -2 & 3 & 1 & 3 & 4 & -1 \\ 0 & 0 & 1 & 5 & -2 & -3 & -1 \\ 0 & 0 & 0 & 0 & 1 & 2 & 3 \\ 0 & 0 & 0 & 0 & 0 & 1 & -1 \end{array}\right].$$

This is row echelon form. The variables $x_2 = t$ and $x_4 = s$ are free.

By back substitution, we get $x_6 = -1$, $x_5 = 3 - 2x_6 = 5$, $x_3 = -1 - 5x_4 + 2x_5 + 3x_6 = -1 - 5s + 10 - 3 = 6 - 5s$, and $x_1 = -1 + 2x_2 - 3x_3 - x_4 - 3x_5 - 4x_6 = -1 + 2t - 18 + 15s - s - 15 + 4 = -30 + 2t + 14s$. In vector form, the solution is

$$\mathsf{x} = \begin{bmatrix} x_1 \\ x_2 \\ x_3 \\ x_4 \\ x_5 \\ x_6 \end{bmatrix} = \begin{bmatrix} -30 + 2t + 14s \\ t \\ 6 - 5s \\ s \\ 5 \\ -1 \end{bmatrix} = \begin{bmatrix} -30 \\ 0 \\ 6 \\ 0 \\ 5 \\ -1 \end{bmatrix} + t\begin{bmatrix} 2 \\ 1 \\ 0 \\ 0 \\ 0 \\ 0 \end{bmatrix} + s\begin{bmatrix} 14 \\ 0 \\ -5 \\ 1 \\ 0 \\ 0 \end{bmatrix}.$$

This is $\mathsf{x}_p + \mathsf{x}_h$ with $\mathsf{x}_p = \begin{bmatrix} -30 \\ 0 \\ 6 \\ 0 \\ 5 \\ -1 \end{bmatrix}$ and $\mathsf{x}_h = t\begin{bmatrix} 2 \\ 1 \\ 0 \\ 0 \\ 0 \\ 0 \end{bmatrix} + s\begin{bmatrix} 14 \\ 0 \\ -5 \\ 1 \\ 0 \\ 0 \end{bmatrix}$.

(c) In part (b), we saw that $A\begin{bmatrix}-30\\0\\6\\0\\5\\-1\end{bmatrix} = \begin{bmatrix}-1\\2\\3\\1\end{bmatrix}$ so, in view of **5.23**, we have

$$\begin{bmatrix}-1\\2\\3\\1\end{bmatrix} = -30\begin{bmatrix}1\\-3\\1\\-2\end{bmatrix} + 0\begin{bmatrix}-2\\6\\-2\\4\end{bmatrix} + 6\begin{bmatrix}3\\-8\\2\\-6\end{bmatrix} + 0\begin{bmatrix}1\\2\\-4\\-2\end{bmatrix} + 5\begin{bmatrix}3\\-11\\6\\-6\end{bmatrix} - \begin{bmatrix}4\\-15\\9\\-7\end{bmatrix}.$$

3. Matrices in general do not commute. It is possible for $AB = 0$ without $A = 0$ or $B = 0$. A nonzero matrix need not be invertible.

4. (a) $AB = I_2 = BA$, so these matrices are inverses.

   (b) $AB \neq I_3$, so A and B are not inverses.

5. $AB = \begin{bmatrix}5 & -4\\1 & -3\end{bmatrix}$, so $(AB)^{-1} = \frac{1}{11}\begin{bmatrix}3 & -4\\1 & -5\end{bmatrix}$. Also, $A^{-1} = \frac{1}{11}\begin{bmatrix}5 & -3\\2 & 1\end{bmatrix}$, $B^{-1} = \begin{bmatrix}1 & -1\\1 & -2\end{bmatrix}$, and

$$B^{-1}A^{-1} = \frac{1}{11}\begin{bmatrix}1 & -1\\1 & -2\end{bmatrix}\begin{bmatrix}5 & -3\\2 & 1\end{bmatrix} = \frac{1}{11}\begin{bmatrix}3 & -4\\1 & -5\end{bmatrix} = (AB)^{-1}.$$

On the other hand, $A^{-1}B^{-1} = \begin{bmatrix}5 & -3\\2 & 1\end{bmatrix}\frac{1}{11}\begin{bmatrix}1 & -1\\1 & -2\end{bmatrix} = \frac{1}{11}\begin{bmatrix}2 & 1\\3 & -4\end{bmatrix} \neq (AB)^{-1}$.

6. The solution is the $t \times n$ matrix $X = BA^{-1}$ since $XA = (BA^{-1})A = BI = B$.

7. No, A cannot be invertible; otherwise, multiplying both sides of the given equation on the left by A^{-1} would give $\begin{bmatrix}1\\2\\3\end{bmatrix} = \begin{bmatrix}0\\0\\0\end{bmatrix}$, which is not true.

8. (a) $AB = \begin{bmatrix}1 & -7 & 1\\2 & -9 & 1\end{bmatrix}\begin{bmatrix}-1 & 3\\0 & 1\\2 & 4\end{bmatrix} = \begin{bmatrix}1 & 0\\0 & 1\end{bmatrix}$;

$$BA = \begin{bmatrix}-1 & 3\\0 & 1\\2 & 4\end{bmatrix}\begin{bmatrix}1 & -7 & 1\\2 & -9 & 1\end{bmatrix} = \begin{bmatrix}5 & -20 & 2\\2 & -9 & 1\\10 & -50 & 6\end{bmatrix}.$$

   (b) A is not invertible since $BA \neq I$. Only square matrices can be invertible.

9. We have $B = A^{-1}C$ and $A^{-1} = \frac{1}{2}\begin{bmatrix}4 & -1\\-2 & 1\end{bmatrix}$, so $B = \frac{1}{2}\begin{bmatrix}4 & -1\\-2 & 1\end{bmatrix}\begin{bmatrix}5 & 3\\2 & 2\end{bmatrix} = \begin{bmatrix}9 & 5\\-4 & -2\end{bmatrix}$.

10. Taking the inverse of each side, we have $X - A = XB^{-1}$, so $X - XB^{-1} = A$, $X(I - B^{-1}) = A$. We are given that $I - B^{-1}$ is invertible, so $X = A(I - B^{-1})^{-1}$.

11. (a) Multiplying $AXB = A + B$ on the left by A^{-1} gives $XB = I + A^{-1}B$ and multiplying this on the right by B^{-1} gives $X = B^{-1} + A^{-1}$.

(b) Multiplying $X^{-1}A = C - X^{-1}B$ on the left by X gives $A = XC - B$, so $XC = A + B$. Thus $X = (A+B)C^{-1}$.

(c) Multiplying $BAX = XABX$ on the right by X^{-1} gives $BA = XAB$. Thus $X = (BA)(AB)^{-1} = BAB^{-1}A^{-1}$.

12. The given system is $A\mathsf{x} = \mathsf{b}$, with A as given, $\mathsf{x} = \begin{bmatrix} x \\ y \\ z \end{bmatrix}$, and $\mathsf{b} = \begin{bmatrix} 12 \\ -1 \\ -8 \end{bmatrix}$. Since A is invertible, the solution is $\mathsf{x} = A^{-1}\mathsf{b} = \frac{1}{9}\begin{bmatrix} 5 & -2 & 10 \\ 10 & -4 & 11 \\ -3 & 3 & -6 \end{bmatrix}\begin{bmatrix} 12 \\ -1 \\ -8 \end{bmatrix} = \begin{bmatrix} -2 \\ 4 \\ 1 \end{bmatrix}$; that is, $x = -2$, $y = 4$, $z = 1$.

13. (a) $[A|I] = \left[\begin{array}{cc|cc} 3 & 1 & 1 & 0 \\ -6 & -3 & 0 & 1 \end{array}\right] \to \left[\begin{array}{cc|cc} 1 & \frac{1}{3} & \frac{1}{3} & 0 \\ 0 & -1 & 2 & 1 \end{array}\right] \to \left[\begin{array}{cc|cc} 1 & 0 & 1 & \frac{1}{3} \\ 0 & 1 & -2 & -1 \end{array}\right]$.

The inverse is $\begin{bmatrix} 1 & \frac{1}{3} \\ -2 & -1 \end{bmatrix}$.

(b) $[A|I] = \left[\begin{array}{ccc|ccc} 2 & 4 & 2 & 1 & 0 & 0 \\ 1 & 2 & 3 & 0 & 1 & 0 \\ 3 & 2 & 1 & 0 & 0 & 1 \end{array}\right] \to \left[\begin{array}{ccc|ccc} 1 & 2 & 3 & 0 & 1 & 0 \\ 2 & 4 & 2 & 1 & 0 & 0 \\ 3 & 2 & 1 & 0 & 0 & 1 \end{array}\right] \to \left[\begin{array}{ccc|ccc} 1 & 2 & 3 & 0 & 1 & 0 \\ 0 & 0 & -4 & 1 & -2 & 0 \\ 0 & -4 & -8 & 0 & -3 & 1 \end{array}\right]$

$\to \left[\begin{array}{ccc|ccc} 1 & 2 & 3 & 0 & 1 & 0 \\ 0 & 1 & 2 & 0 & \frac{3}{4} & -\frac{1}{4} \\ 0 & 0 & 1 & -\frac{1}{4} & \frac{1}{2} & 0 \end{array}\right] \to \left[\begin{array}{ccc|ccc} 1 & 0 & -1 & 0 & -\frac{1}{2} & \frac{1}{2} \\ 0 & 1 & 2 & 0 & \frac{3}{4} & -\frac{1}{4} \\ 0 & 0 & 1 & -\frac{1}{4} & \frac{1}{2} & 0 \end{array}\right]$

$\to \left[\begin{array}{ccc|ccc} 1 & 0 & 0 & -\frac{1}{4} & 0 & \frac{1}{2} \\ 0 & 1 & 0 & \frac{1}{2} & -\frac{1}{4} & -\frac{1}{4} \\ 0 & 0 & 1 & -\frac{1}{4} & \frac{1}{2} & 0 \end{array}\right]$. The inverse is $\begin{bmatrix} -\frac{1}{4} & 0 & \frac{1}{2} \\ \frac{1}{2} & -\frac{1}{4} & -\frac{1}{4} \\ -\frac{1}{4} & \frac{1}{2} & 0 \end{bmatrix}$.

(c) $[A|I] = \left[\begin{array}{ccc|ccc} 0 & -1 & 2 & 1 & 0 & 0 \\ 2 & 1 & 4 & 0 & 1 & 0 \\ 1 & -1 & 5 & 0 & 0 & 1 \end{array}\right] \to \left[\begin{array}{ccc|ccc} 1 & -1 & 5 & 0 & 0 & 1 \\ 0 & -1 & 2 & 1 & 0 & 0 \\ 2 & 1 & 4 & 0 & 1 & 0 \end{array}\right]$

$\to \left[\begin{array}{ccc|ccc} 1 & -1 & 5 & 0 & 0 & 1 \\ 0 & 1 & -2 & -1 & 0 & 0 \\ 0 & 3 & -6 & 0 & 1 & -2 \end{array}\right] \to \left[\begin{array}{ccc|ccc} 1 & 0 & 3 & -1 & 0 & 1 \\ 0 & 1 & -2 & -1 & 0 & 0 \\ 0 & 0 & 0 & 3 & 1 & -2 \end{array}\right]$.

There is no inverse.

14. In cases (a) and (b), the columns are are linearly independent. The columns of the matrix in (c) are linearly dependent. See Examples 7.3 and 7.4.

15. $\begin{bmatrix} 1 & 2 & 0 \\ 3 & -1 & 2 \\ -2 & 3 & -2 \end{bmatrix} \to \begin{bmatrix} 1 & 2 & 0 \\ 0 & -7 & 2 \\ 0 & 7 & -2 \end{bmatrix} \to \begin{bmatrix} 1 & 2 & 0 \\ 0 & 1 & -\frac{2}{7} \\ 0 & 0 & 0 \end{bmatrix} \to \begin{bmatrix} 1 & 0 & \frac{4}{7} \\ 0 & 1 & -\frac{2}{7} \\ 0 & 0 & 0 \end{bmatrix}$.

Week 8—Test Yourself

1. (a) The second row of EA is the sum of the second row and four times the third row of A; all other rows of EA are the same as those of A.

   (b) E is called an elementary matrix.

   (c) Any elementary matrix is invertible, its inverse being that elementary matrix that "undoes" E. Here, $E^{-1} = \begin{bmatrix} 1 & 0 & 0 \\ 0 & 1 & -4 \\ 0 & 0 & 1 \end{bmatrix}$.

2. (a) EA was formed by the operation $R3 \to R3 - 6R1$, so $E = \begin{bmatrix} 1 & 0 & 0 \\ 0 & 1 & 0 \\ -6 & 0 & 1 \end{bmatrix}$.

   (b) EA was formed by the operation $R2 \to R2 - 4R1$, so $E = \begin{bmatrix} 1 & 0 & 0 \\ -4 & 1 & 0 \\ 0 & 0 & 1 \end{bmatrix}$.

   (c) EA was formed by the operation $R3 \to R3 - \frac{3}{2}R2$, so $E = \begin{bmatrix} 1 & 0 & 0 \\ 0 & 1 & 0 \\ 0 & -\frac{3}{2} & 1 \end{bmatrix}$.

   (d) EA was formed by interchanging rows two and three of A, so $E = \begin{bmatrix} 1 & 0 & 0 \\ 0 & 0 & 1 \\ 0 & 1 & 0 \end{bmatrix}$.

3. (a) A is elementary, so its inverse is the elementary matrix that "undoes" A: $A^{-1} = \begin{bmatrix} 1 & -3 \\ 0 & 1 \end{bmatrix}$.

   (b) A is elementary, so its inverse is the elementary matrix that "undoes" A: $A^{-1} = \begin{bmatrix} 1 & 0 \\ 5 & 1 \end{bmatrix}$.

   (c) A is a elementary, so its inverse is the elementary matrix that "undoes" A: $A^{-1} = A$.

4. $\begin{bmatrix} 2 & 5 & 1 \\ 4 & x & 1 \\ 0 & 1 & -1 \end{bmatrix} \to \begin{bmatrix} 2 & 5 & 1 \\ 0 & x-10 & -1 \\ 0 & 1 & -1 \end{bmatrix}$, so if $x = 10$, it will be necessary to interchange rows two and three.

5. Yes, because an elementary matrix is invertible and the product of invertible matrices is invertible (if A and B are invertible, the inverse of AB is $B^{-1}A^{-1}$).

6. No. The previous exercise showed that an interchange of rows is required when moving A to a row echelon matrix.

7. An LU factorization of a matrix A is a factorization $A = LU$ of A where L is a lower triangular matrix with 1s on the diagonal and U is a row echelon (and hence upper triangular) matrix that is the same size as A.

8. Let A and U be $m \times n$. If order for the product LU to be defined, L has to be $k \times m$, and then LU is $k \times n$. But $LU = A$ is $m \times n$, so $k = m$ which says that L is square.

9. Each step in the Gaussian elimination $A \to U$ can be achieved by multiplying the current matrix on the left by the elementary matrix that corresponds to an elementary row operation of types two or three. Such elementary matrices are lower triangular and the product of lower triangular matrices is lower triangular. Thus we have $U = E_n E_{n-1} \cdots E_2 E_1 A$. The matrix L is the lower triangular matrix $(E_n E_{n-1} \cdots E_2 E_1)^{-1}$.

10. We have $A = \begin{bmatrix} -1 & 1 & -2 \\ 2 & 1 & 7 \end{bmatrix} \to \begin{bmatrix} -1 & 1 & -2 \\ 0 & 3 & 3 \end{bmatrix} = EA = U$ with $E = \begin{bmatrix} 1 & 0 \\ 2 & 1 \end{bmatrix}$. Thus $A = E^{-1}U = LU$ with $L = E^{-1} = \begin{bmatrix} 1 & 0 \\ -2 & 1 \end{bmatrix}$.

11. We bring A to a row echelon matrix using only the third elementary row operation placing the multpliers in a lower triangular matrix as we proceed. $A = \begin{bmatrix} -2 & 1 & 3 & 4 \\ 4 & 0 & -9 & -7 \\ -6 & 10 & 4 & 15 \end{bmatrix} \to \begin{bmatrix} -2 & 1 & 3 & 4 \\ 0 & 2 & -3 & 1 \\ 0 & 7 & -5 & 3 \end{bmatrix} \to \begin{bmatrix} -2 & 1 & 3 & 4 \\ 0 & 2 & -3 & 1 \\ 0 & 0 & \frac{11}{2} & -\frac{1}{2} \end{bmatrix} = U$ with $L = \begin{bmatrix} 1 & 0 & 0 \\ -2 & 1 & 0 \\ 3 & \frac{7}{2} & 1 \end{bmatrix}$.

12. (a) First we solve $L\begin{bmatrix} y_1 \\ y_2 \end{bmatrix} = \begin{bmatrix} -2 \\ 9 \end{bmatrix}$. Using forward substitution, we have $2y_1 = -2$, so $y_1 = -1$, and then $6y_1 + 5y_2 = 9$, so $y_2 = 3$. Thus $\begin{bmatrix} c_1 \\ c_2 \end{bmatrix} = \begin{bmatrix} -1 \\ 3 \end{bmatrix}$. Now we solve $U\begin{bmatrix} x \\ y \end{bmatrix} = \begin{bmatrix} -1 \\ 3 \end{bmatrix}$. Using back substitution, we obtain $y = 3$; then, since $x + \frac{1}{2}y = -1$, we get $x = -\frac{5}{2}$. Thus $\begin{bmatrix} x \\ y \end{bmatrix} = \begin{bmatrix} -\frac{5}{2} \\ 3 \end{bmatrix}$.

(b) We have just shown that $A\begin{bmatrix} -\frac{5}{2} \\ 3 \end{bmatrix} = \begin{bmatrix} -2 \\ 9 \end{bmatrix}$, thus $\begin{bmatrix} -2 \\ 9 \end{bmatrix} = -\frac{5}{2}\begin{bmatrix} 1 \\ 6 \end{bmatrix} + 3\begin{bmatrix} 1 \\ 8 \end{bmatrix}$.

13. A is a permutation matrix, so $A^{-1} = A^T = \begin{bmatrix} 0 & 1 & 0 \\ 0 & 0 & 1 \\ 1 & 0 & 0 \end{bmatrix}$.

14. Replace rows one, two and three by rows three, one and two, respectively: $P = \begin{bmatrix} 0 & 0 & 1 \\ 1 & 0 & 0 \\ 0 & 1 & 0 \end{bmatrix}$.

Week 9—Test Yourself

1. (a) i. $M = \begin{bmatrix} -26 & -12 & 4 \\ -13 & -6 & 2 \\ 13 & 6 & -2 \end{bmatrix}$, $C = \begin{bmatrix} -26 & 12 & 4 \\ 13 & -6 & -2 \\ 13 & -6 & -2 \end{bmatrix}$,

$$AC^T = \begin{bmatrix} 2 & 3 & 4 \\ 0 & -1 & 3 \\ 4 & 7 & 5 \end{bmatrix}\begin{bmatrix} -26 & 13 & 13 \\ 12 & -6 & -6 \\ 4 & -2 & -2 \end{bmatrix} = \begin{bmatrix} 0 & 0 & 0 \\ 0 & 0 & 0 \\ 0 & 0 & 0 \end{bmatrix} = C^T A.$$

ii. Since $AC^T = (\det A)I$, we must have $\det A = 0$ in this case.

iii. A is not invertible for lots or reasons; for instance, $AC^T = 0$ but $C^T \neq 0$. See Example 10.15.

2. (a) Expanding by cofactors of the third row gives

$$\det A = 2\begin{vmatrix} -1 & 2 \\ 1 & 1 \end{vmatrix} + \begin{vmatrix} 1 & 2 \\ 3 & 1 \end{vmatrix} + 3\begin{vmatrix} 1 & -1 \\ 3 & 1 \end{vmatrix} = 2(-3) + (-5) + 3(4) = 1.$$

(b) Expanding by cofactors of the second column gives

$$\det A = \begin{vmatrix} 3 & 1 \\ 2 & 3 \end{vmatrix} + \begin{vmatrix} 1 & 2 \\ 2 & 3 \end{vmatrix} + \begin{vmatrix} 1 & 2 \\ 3 & 1 \end{vmatrix} = 7 - 1 - 5 = 1.$$

3. (a) $c_{13} = \begin{vmatrix} 2 & 1 \\ 1 & 2 \end{vmatrix} = 3$, $c_{21} = -\begin{vmatrix} 0 & 1 \\ 2 & 1 \end{vmatrix} = -(-2) = 2$,

$c_{32} = -\begin{vmatrix} 1 & 1 \\ 2 & 1 \end{vmatrix} = -(-1) = 1.$

(b) Expanding by cofactors of the first row, $\det A = 1(-1) + 0(-1) + 1(3) = 2$.

(c) A is invertible since $\det A \neq 0$.

(d) $A^{-1} = \dfrac{1}{\det A} C^T = \dfrac{1}{2}\begin{bmatrix} -1 & 2 & -1 \\ -1 & 0 & 1 \\ 3 & -2 & 1 \end{bmatrix}$.

4. Not possible. Only a square matrix has a determinant.

Week 10—Test Yourself

1. (a) 0 (two equal rows)

   (b) 236 (third row has been multiplied by 2)

2. The determinant of a triangular matrix is the product of its diagonal entries, so $\det A = 60$. Also, $\det A^{-1} = \dfrac{1}{\det A} = \frac{1}{60}$ and $\det A^2 = (\det A)^2 = 3600$.

3. $\det AB = \det A \det B$.

4. $\det A = -15 \neq 0$ (easy because A is triangular), so use Proposition 10.17.

5. A singular matrix is a square matrix that is not invertible; equivalently, a square matrix with determinant 0.

6. (a) A matrix is not invertible if and only if its determinant is 0. The determinant of the given matrix is $x^2 + 3x + 2 = 0$, so the matrix is singular if and only if $x = -1$ or $x = -2$.

   (b) With a Laplace expansion along the second row, we see that the matrix has determinant $-3\begin{vmatrix} 1 & x \\ 4 & 7 \end{vmatrix} = -3(7 - 4x)$, so the matrix is singular if and only if $x = \frac{7}{4}$.

7. (a) $\det A = \begin{vmatrix} -1 & -1 & 1 & 0 \\ 2 & 1 & 1 & 3 \\ 0 & 1 & 1 & 2 \\ 1 & 3 & -1 & 2 \end{vmatrix} = \begin{vmatrix} -1 & -1 & 1 & 0 \\ 0 & -1 & 3 & 3 \\ 0 & 1 & 1 & 2 \\ 0 & 2 & 0 & 2 \end{vmatrix} = \begin{vmatrix} -1 & -1 & 1 & 0 \\ 0 & -1 & 3 & 3 \\ 0 & 0 & 4 & 5 \\ 0 & 0 & 6 & 8 \end{vmatrix}$

$$= 4\begin{vmatrix} -1 & -1 & 1 & 0 \\ 0 & -1 & 3 & 3 \\ 0 & 0 & 1 & \frac{5}{4} \\ 0 & 0 & 6 & 8 \end{vmatrix} = 4\begin{vmatrix} -1 & -1 & 1 & 0 \\ 0 & -1 & 3 & 3 \\ 0 & 0 & 1 & \frac{5}{4} \\ 0 & 0 & 0 & \frac{1}{2} \end{vmatrix} = 4(\tfrac{1}{2}) = 2.$$

(b) Since $\det A \neq 0$, the columns of A are linearly independent.

Week 11—Test Yourself

1. v_1 is not an eigenvector: an eigenvector is, by definition, **not zero.**

   $Av_2 = \begin{bmatrix} 5 \\ 4 \\ 4 \end{bmatrix}$ is not λv_2 for any scalar λ, so v_2 is not an eigenvector.

   $Av_3 = v_3$, so v_3 is an eigenvector of A corresponding to $\lambda = 1$.

   $Av_4 = -2v_4$, so v_4 is an eigenvector of A corresponding to $\lambda = -2$.

   $Av_5 = 5v_5$, so v_5 is an eigenvector of A corresponding to $\lambda = 5$.

   $Av_6 = \begin{bmatrix} 12 \\ 8 \\ 6 \end{bmatrix}$ is not λv_6 for any scalar λ, so v_6 is not an eigenvector.

2. We have $A - \lambda I = \begin{bmatrix} 5-\lambda & -2 \\ 4 & -1-\lambda \end{bmatrix}$, so $\det(A - \lambda I) = \lambda^2 - 4\lambda + 3 = (\lambda - 1)(\lambda - 3)$, so $\lambda = 1$ and $\lambda = 3$ are eigenvalues, while -2, -1, and 4 are not.

3. (a) The characteristic polynomial of A is

   $$\det(A - \lambda I) = \begin{vmatrix} 5-\lambda & 1 \\ -1 & 3-\lambda \end{vmatrix} = (5-\lambda)(3-\lambda) + 1 = \lambda^2 - 8\lambda + 16 = (\lambda - 4)^2.$$

   (b) The only eigenvalue is $\lambda = 4$.

   (c) $A\begin{bmatrix} -3 \\ 3 \end{bmatrix} = \begin{bmatrix} -12 \\ 12 \end{bmatrix} = 4\begin{bmatrix} -3 \\ 3 \end{bmatrix}$, so $\begin{bmatrix} -3 \\ 3 \end{bmatrix}$ is an eigenvector corresponding to $\lambda = 4$.

4. The characteristic polynomial of A is $\det(A - \lambda I)$

   $$= \begin{vmatrix} 4-\lambda & 2 & 2 \\ 4 & 2-\lambda & -4 \\ -2 & 0 & -4-\lambda \end{vmatrix} = -\lambda^3 + 2\lambda^2 + 20\lambda + 24.$$

   By inspection, $\lambda = 6$ is a root; long division then gives $-\lambda^3 + 2\lambda^2 + 20\lambda + 24 = (6 - \lambda)(\lambda + 2)^2$ so the eigenvalues are $\lambda = 6$ and $\lambda = -2$. To find the eigenspace for $\lambda = 6$, we solve the homogeneous system $(A - \lambda I)x = 0$ for $x = \begin{bmatrix} x_1 \\ x_2 \\ x_3 \end{bmatrix}$ with

$\lambda = 6$. We have

$$A - \lambda I = \begin{bmatrix} -2 & 2 & 2 \\ 4 & -4 & -4 \\ -2 & 0 & -10 \end{bmatrix} \to \begin{bmatrix} 1 & -1 & -1 \\ 0 & -2 & -12 \\ 0 & 0 & 0 \end{bmatrix} \to \begin{bmatrix} 1 & -1 & -1 \\ 0 & 1 & 6 \\ 0 & 0 & 0 \end{bmatrix}.$$

The solutions are $x_3 = t, x_2 = -6t, x_1 = -5t$. The eigenspace consists of vectors of the form $\begin{bmatrix} -5t \\ -6t \\ t \end{bmatrix} = t\begin{bmatrix} -5 \\ -6 \\ 1 \end{bmatrix}$. To find the eigenspace for $\lambda = -2$, we solve the homogeneous system $(A - \lambda I)\mathsf{x} = 0$ for $\mathsf{x} = \begin{bmatrix} x_1 \\ x_2 \\ x_3 \end{bmatrix}$ with $\lambda = -2$. We have

$$A - \lambda I = \begin{bmatrix} 6 & 2 & 2 \\ 4 & 4 & -4 \\ -2 & 0 & -2 \end{bmatrix} \to \begin{bmatrix} 1 & 0 & 1 \\ 1 & 1 & -1 \\ 3 & 1 & 1 \end{bmatrix} \to \begin{bmatrix} 1 & 0 & 1 \\ 0 & 1 & -2 \\ 0 & 1 & -2 \end{bmatrix} \to \begin{bmatrix} 1 & 0 & 1 \\ 0 & 1 & -2 \\ 0 & 0 & 0 \end{bmatrix}.$$

The solutions are $x_3 = t, x_2 = 2t, x_1 = -t$. The eigenspace consists of vectors of the form $\begin{bmatrix} -t \\ 2t \\ t \end{bmatrix} = t\begin{bmatrix} -1 \\ 2 \\ 1 \end{bmatrix}$.

5. If A is similar to I, then there exists an invertible matrix P such that $A = P^{-1}IP = I$. So $A = I$.

6. The sum of the entries on the main diagonal.

7. The characteristic polynomial of A is $f(\lambda) = \det(A - \lambda I)$. The characteristic polynomial of B is $g(\lambda) = \det(B - \lambda I)$. Now there exists an invertible matrix P such that $B = P^{-1}AP$. Thus $B - \lambda I = P^{-1}AP - \lambda I = P^{-1}(A - \lambda I)P$. Since $\det P^{-1}XP = \det X$ for any X, we have $\det P^{-1}(A - \lambda I)P = \det(A - \lambda I)$, so $g(\lambda) = f(\lambda)$.

8. (a) $\det(A - \lambda I) = \begin{vmatrix} -\lambda & 1 \\ 0 & -\lambda \end{vmatrix} = \lambda^2$ so $\lambda = 0$ is the only eigenvalue. Corresponding eigenvectors are obtained by solving $(A - \lambda I)\mathsf{x} = 0$ for $\mathsf{x} = \begin{bmatrix} x_1 \\ x_2 \end{bmatrix}$ with $\lambda = 0$. Since A is already in row echelon form, we have $x_1 = t$ is free and $x_2 = 0$, so $\mathsf{x} = t\begin{bmatrix} 1 \\ 0 \end{bmatrix}$.

   (b) If $P^{-1}AP = D$, the columns of P are eigenvectors. The only possibility for P here is a matrix of the form $\begin{bmatrix} t & s \\ 0 & 0 \end{bmatrix}$ and such a matrix is not invertible.

9. We have an invertible matrix P with $P^{-1}AP = D$ a diagonal matrix.So $AP = PD$ and, as shown this week, this means that the columns of P are eigenvectors of A. These are linearly independent because P is invertible and there are n of them because P is $n \times n$.

10. (a) The characteristic polynomial of A is $\begin{vmatrix} 1-\lambda & 0 \\ 2 & 3-\lambda \end{vmatrix} = (1-\lambda)(3-\lambda)$, so A has two distinct eigenvalues and hence is diagonalizable. For $\lambda = 1$, we find that $\mathsf{x} = \begin{bmatrix} 1 \\ -1 \end{bmatrix}$ is an eigenvector and, for $\lambda = 3$, $\mathsf{x} = \begin{bmatrix} 0 \\ 1 \end{bmatrix}$. For $P = \begin{bmatrix} 1 & 0 \\ -1 & 1 \end{bmatrix}$, we have $P^{-1}AP = D = \begin{bmatrix} 1 & 0 \\ 0 & 3 \end{bmatrix}$.

(b) The characteristic polynomial of A is

$$\begin{vmatrix} -1-\lambda & 3 & 0 \\ 0 & 2-\lambda & 0 \\ 2 & 1 & -1-\lambda \end{vmatrix} = (2-\lambda)\begin{vmatrix} -1-\lambda & 0 \\ 2 & -1-\lambda \end{vmatrix} = (2-\lambda)(\lambda+1)^2.$$

There are two eigenvalues, $\lambda = -1$ and $\lambda = 2$. For $\lambda = 2$, the eigenspace is spanned by $\begin{bmatrix} 1 \\ 1 \\ 1 \end{bmatrix}$. For $\lambda = -1$, the eigenspace is spanned by $\begin{bmatrix} 0 \\ 0 \\ 1 \end{bmatrix}$. There are just two linearly independent eigenvectors. The matrix is not diagonalizable.

11. Similar matrices have the same characteristic polynomial, so the characteristic polynomial of A is the same as that of the given matrix. The given matrix is triangular; its characteristic polynomial is $(-1-\lambda)(1-\lambda)(-3-\lambda)^2$.

12. If $A = P^{-1}(5I)P$, then $A = 5I$.

13. This matrix is $A = 2I$, so $A\mathsf{x} = 2\mathsf{x}$ for every x. Every vector in R^4 is an eigenvector corresponding to the eigenvalue $\lambda = 2$.

14. (a) The characteristic polynomial of A is

$$\begin{vmatrix} 5-\lambda & 3 \\ 2 & 4-\lambda \end{vmatrix} = 20 - 9\lambda + \lambda^2 - 6 = \lambda^2 - 9\lambda + 14 = (\lambda-2)(\lambda-7).$$

Since the 2×2 matrix A has two distinct eigenvalues, it is diagonalizable by Theorem 11.19.

(b) It is similar to either of the two matrices whose diagonal entries are eigenvalues: $\begin{bmatrix} 2 & 0 \\ 0 & 7 \end{bmatrix}$ and $\begin{bmatrix} 7 & 0 \\ 0 & 2 \end{bmatrix}$.

Glossary

If you do not know what the words mean, it is impossible to read anything with understanding. This fact, which is completely obvious to students of German or Russian, is often lost on mathematics students, but it is just as applicable. What follows is a vocabulary list of the technical terms discussed in these notes. In most cases, each definition is followed by an example.

Angle between vectors: If u and v are nonzero vectors in R^n, the angle between u and v is that angle θ, $0 \le \theta \le \pi$, whose cosine satisfies $\cos\theta = \dfrac{\mathsf{u}\cdot\mathsf{v}}{\|\mathsf{u}\|\,\|\mathsf{v}\|}$. For example, if $\mathsf{u} = \begin{bmatrix}1\\2\\1\\0\\3\end{bmatrix}$ and $\mathsf{v} = \begin{bmatrix}1\\-1\\1\\1\\1\end{bmatrix}$, then $\cos\theta = \dfrac{\mathsf{u}\cdot\mathsf{v}}{\|\mathsf{u}\|\,\|\mathsf{v}\|} = \dfrac{3}{\sqrt{15}\sqrt{5}} = \dfrac{\sqrt{3}}{5} \approx .346$, so $\theta \approx 1.217$ rads $\approx 70°$.

Cofactor: See *Minor.*

Component (of a vector): See *Vector.*

Cross product: The cross product of vectors $\mathsf{u} = \begin{bmatrix}u_1\\u_2\\u_3\end{bmatrix}$ and $\mathsf{v} = \begin{bmatrix}v_1\\v_2\\v_3\end{bmatrix}$ is a vector whose calculation we remember by the expression

$$\mathsf{u}\times\mathsf{v} = \begin{vmatrix}\mathsf{i} & \mathsf{j} & \mathsf{k}\\ u_1 & u_2 & u_3\\ v_1 & v_2 & v_3\end{vmatrix}.$$

Thus

$$\mathsf{u}\times\mathsf{v} = \begin{vmatrix}u_2 & u_3\\ v_2 & v_3\end{vmatrix}\mathsf{i} - \begin{vmatrix}u_1 & u_3\\ v_1 & v_3\end{vmatrix}\mathsf{j} + \begin{vmatrix}u_1 & u_2\\ v_1 & v_2\end{vmatrix}\mathsf{k}.$$

For example, let $\mathsf{u} = \begin{bmatrix} 1 \\ 3 \\ 2 \end{bmatrix}$ and $\mathsf{v} = \begin{bmatrix} 0 \\ 2 \\ -1 \end{bmatrix}$. We have

$$\mathsf{u} \times \mathsf{v} = \begin{vmatrix} \mathsf{i} & \mathsf{j} & \mathsf{k} \\ 1 & 3 & 2 \\ 0 & -2 & 1 \end{vmatrix} = \begin{vmatrix} 3 & 2 \\ -2 & 1 \end{vmatrix} \mathsf{i} - \begin{vmatrix} 1 & 2 \\ 0 & 1 \end{vmatrix} \mathsf{j} + \begin{vmatrix} 1 & 3 \\ 0 & -2 \end{vmatrix} \mathsf{k}$$

$$= 7\mathsf{i} - 1\mathsf{j} + (-2)\mathsf{k} = \begin{bmatrix} 7 \\ -1 \\ -2 \end{bmatrix}.$$

p. 33

Determinant: The determinant of the 2×2 matrix $A = \begin{bmatrix} a & b \\ c & d \end{bmatrix}$ is the number $ad - bc$. It is denoted $\det A$ or $\begin{vmatrix} a & b \\ c & d \end{vmatrix}$. Thus

$$\det A = \begin{vmatrix} a & b \\ c & d \end{vmatrix} = ad - bc.$$

For $n > 2$, the determinant of an $n \times n$ matrix A is defined by the equation $AC^T = (\det A)I$, where C denotes the matrix of cofactors of A.

Diagonal: See *Main diagonal.*

Diagonalizable: A matrix A is diagonalizable if it is similar to a diagonal matrix; that is, if there exists an invertible matrix P and a diagonal matrix D such that $P^{-1}AP = D$. For example, $A = \begin{bmatrix} -11 & 18 \\ -6 & 10 \end{bmatrix}$ is diagonalizable because $P^{-1}AP = D$, with $P = \begin{bmatrix} 2 & 3 \\ 1 & 2 \end{bmatrix}$ and $D = \begin{bmatrix} -2 & 0 \\ 0 & 1 \end{bmatrix}$. A real $n \times n$ matrix A is orthogonally diagonalizable if there exists an orthogonal matrix Q such that $Q^{-1}AQ = D$ is a diagonal matrix. For example, let $A = \begin{bmatrix} 1 & -2 \\ -2 & -2 \end{bmatrix}$. The matrix $Q = \begin{bmatrix} \frac{1}{\sqrt{5}} & -\frac{2}{\sqrt{5}} \\ \frac{2}{\sqrt{5}} & \frac{1}{\sqrt{5}} \end{bmatrix}$ is orthogonal and $Q^{-1}AQ = Q^TAQ = \begin{bmatrix} -3 & 0 \\ 0 & 2 \end{bmatrix} = D$.

Dot product: The dot product of n-dimensional vectors $\mathsf{x} = \begin{bmatrix} x_1 \\ x_2 \\ \vdots \\ x_n \end{bmatrix}$ and $\mathsf{y} = \begin{bmatrix} y_1 \\ y_2 \\ \vdots \\ y_n \end{bmatrix}$ is the number $\mathsf{x} \cdot \mathsf{y} = x_1y_1 + x_2y_2 + \cdots + x_ny_n$. For example, with $\mathsf{x} = \begin{bmatrix} -1 \\ 2 \end{bmatrix}$ and $\mathsf{y} = \begin{bmatrix} 4 \\ -3 \end{bmatrix}$, we have $\mathsf{x} \cdot \mathsf{y} = -1(4) + 2(-3) = -10$. If $\mathsf{x} = \begin{bmatrix} -1 \\ 0 \\ 2 \end{bmatrix}$ and $\mathsf{y} = \begin{bmatrix} 1 \\ 2 \\ 3 \end{bmatrix}$, then $\mathsf{x} \cdot \mathsf{y} = -1(1) + 0(2) + 2(3) = 5$. The complex dot product of vectors $\mathsf{z} = \begin{bmatrix} z_1 \\ \vdots \\ z_n \end{bmatrix}$ and $\mathsf{w} = \begin{bmatrix} w_1 \\ \vdots \\ w_n \end{bmatrix}$ in C^n is $\mathsf{z} \cdot \mathsf{w} = \bar{z}_1w_1 + \bar{z}_2w_2 + \cdots + \bar{z}_nw_n$ For example, if $\mathsf{z} = \begin{bmatrix} 1 \\ i \end{bmatrix}$ and $\mathsf{w} = \begin{bmatrix} 2+i \\ 3-i \end{bmatrix}$ are vectors in C^2, then $\mathsf{z} \cdot \mathsf{w} = 1(2+i) - i(3-i) = 2 + i - 3i - 1 = 1 - 2i$ and $\mathsf{w} \cdot \mathsf{z} = (2-i)(1) + (3+i)(i) = 2 - i + 3i - 1 = 1 + 2i$.

Eigenspace; Eigenvalue; Eigenvector: An eigenvalue of a (square) matrix is a real number λ with the property that $A\mathsf{x} = \lambda\mathsf{x}$ for some nonzero vector x. The vector x is called an eigenvector of A corresponding to λ. The set of all solutions to $A\mathsf{x} = \lambda\mathsf{x}$ is called the eigenspace of A corresponding to λ. For example, for $A = \begin{bmatrix} 1 & 3 \\ 2 & -4 \end{bmatrix}$, we have

$$A\begin{bmatrix} 3 \\ 1 \end{bmatrix} = \begin{bmatrix} 1 & 3 \\ 2 & -4 \end{bmatrix}\begin{bmatrix} 3 \\ 1 \end{bmatrix} = \begin{bmatrix} 6 \\ 2 \end{bmatrix} = 2\mathsf{x},$$

so $\begin{bmatrix} 3 \\ 1 \end{bmatrix}$ is an eigenvector of A corresponding to the eigenvalue $\lambda = 2$. The eigenspace corresponding to $\lambda = 2$ is the set of multiples of $\begin{bmatrix} 3 \\ 1 \end{bmatrix}$. You find this by solving the homogeneous system $(A - 2I)\mathsf{x} = 0$, which is a useful way to rewrite $A\mathsf{x} = 2\mathsf{x}$.

Elementary matrix: An elementary matrix is a (square) matrix obtained from the identity matrix by a single elementary row operation. For example, $E = \begin{bmatrix} 0 & 1 & 0 \\ 1 & 0 & 0 \\ 0 & 0 & 1 \end{bmatrix}$ is an elementary matrix, obtained from the 3×3 identity matrix I by interchanging rows one and two. The matrix $E = \begin{bmatrix} 1 & 0 & 0 \\ 0 & 3 & 0 \\ 0 & 0 & 1 \end{bmatrix}$ is elementary; it is obtained from I by multiplying row two by 3.

Elementary row operations: The three elementary row operations on a matrix are

1. the interchange of two rows,
2. the multiplication of a row by a nonzero scalar, and
3. replacement of a row by that row minus a multiple of another row.

Euclidean n-space: Euclidean n-space is the set of all n-dimensional vectors. It is denoted R^n. Euclidean 2-space, $\mathsf{R}^2 = \{\begin{bmatrix} x \\ y \end{bmatrix} \mid x, y \in \mathsf{R}\}$ is more commonly called the Euclidean plane and Euclidean 3-space $\mathsf{R}^3 = \{\begin{bmatrix} x \\ y \\ z \end{bmatrix} \mid x, y, z \in \mathsf{R}\}$ is often called simply 3-*space.*

Homogeneous system: A system of linear equations is homogeneous if it has the form $A\mathsf{x} = 0$, that is, the vector of constants to the right of = is the zero vector. For example, the system

$$\begin{aligned} 3x_1 + 2x_2 - x_3 &= 0 \\ 2x_1 - 5x_2 + 7x_3 &= 0, \end{aligned}$$

which is $A\mathsf{x} = 0$ with $A = \begin{bmatrix} 3 & 2 & -1 \\ 2 & -5 & 7 \end{bmatrix}$ and $\mathsf{x} = \begin{bmatrix} x_1 \\ x_2 \end{bmatrix}$, is homogeneous.

Identity matrix: An identity matrix is a square matrix with 1s on the diagonal and 0s everywhere else. For example, $\begin{bmatrix} 1 & 0 \\ 0 & 1 \end{bmatrix}$ and $\begin{bmatrix} 1 & 0 & 0 \\ 0 & 1 & 0 \\ 0 & 0 & 1 \end{bmatrix}$ are identity matrices, the first being the 2×2 identity matrix and the second the 3×3 identity matrix.

Inconsistent: See *Linear equation/Linear system.*

Inverse: See *invertible matrix.*

Invertible matrix: A square matrix A is invertible (or has an inverse) if there is another matrix B such that $AB = I$ and $BA = I$. The matrix B is called the inverse of A and we write $B = A^{-1}$. For example, the matrix $A = \begin{bmatrix} 1 & -2 \\ 2 & -3 \end{bmatrix}$ has an inverse, namely, $B = \begin{bmatrix} -3 & 2 \\ -2 & 1 \end{bmatrix}$, since

$$AB = \begin{bmatrix} 1 & -2 \\ 2 & -3 \end{bmatrix}\begin{bmatrix} -3 & 2 \\ -2 & 1 \end{bmatrix} = \begin{bmatrix} 1 & 0 \\ 0 & 1 \end{bmatrix} = I$$

and

$$BA = \begin{bmatrix} -3 & 2 \\ -2 & 1 \end{bmatrix}\begin{bmatrix} 1 & -2 \\ 2 & -3 \end{bmatrix} = \begin{bmatrix} 1 & 0 \\ 0 & 1 \end{bmatrix} = I.$$

From theory, we know that if $AB = I$ with A and B square, then BA must also be I so, in point of fact, we had only to compute one of the above two products to be sure that A and B are invertible. The matrix $A = \begin{bmatrix} 1 & 2 \\ 2 & 4 \end{bmatrix}$ is not invertible because, if we let $B = \begin{bmatrix} x & y \\ z & w \end{bmatrix}$, then $AB = \begin{bmatrix} x+2z & y+2w \\ 2x+4z & 2y+4w \end{bmatrix}$. Since the second row of AB is twice the first, AB can never be I.

Length: The length of the n-dimensional vector $\mathsf{x} = \begin{bmatrix} x_1 \\ x_2 \\ \vdots \\ x_n \end{bmatrix}$ is the number $\sqrt{x_1^2 + x_2^2 + \cdots + x_n^2}$, which is $\sqrt{\mathsf{x} \cdot \mathsf{x}}$. It's denoted $\|\mathsf{x}\|$. For example, if $\mathsf{x} = \begin{bmatrix} 2 \\ 1 \\ 2 \end{bmatrix}$, then $\|\mathsf{x}\| = \sqrt{4+1+4} = 3$. The rule is the same for complex vectors (but you must use the complex dot product). For instance, if $\mathsf{z} = \begin{bmatrix} 1 \\ i \end{bmatrix}$ and $\mathsf{w} = \begin{bmatrix} 2+i \\ 3-i \end{bmatrix}$, $\|\mathsf{z}\|^2 = \mathsf{z} \cdot \mathsf{z} = 1(1) - i(i) = 1 - i^2 = 2$, so $\|\mathsf{z}\| = \sqrt{2}$, and $\|\mathsf{w}\|^2 = \mathsf{w} \cdot \mathsf{w} = (2-i)(2+i) + (3+i)(3-i) = 4+1+9+1 = 15$, so $\|\mathsf{w}\| = \sqrt{15}$.

Linear combination: A linear combination of k vectors $\mathsf{u}_1, \mathsf{u}_2, \ldots, \mathsf{u}_k$ is a vector of the form $c_1\mathsf{u}_1 + c_2\mathsf{u}_2 + \cdots + c_k\mathsf{u}_k$, where $c_1, \ldots, c_k$ are scalars. For example, $\begin{bmatrix} -5 \\ 9 \end{bmatrix}$ is a linear combination of $\mathsf{u} = \begin{bmatrix} -2 \\ 3 \end{bmatrix}$ and $\mathsf{v} = \begin{bmatrix} -1 \\ 1 \end{bmatrix}$ since $\begin{bmatrix} -5 \\ 9 \end{bmatrix} = 4\begin{bmatrix} -2 \\ 3 \end{bmatrix} - 3\begin{bmatrix} -1 \\ 1 \end{bmatrix} = 4\mathsf{u} - 3\mathsf{v}$ and $\begin{bmatrix} 2 \\ -6 \end{bmatrix}$ is a linear combination of $\mathsf{u}_1 = \begin{bmatrix} -2 \\ 3 \end{bmatrix}$, $\mathsf{u}_2 = \begin{bmatrix} 6 \\ -5 \end{bmatrix}$ and $\mathsf{u}_3 = \begin{bmatrix} 4 \\ 5 \end{bmatrix}$ since $\begin{bmatrix} 2 \\ -6 \end{bmatrix} = 3\begin{bmatrix} -2 \\ 3 \end{bmatrix} + 2\begin{bmatrix} 6 \\ -5 \end{bmatrix} - \begin{bmatrix} 4 \\ 5 \end{bmatrix} = 3\mathsf{u}_1 + 2\mathsf{u}_2 + (-1)\mathsf{u}_3$. A linear combination of matrices $A_1, A_2, \ldots, A_k$ (all of the same size) is a matrix of the form $c_1A_1 + c_2A_2 + \cdots + c_kA_k$, where $c_1, c_2, \ldots, c_k$ are scalars.

Linear equation/Linear system: A linear equation is an equation of the form

$$a_1x_1 + a_2x_2 + \cdots + a_nx_n = b,$$

where $a_1, a_2, \ldots, a_n$ and b are real numbers and $x_1, x_2, \ldots, x_n$ are variables. A set of one or more linear equations is called a linear system. To solve a linear system means to find values of the variables that make each equation true. A system that has no solution is called inconsistent.

Linearly dependent: See *Linearly independent.*

Linearly independent: A set of vectors $\mathsf{x}_1, \mathsf{x}_2, \ldots, \mathsf{x}_n$ is linearly independent if the only linear combination of them that equals the zero vector is the trivial one, where all the coefficients are 0:

$$c_1\mathsf{x}_1 + c_2\mathsf{x}_2 + \cdots + c_n\mathsf{x}_n = \mathbf{0} \quad \text{implies} \quad c_1 = c_2 = \cdots = c_n = 0.$$

Vectors that are not linearly independent are linearly dependent; that is, there exists some linear combination $c_1\mathsf{x}_1 + c_2\mathsf{x}_2 + \cdots + c_n\mathsf{x}_n = \mathbf{0}$ without **all** the coefficients $c_1, c_2, \ldots, c_n$ equal to 0. The vectors $\mathsf{x}_1 = \begin{bmatrix} 1 \\ 2 \end{bmatrix}$ and $\mathsf{x}_2 = \begin{bmatrix} -1 \\ 0 \end{bmatrix}$ are linearly independent (solve the equation $c_1\mathsf{x}_1 + c_2\mathsf{x}_2 = \mathbf{0}$ and obtain $c_1 = c_2 = 0$) whereas the vectors $\mathsf{u}_1 = \begin{bmatrix} 1 \\ 2 \end{bmatrix}$, $\mathsf{u}_2 = \begin{bmatrix} 1 \\ 0 \end{bmatrix}$ and $\mathsf{u}_3 = \begin{bmatrix} 2 \\ 4 \end{bmatrix}$ are linearly dependent because $2\mathsf{u}_1 + 0\mathsf{u}_2 - 1\mathsf{u}_3 = \mathbf{0}$.

LU Factorization: An LU factorization of a matrix A is the representation of $A = LU$ as the product of a lower triangular matrix L with 1s on the diagonal and a row echelon matrix U. For example, $\begin{bmatrix} 1 & 2 \\ 4 & 6 \end{bmatrix} = \begin{bmatrix} 1 & 0 \\ 4 & 1 \end{bmatrix} \begin{bmatrix} 1 & 2 \\ 0 & -2 \end{bmatrix}$ is an LU factorization of $A = \begin{bmatrix} 1 & 2 \\ 4 & 6 \end{bmatrix}$.

Main diagonal: The main diagonal of an $m \times n$ matrix $A = [a_{ij}]$ is the list of elements $a_{11}, a_{22}, a_{33}, \ldots$. For instance, the main diagonal of $\begin{bmatrix} 1 & 2 & 3 \\ 4 & 5 & 6 \\ 7 & 8 & 9 \end{bmatrix}$ is 1, 5, 9 and the main diagonal of $\begin{bmatrix} 7 & 9 & 3 \\ 6 & 5 & 4 \end{bmatrix}$ is 7, 5. A matrix is diagonal if and only if its only nonzero entries lie on the main diagonal. For example, the matrix $\begin{bmatrix} 1 & 0 & 0 \\ 0 & -7 & 0 \\ 0 & 0 & 2 \end{bmatrix}$ is diagonal. So is $\begin{bmatrix} -2 & 0 & 0 \\ 0 & 7 & 0 \end{bmatrix}$.

Matrix: A matrix is a rectangular array of numbers enclosed in square brackets. If there are m rows and n columns in the array, the matrix is called $m \times n$ (read "m by n") and said to have *size* $m \times n$. For example, $A = \begin{bmatrix} 2 & 1 & -1 \\ 3 & -2 & 6 \\ 1 & 0 & -5 \end{bmatrix}$ is a 3×3 ("three by three")matrix, while $B = \begin{bmatrix} 1 & 2 & 3 \\ 4 & 5 & 6 \end{bmatrix}$ is 2×3 and $\mathsf{x} = \begin{bmatrix} x_1 \\ x_2 \\ x_3 \end{bmatrix}$ is 3×1. The numbers in the matrix are its entries. We always use a capital letter to denote a matrix and the corresponding lower case letter, with subscripts, for its entries. Thus if $A = [a_{ij}]$, the $(2,3)$ entry is a_{23}. If $A = \begin{bmatrix} 2 & 1 & 3 \\ 3 & 2 & 8 \\ 9 & 0 & 1 \end{bmatrix}$, then $a_{32} = 0$. If $A = [2i - j]$ is a 3×4 matrix, the $(1,3)$ entry of A is $2(1) - 3 = -1$. *p. 56*

Minor: If A is a square matrix, the (i,j) minor, denoted m_{ij}, is the determinant of the $(n-1) \times (n-1)$ matrix obtained from A by deleting row i and column j. The (i,j) cofactor of A is $(-1)^{i+j}m_{ij}$ and is denoted c_{ij}. For example, let $A = \begin{bmatrix} 1 & 2 & 3 \\ 0 & -2 & 2 \\ 3 & 7 & -4 \end{bmatrix}$. The $(1,1)$ minor is the determinant of the matrix $\begin{bmatrix} -2 & 2 \\ 7 & -4 \end{bmatrix}$, which is what remains when we remove row one and column one of A. So $m_{11} = -2(-4) - 2(7) = -6$ and $c_{11} = (-1)^{1+1}m_{11} = (-1)^2(-6) = -6$. The $(-1)^{i+j}$ in the definition of cofactor has the affect of either leaving the minor alone or changing its sign according to the pattern $\begin{bmatrix} + & - & + & - & \cdots \\ - & + & - & + & \cdots \\ \vdots & \vdots & & & \end{bmatrix}$. Thus the $(2,3)$ minor of A is the determinant of $\begin{bmatrix} 1 & 2 \\ 3 & 7 \end{bmatrix}$, which is what remains after removing row two and column three of A, $m_{23} = 1$, while $c_{23} = -1$ since there is a $-$ in the $(2,3)$ position of the pattern.

Opposite direction: Vectors u and v have opposite direction if $\mathsf{u} = c\mathsf{v}$ for some scalar $c < 0$. For example, $\mathsf{u} = \begin{bmatrix} 2 \\ -3 \end{bmatrix}$ and $\mathsf{v} = \begin{bmatrix} -3 \\ \frac{9}{2} \end{bmatrix}$ have opposite direction because $\mathsf{u} = -\frac{2}{3}\mathsf{v}$ with the scalar $-\frac{2}{3} < 0$.

Orthogonal: A set $\{\mathsf{f}_1, \mathsf{f}_2, \ldots, \mathsf{f}_k\}$ of vectors is orthogonal if and only if the vectors are pairwise orthogonal: $\mathsf{f}_i \cdot \mathsf{f}_j = 0$ if $i \neq j$. If, in addition, each f_i has norm 1, then the set is called orthonormal. For example, any subset of the set of standard basis vectors $\mathsf{e}_1, \mathsf{e}_2, \ldots, \mathsf{e}_n$ is an orthonormal set in R^n, say, $\{\begin{bmatrix} 0 \\ 1 \\ 0 \end{bmatrix}, \begin{bmatrix} 0 \\ 0 \\ 1 \end{bmatrix}\}$ in R^3. The vectors $\mathsf{u} = \begin{bmatrix} 3 \\ 2 \\ -1 \\ 4 \end{bmatrix}$ and $\mathsf{v} = \begin{bmatrix} 1 \\ -1 \\ 1 \\ 0 \end{bmatrix}$ are orthogonal vectors in R^4. The four vectors $\mathsf{f}_1 = \begin{bmatrix} 1 \\ 1 \\ 1 \\ -1 \end{bmatrix}$, $\mathsf{f}_2 = \begin{bmatrix} 1 \\ 0 \\ 1 \\ 2 \end{bmatrix}$, $\mathsf{f}_3 = \begin{bmatrix} -1 \\ 0 \\ 1 \\ 0 \end{bmatrix}$, $\mathsf{f}_4 = \begin{bmatrix} -1 \\ 3 \\ -1 \\ 1 \end{bmatrix}$ form an orthogonal set in R^4.

Orthogonal matrix: An orthogonal matrix is a square matrix with orthonormal columns. For example, the rotation matrix $Q = \begin{bmatrix} \cos\theta & -\sin\theta \\ \sin\theta & \cos\theta \end{bmatrix}$ is orthogonal for any angle θ. Any permutation matrix is orthogonal because its columns are just the standard basis vectors $\mathsf{e}_1, \mathsf{e}_2, \ldots, \mathsf{e}_n$ in some order. One such matrix is $P = \begin{bmatrix} 0 & 0 & 1 \\ 1 & 0 & 0 \\ 0 & 1 & 0 \end{bmatrix}$.

Orthonormal: See *Orthogonal.*

Pairwise orthogonal: See *Orthogonal.*

Parallel: Vectors u and v are parallel if one is a scalar multiple of the other, that is, if $\mathsf{u} = c\mathsf{v}$ or $\mathsf{v} = c\mathsf{u}$ for some scalar c. For example, $\mathsf{u} = \begin{bmatrix} -2 \\ 1 \end{bmatrix}$ and $\mathsf{v} = \begin{bmatrix} 6 \\ -3 \end{bmatrix}$ are parallel because $\mathsf{v} = -3\mathsf{u}$. Any vector u is parallel to the zero vector because $\mathsf{u} = 15(0)$.

Permutation matrix: A permutation matrix is a matrix whose rows are the standard basis vectors, in some order; equivalently, a matrix whose columns are the standard basis vectors, in some order. For example, $\begin{bmatrix} 0 & 1 \\ 1 & 0 \end{bmatrix}$, $\begin{bmatrix} 0 & 1 & 0 \\ 0 & 0 & 1 \\ 1 & 0 & 0 \end{bmatrix}$ and $\begin{bmatrix} 0 & 0 & 1 & 0 \\ 0 & 1 & 0 & 0 \\ 0 & 0 & 0 & 1 \\ 1 & 0 & 0 & 0 \end{bmatrix}$ are all permutation matrices.

Pivot: A pivot of a row echelon matrix is the first nonzero entry in a nonzero row and a pivot column is a column containing a pivot. For example, the pivots of $\begin{bmatrix} 0 & -3 & 1 & 2 & 1 \\ 0 & 0 & 2 & 0 & -3 \\ 0 & 0 & 0 & 0 & 8 \end{bmatrix}$ are -3, 2 and 8 and the pivot columns are two, three and five. If U is a row echelon form of a matrix A, the pivot columns of A are the pivot columns of U. If U was obtained without multiplying any row by a scalar, then the pivots of A are the pivots of U. For example,

$$A = \begin{bmatrix} -2 & 4 & 3 \\ 3 & -6 & 0 \\ 4 & -8 & 1 \end{bmatrix} \to \begin{bmatrix} -2 & 4 & 3 \\ 0 & 0 & \frac{9}{2} \\ 0 & 0 & 7 \end{bmatrix} \to \begin{bmatrix} -2 & 4 & 3 \\ 0 & 0 & \frac{9}{2} \\ 0 & 0 & 0 \end{bmatrix} = U$$

so the pivots of A are -2 and $\frac{9}{2}$, and the pivot columns of A are columns one and three.

Projection: The projection of a vector u on the line with direction the (nonzero) vector v is the vector $\mathsf{p} = \operatorname{proj}_\ell \mathsf{u}$ (also written $\operatorname{proj}_\mathsf{v} \mathsf{u}$) with the property that $\mathsf{u} - \mathsf{p}$ is orthogonal to v. One can show that $\operatorname{proj}_\mathsf{v} \mathsf{u} = \dfrac{\mathsf{u} \cdot \mathsf{v}}{\mathsf{v} \cdot \mathsf{v}} \mathsf{v}$. For example, if $\mathsf{u} = \begin{bmatrix} 1 \\ -3 \\ 4 \end{bmatrix}$ and $\mathsf{v} = \begin{bmatrix} -3 \\ 1 \\ 2 \end{bmatrix}$,

$\operatorname{proj}_\mathsf{v} \mathsf{u} = \frac{1}{7}\mathsf{v} = \begin{bmatrix} -\frac{3}{7} \\ \frac{1}{7} \\ \frac{2}{7} \end{bmatrix}$.

The projection of a vector w onto a plane π is the vector $\mathsf{p} = \operatorname{proj}_\pi \mathsf{w}$ in π with the property that $\mathsf{w} - \mathsf{p}$ is orthogonal to every vector in π. If the plane π is spanned by orthogonal vectors e and f, one can show that $\operatorname{proj}_\pi \mathsf{w} = \frac{\mathsf{w}\cdot\mathsf{e}}{\mathsf{e}\cdot\mathsf{e}} \mathsf{e} + \frac{\mathsf{w}\cdot\mathsf{f}}{\mathsf{f}\cdot\mathsf{f}} \mathsf{f}$. For example, the plane π with equation $3x - 2y + z = 0$ is spanned by the orthogonal vectors $\mathsf{e} = \begin{bmatrix} 5 \\ 6 \\ -3 \end{bmatrix}$ and $\mathsf{f} = \begin{bmatrix} 0 \\ 1 \\ 2 \end{bmatrix}$. If $\mathsf{w} = \begin{bmatrix} 5 \\ -3 \\ -7 \end{bmatrix}$,

$$\begin{aligned} \operatorname{proj}_\pi \mathsf{w} &= \frac{\mathsf{w} \cdot \mathsf{e}}{\mathsf{e} \cdot \mathsf{e}} \mathsf{e} + \frac{\mathsf{w} \cdot \mathsf{f}}{\mathsf{f} \cdot \mathsf{f}} \mathsf{f} \\ &= \frac{28}{70} \begin{bmatrix} 5 \\ 6 \\ -3 \end{bmatrix} - \frac{17}{5} \begin{bmatrix} 0 \\ 1 \\ 2 \end{bmatrix} = \frac{2}{5} \begin{bmatrix} 5 \\ 6 \\ -3 \end{bmatrix} - \frac{17}{5} \begin{bmatrix} 0 \\ 1 \\ 2 \end{bmatrix} = \begin{bmatrix} 2 \\ -1 \\ -8 \end{bmatrix}. \end{aligned}$$

In general, the projection of a vector w on a subspace U is the vector $\mathsf{p} = \operatorname{proj}_U \mathsf{w}$ with the property that $\mathsf{w} - \mathsf{p}$ is orthogonal to U. If $\mathsf{f}_1, \mathsf{f}_2, \ldots, \mathsf{f}_n$ is an orthogonal basis for U, then

$$\operatorname{proj}_U \mathsf{w} = \frac{\mathsf{w} \cdot \mathsf{f}_1}{\mathsf{f}_1 \cdot \mathsf{f}_1} \mathsf{f}_1 + \frac{\mathsf{w} \cdot \mathsf{f}_2}{\mathsf{f}_2 \cdot \mathsf{f}_2} \mathsf{f}_2 + \cdots + \frac{\mathsf{w} \cdot \mathsf{f}_n}{\mathsf{f}_n \cdot \mathsf{f}_n} \mathsf{f}_n.$$

Reduced row echelon form: See *Row echelon form.*

Row echelon form: A matrix is in row echelon form if and only if it is upper triangular with the following properties:

1. all rows consisting entirely of 0s are at the bottom,
2. the leading nonzero entries of the nonzero rows step from left to right as you read down the matrix, and
3. every entry below a leading nonzero entry is a 0.

For example, $\begin{bmatrix} 7 & 1 & 3 & 4 & 5 \\ 0 & 0 & 5 & 3 & 1 \\ 0 & 0 & 0 & 0 & 1 \end{bmatrix}$ and $\begin{bmatrix} 0 & 0 & 1 & 1 & 5 \\ 0 & 0 & 0 & 0 & 1 \\ 0 & 0 & 0 & 0 & 0 \end{bmatrix}$ are in row echelon form, while $\begin{bmatrix} 1 & 0 & 3 & 2 \\ 0 & -1 & 2 & 4 \\ 0 & 1 & 0 & 1 \end{bmatrix}$ and $\begin{bmatrix} 1 & 0 & 3 & 2 \\ 0 & 0 & 1 & 4 \\ 0 & 1 & 0 & 2 \end{bmatrix}$ are not. A matrix is in reduced row echelon form if it is in row echelon form, each pivot (that is, each nonzero leading entry in a nonzero row) is a 1, and all other entries in a column containing a leading 1 are 0. For example, $\begin{bmatrix} 1 & 0 \\ 0 & 1 \end{bmatrix}$, $\begin{bmatrix} 1 & 0 & -2 \\ 0 & 1 & 1 \end{bmatrix}$ and $\begin{bmatrix} 1 & 0 & 0 & 0 \\ 0 & 0 & 1 & 0 \\ 0 & 0 & 0 & 1 \end{bmatrix}$ are all in reduced row echelon form.

Same direction: Vectors u and v have the same direction if $\mathsf{u} = c\mathsf{v}$ with $c > 0$. For example, $\mathsf{u} = \begin{bmatrix} -2 \\ 4 \\ 6 \end{bmatrix}$ and $\mathsf{v} = \begin{bmatrix} -1 \\ 2 \\ 3 \end{bmatrix}$ have the same direction because $\mathsf{v} = \frac{1}{2}\mathsf{u}$ with the scalar $\frac{1}{2} > 0$.

Similar: Matrices A and B are similar if there is an invertible matrix P such that $B = P^{-1}AP$. For example, $A = \begin{bmatrix} 1 & 2 \\ 3 & 4 \end{bmatrix}$ and $B = \begin{bmatrix} -55 & -97 \\ 34 & 60 \end{bmatrix}$ are similar since

$$B = \begin{bmatrix} -55 & -97 \\ 34 & 60 \end{bmatrix} = \begin{bmatrix} 2 & -5 \\ -1 & 3 \end{bmatrix}\begin{bmatrix} 1 & 2 \\ 3 & 4 \end{bmatrix}\begin{bmatrix} 3 & 5 \\ 1 & 2 \end{bmatrix} = P^{-1}AP$$

with $P = \begin{bmatrix} 3 & 5 \\ 1 & 2 \end{bmatrix}$.

Singular matrix: A matrix is singular if it is not invertible, a condition equivalent to having determinant 0. For example, $A = \begin{bmatrix} 2 & 3 \\ 4 & 6 \end{bmatrix}$ is singular, because it has determinant 0.

Size: See *Matrix*.

Span, spanned by: The span of vectors $\mathsf{u}_1, \mathsf{u}_2, \ldots, \mathsf{u}_n$ is the set of all linear combinations of them. It is denoted $\operatorname{sp}\{\mathsf{u}_1, \ldots, \mathsf{u}_n\}$. We say that the space $\operatorname{sp}\{\mathsf{u}_1, \ldots, \mathsf{u}_n\}$ is spanned by the vectors $\mathsf{u}_1, \ldots, \mathsf{u}_n$. For example, the span of i and j in R^2 is all R^2 since any vector $\mathsf{u} = \begin{bmatrix} a \\ b \end{bmatrix}$ in R^2 is $a\mathsf{i} + b\mathsf{j}$. We say that R^2 is spanned by i and j. On the other hand, the span of $\mathsf{u} = \begin{bmatrix} -1 \\ 2 \end{bmatrix}$ and $\mathsf{v} = \begin{bmatrix} -3 \\ 6 \end{bmatrix}$ is the set of scalar multiples of u since $a\mathsf{u} + b\mathsf{v} = \begin{bmatrix} -a - 3b \\ 2a + 6b \end{bmatrix} = (a + 3b)\begin{bmatrix} -1 \\ 2 \end{bmatrix}$. The span of two nonparallel vectors is a plane. For example, the span of $\begin{bmatrix} 1 \\ 2 \end{bmatrix}$ and $\begin{bmatrix} -3 \\ 4 \end{bmatrix}$ is the xy-plane. Any plane through the origin in R^3 is spanned by any two nonparallel vectors it contains. For example, the plane with equation $2x - y + 3z = 0$ is spanned by $\mathsf{u} = \begin{bmatrix} 1 \\ 2 \\ 0 \end{bmatrix}$ and $\mathsf{v} = \begin{bmatrix} 0 \\ 3 \\ 1 \end{bmatrix}$.

Standard basis vectors: The standard basis vectors in R^n are the n-dimensional vectors $\mathsf{e}_1, \mathsf{e}_2, \ldots, \mathsf{e}_n$ where e_i has ith component 1 and all other components 0; that is,

$$\mathsf{e}_1 = \begin{bmatrix} 1 \\ 0 \\ 0 \\ 0 \\ \vdots \\ 0 \end{bmatrix}, \quad \mathsf{e}_2 = \begin{bmatrix} 0 \\ 1 \\ 0 \\ 0 \\ \vdots \\ 0 \end{bmatrix}, \quad \mathsf{e}_3 = \begin{bmatrix} 0 \\ 0 \\ 1 \\ 0 \\ \vdots \\ 0 \end{bmatrix}, \quad \ldots, \quad \mathsf{e}_n = \begin{bmatrix} 0 \\ 0 \\ 0 \\ \vdots \\ 0 \\ 1 \end{bmatrix}.$$

In R^2, the standard basis vectors $\mathsf{e}_1 = \begin{bmatrix} 1 \\ 0 \end{bmatrix}$ and $\mathsf{e}_2 = \begin{bmatrix} 0 \\ 1 \end{bmatrix}$ are also denoted i and j, respectively, and, in R^3, the standard basis vectors $\mathsf{e}_1 = \begin{bmatrix} 1 \\ 0 \\ 0 \end{bmatrix}$, $\mathsf{e}_2 = \begin{bmatrix} 0 \\ 0 \\ 1 \end{bmatrix}$ and $\mathsf{e}_3 = \begin{bmatrix} 0 \\ 0 \\ 1 \end{bmatrix}$ are also denoted i, j, and k, respectively.

Symmetric: A matrix A is symmetric if it equals its transpose: $A = A^T$. For example, $\begin{bmatrix} 1 & 2 \\ 2 & 4 \end{bmatrix}$ is a symmetric matrix, as is $\begin{bmatrix} 1 & 2 & 3 \\ 2 & 7 & -8 \\ 3 & -8 & 5 \end{bmatrix}$.

Transpose: The transpose of an $m \times n$ **real** matrix A is the $n \times m$ matrix whose rows are the columns of A in the same order. The transpose of A is denoted A^T. For example, if $A = \begin{bmatrix} 1 & 2 & 3 \\ 4 & 5 & 6 \\ 7 & 8 & 9 \end{bmatrix}$, then $A^T = \begin{bmatrix} 1 & 4 & 7 \\ 2 & 5 & 8 \\ 3 & 6 & 9 \end{bmatrix}$ and if $A = \begin{bmatrix} -1 & 2 \\ 3 & 4 \\ 0 & -5 \end{bmatrix}$, then $A^T = \begin{bmatrix} -1 & 3 & 0 \\ 2 & 4 & -5 \end{bmatrix}$.

Triangular: A matrix is upper triangular if all entries below the main diagonal are 0 and lower triangular if all entries above the main diagonal are 0. For example, $U = \begin{bmatrix} 1 & -1 & 2 \\ 0 & 1 & 1 \end{bmatrix}$ is upper triangular, while $L = \begin{bmatrix} -1 & 0 \\ 2 & 3 \end{bmatrix}$ is lower triangular. A matrix is triangular if it is either upper triangular or lower triangular.

Unit vector: A unit vector is a vector of norm 1. For example, the vector $\begin{bmatrix} \frac{1}{\sqrt{2}} \\ \frac{1}{\sqrt{2}} \end{bmatrix}$ is a unit two-dimensional vector and $\begin{bmatrix} \frac{1}{\sqrt{6}} \\ -\frac{2}{\sqrt{6}} \\ \frac{1}{\sqrt{6}} \end{bmatrix}$ is a unit three-dimensional vector.

Vector: An n-dimensional vector ($n \geq 1$) is a column of n numbers enclosed in brackets. The numbers are called the components of the vector. For example, $\begin{bmatrix} 1 \\ 2 \end{bmatrix}$ is a two-dimensional vector with components 1 and 2, and $\begin{bmatrix} -1 \\ 2 \\ 2 \\ 5 \end{bmatrix}$ is a four-dimensional vector with components -1, 2, 2, and 5.

Index